W9-ADU-840

Limnological Methods

Limnological Methods

By PAUL S. WELCH, Ph.D.
Professor of Zoology,
University of Michigan

NEW YORK · TORONTO · LONDON

McGRAW-HILL BOOK COMPANY, INC.

V

69175

Preface

One of the characteristic features of limnology is its dependence upon numerous, diverse, technical methods. Many of them are wholly limnological in origin, some are special modifications and adaptations of procedures from other fields of science, and still others are taken bodily from outside sources. Descriptions of these methods are so scattered throughout the literature that the less experienced worker is often perplexed or discouraged. Since no single source of information now exists, the need for an assembly of fundamental limnological methods seems urgent enough to justify the preparation of a book of this kind.

The purpose of this volume is to present the essentials of those *basic* methods necessary for (a) entry into the subject of limnology, (b) limnological surveys of lakes and streams, and (c) fundamental information upon which specialized researches depend. No attempt is made to assemble all of the different methods in current use. The writer has deliberately selected certain of those which, for the most part, long have been used in his field and laboratory programs and which provide, when properly used, satisfactory results.

Certain items relating to the choice and treatment of materials require comment. Each method is presented in the form of simple basic directions for procedure, with supplemental information, directions, and additions assembled at the end of each section. Lake and stream mapping is stressed because of (a) its fundamental importance in limnology, (b) the common failure of limnologists to secure training in surveying, and (c) the possibility of acquiring some skill in hydrographic mapping in the absence of training in professional surveying. Application of statistical methods to many limnological processes is still too undeveloped to justify consideration here. A few simple methods and descriptions of improvised instruments are included for the benefit of those whose facilities are limited. Forms for records are presented in a few instances as samples to show how these useful devices can be constructed. Only those chemical methods commonly used in routine work have been included since authoritative manuals on water analysis are easily available. In some instances it has not been possible to indicate credit for a method since the original authorship has become uncertain or even lost. References listed at the ends of chapters were selected on the basis of outstanding utility; completeness is not intended. English units of measure are so firmly established in the construction of many instruments and in the descriptions of so many materials and procedures that to convert all values into metric units for sake of consistency becomes a futile gesture; therefore the writer has yielded, reluctantly, to the necessity of intermingling the two systems.

For permission to quote or otherwise use materials, some of which are copyrighted, the author acknowledges indebtedness to: The American Public Health Association, New York; Eugene Dietzgen Co., Chicago, Ill.; Evershed and Vignoles, London, England; Gray Printing Company, DuBois, Pa.; W. and L. E. Gurley, Troy, N. Y.; Hellige, Inc., Long Island City, N. Y.; Leupold and Stevens Instruments, Portland, Ore.; Negretti and Zambra, London, England; Precision Scientific Co., Chicago, Ill.; Prentice-Hall, Inc., New York; John Wiley and Sons, Inc., New York; and Yoder Instruments, East Palestine, Ohio.

The ink work on all line drawings was done by Mr. William L. Cristanelli.

PAUL S. WELCH

Ann Arbor, Michigan

Contents

Introductory Considerations

Choice and Application of Methods

For the execution of some particular piece of limnological work different methods representing various grades of precision may be available or some one standard method may be so flexible that it can be adapted or modified to meet the needs of the task. The choice of a method or its modification should be determined largely by the *accuracy required* and, at the outset, the operator must ascertain this requirement as definitely as circumstances will permit. Highly precise methods applied to preliminary or exploratory work may be ill-suited or unnecessarily expensive in time and effort. Often valuable time is wasted in striving for a precision which is not needed. On the other hand, use of some gross method for a precision task is likely to result in worthless data.

In order that the operator may select methods intelligently, he must not only determine what accuracy is needed for his purposes, but he must also know what precision the available methods provide. The decision as to the *accuracy needed* requires a study of the task to be undertaken. Sometimes the decision is obvious, sometimes it must be based wholly upon sound judgment, sometimes it is arrived at only through preliminary tests or examinations. Each case must be determined on the basis of its own requirements.

Once the needed accuracy is known, suitable methods must be sought. The precision of a given method may depend upon several features, prominent among which are the *inherent limitations* of instruments and procedures. By inherent limitations is meant the fixed capacity of an instrument or method. For example, on an instrument designed for linear measurements, if the finest unit of graduation is 1 mm., then the inherent limitation of that instrument is 1 mm.; no smaller values can be secured with it except by estimation. These limitations are often built into the instrument or method in such a way that they cannot be changed.

Methods may be described as possessing a certain specified precision when operated under the best conditions, but it must be realized that unless these ideal conditions are met the accuracy will be less, to various extents, than the stated one. Such stated values are usually regarded as maximum precisions, unless otherwise qualified. The inexperienced operator, in particular, must be warned that in his hands methods will probably not supply the accuracy claimed for them in printed descriptions.

Complex methods often involve a series of separate operations in a definite sequence. Each operation may be sufficiently independent to have a determinable precision of its own which may differ from those of the other operations. Consequently the accuracy of the method as a whole is a composite of all the precision values of the separate operations.

Errors

In the operation and the results of any method, existence of error should be regarded as an ever-present possibility. Unchecked work cannot escape the threat of unreliability. Therefore, the *testing* of procedures and results and the proper understanding and treatment of errors are essential to dependability.

Errors may be classified as follows: inherent errors, natural errors, and personal errors.

Inherent errors arise from the failure of instruments and methods to accomplish their expected precision up to their inherent limitations. For example, inherent errors may occur in instruments because of faulty construction, worn parts, poor adjustment, or damage from accidents. Commonly, results affected by such errors can be improved by the application of correction values. *Natural errors* arise from the effects of such features as change of temperature, changes in light intensity, action of wind, light refraction, and local-attraction effects upon compass needles. *Personal errors* are those for which the operator is wholly responsible. They may arise from inexperience, lack of manual skill, carelessness, haste, faulty procedures, poor judgment, mistakes in recording data, and many other sources.

Correction and Calibration of Equipment

At the outset the unfailing habit of scrutinizing all equipment for accuracy and good condition must be acquired. It should never be assumed that apparatus, new or old, is accurate, even to that degree of precision for which it is supposed to be constructed. In general, cheaper equipment is more likely to require correction than costly items intended for the same purpose; also, cheaper articles are more apt to vary among themselves. Nevertheless, even expensive instruments inadvertently may be defective in construction, may have suffered injury in transportation, may be affected by use and misuse, or may otherwise fail to function properly. Simple commonplace laboratory materials, such as rulers, meter sticks, cross-section paper, graduated glassware, weights, and balances, are sometimes inaccurate to significant degrees. The name of a well-known instrument maker does not necessarily insure against the possible failure of his products to meet fully the specifications. Too often instruments, simple or complex, new or old, which are supplied by a laboratory, are accepted unquestioningly as satisfactory when actually they may be in need of correction, adjustment, or reconditioning. In limnological work, measurements of some kind are ever-present features and of great importance. Such measurements cannot be dependable unless, among other precautions, all instruments are made to function properly and to the full extent within their individual inherent limitations. Deviation from the rule of constant inspection will lead sooner or later to serious error.

Certain general standards must be available. Properly chosen, they may serve, directly or indirectly, most ordinary needs. The professional limnologist will probably insist upon standards certified by the U.S. Bureau of Standards or by similar agencies. If such certified standards cannot be provided, it will be necessary to select a certain item of equipment as the future standard, have it compared under known conditions with already certified standards, and have it corrected if necessary. Every well-equipped physics or chemistry laboratory will possess certified standards with which comparisons can be made. The first standards which are likely to be required are the following:

LINEAR STANDARDS. Standards for testing and correcting equipment used in making linear measurements should be of at least two general sorts, namely, macroscopic and microscopic. A *macroscopic standard* should be not less than 10 cm. long and have a minimum graduation of 1 mm. The convenience of such a standard is increased if it also possesses graduations in feet, inches, and fractions of inches (minimum graduation of 1/32 in.) since the English system is frequently used. A *microscopic standard* for general limnological work should have a scale about 2 mm. in length graduated to 0.01 mm.

When necessary the advice of an official bureau of standards can be sought in the selection of articles suitable for use as standards.

If certified standards cannot be provided and if "compared with certified standard" articles are used, some appropriate modification of the following suggestions must be adopted. For *macroscopic standards*, select some good quality steel, brass, or boxwood ruler, not less than 10 cm. long and graduated to 1 mm. (also 1/32 in.); have it compared critically with some certified standard by an experienced person; note carefully the physical conditions, particularly temperature, under which the comparison was conducted and make permanent records of the same; record the corrections necessary to make the selected ruler accurate. Label ruler so that it is clearly and permanently distinguishable from all other similar objects, take it out of general use, and retain exclusively as a future reference standard. A carefully selected, well-made meter stick, when so compared with standard, is a very convenient item of limnological equipment.

For a *microscopic standard* suitable for general limnological work, select a well-made microscope stage micrometer of the ordinary type having a total length of scale of 2 mm. and graduated to 0.01 mm.; have it compared with an already certified standard in a fashion similar to that just described for a macroscopic standard; make the necessary records of corrections; mark it distinctively; and reserve it as a reference standard. Similar rulings on a metal plate if available may also be used.

Standard Weights. At least one set of weights, with 1 mg. as the smallest value, should be certified by a standardizing agency. However, if this cannot be done, a set of weights of the type just mentioned should be compared on a sensitive balance with weights of certified accuracy and the necessary corrections determined, the results entered on a permanent record, the set marked in a distinctive way, and reserved as a standard.

Volume Standards. Standards for volume may differ in form with various limnological needs. For general purposes, three articles are convenient; viz., a serological pipet graduated to 0.01 cc. (or ml.), a 50-cc. or 100-cc. buret graduated to 0.1 cc., and a volumetric flask of 1000-cc. capacity. They should be compared with certified standards under known conditions, marked distinctively, and removed from general laboratory use.

Temperature Standards. Since instruments for measuring temperature are so constantly used in limnological work, provision for correction is a necessity. Therefore, it is convenient to have an instrument designed for subsurface temperatures which either has been certified by some bureau of standards or has been very carefully compared with an already certified instrument. Further comments on the correction of temperature-recording apparatus will appear later in the chapter dealing with temperature.

General Considerations. Correction tables of convenient form and detail should be prepared and entered in a set of permanent records for future reference. Make certain that such correction tables include, among other things, the following information: date; name of person making corrections; method of making corrections; standard used; identification number and name of instrument corrected; and physical conditions under which corrections were made.

Special directions for correcting or calibrating equipment will be given later in the various sections on individual methods.

Marking of Equipment

Each piece of equipment should bear an individual identification number or a symbol of some sort, stamped on or otherwise associated with the item in as permanent a way as possible. It should be entered on some prominent part of the equipment at the time it is first secured. Such identification characters are the surest means of relating a piece of equipment with certainty to records, correction tables, or other data. In addition they are useful in the maintenance of inventories. A consistent system of identification characters should be formulated in advance in order that it may be extended at any time to new equipment as it is acquired. Some items will bear a maker's number or symbol which serves as a means of identification and may be so used by the operator. However, it is likely that he will find it more convenient to establish his own identification characters. Whatever the system employed, it must also provide adequate means for marking all *duplicate* items of equipment so that each individual piece is definitely designated. If identification characters can be carried only on appended tags, particular care must be taken that their attachment is as permanent as possible and that in case of loss replacement is made at once. Obviously, identification characters must be on the equipment itself, not on the case or kit in which it is stored.

Reading of Instruments

The chance of error lurks in the operation of reading all kinds of instruments. Familiarity and practice with any instrument tend to reduce this hazard but never eliminate it completely. Therefore, any means which offer some safeguard are deserving of serious consideration. Those which clearly present some insurance against errors should be made a regular part of the procedures in every method. Such safeguards will vary in their detail with different methods and instruments and no complete listing will be attempted here. However, the following means have general value.

1. If the operator is working alone and consequently keeping his own records, he should proceed as follows: Make reading on instrument *aloud;* enter reading on record sheet; reinspect instrument for a second reading and verify first record.

2. If operator has the aid of another person who is acting as recorder, the procedure should be as follows: instrumentman announces reading clearly, with an extra margin of loudness, and then *glances away from the instrument;* recorder repeats distinctly the announced value and enters it in record. Instrumentman again directs his eyes to instrument and announces second reading; if error was made in first reading, instrumentman should call "correction" and then state revised reading, and recorder should in turn announce "correction" followed by statement of new value. If no error occurs in original reading, instrumentman again announces second reading, and recorder responds with the word "check." These procedures should be adhered to rigidly until they become a fixed habit. Many an error either of instrument reading, of recording, or both, will be caught in this way.

3. Avoid haste both in reading instruments and in making records.

Even under pressure of unusual circumstances, resist the temptation to make hasty records or instrument readings.

4. If possible, avoid making readings on instruments in inadequate light.

5. When making readings on finely divided graduations, use a lens of adequate form and magnification.

6. Avoid use of instruments on which graduation marks have become obscure. In most instances the distinctness of graduation can be restored by means well known in any laboratory.

Recording of Data

It is a self-evident fact that the value and usableness of any piece of work depends upon the accuracy, completeness, and clarity of the records. Therefore, the adequate recording of data is of preëminent importance and every care must be exercised to meet its exacting requirements. Good habits of record making, acquired by meticulous care and repetition, eventually become fixed. The following suggestions will probably be found applicable in most instances:

1. Whenever possible, use detailed *tabular forms,* preferably printed or duplicated. Each method may require its own particular set of items in such a form, but the end will justify the expense and effort. Forms constitute one of the best means of insuring completeness, brevity, and orderly arrangement of data.

2. Do not use loose pieces of memorandum paper; enter all records and all information on regular, safeguarded sheets of the record book.

3. Draw one diagonal line through items thought to be in error; do not erase, and do not pull out and throw away any material considered to be inaccurate.

4. Enter all field records with a good-quality pencil; avoid using ink because of danger of wetting.

5. Enter all information in clear, concise, direct language; avoid use of extra words which add nothing essential to the records.

6. Record all data at the moment when they are secured. Never make records from memory.

7. Legibility of all records is a prime necessity.

8. Make all computations on regular sheets of the record book and preserve them.

9. Original records must be carefully preserved, even though for some special reason copies have been made. The original records are always the court of last appeal. Errors in copying are notoriously common and correction can be made only by reference to the originals.

10. Records must always contain the identification numbers or symbols of all instruments used.

Care of Equipment

Since limnological work requires considerable quantities of diverse equipment, the proper care of instruments and other properties must always be a serious consideration.

The operator must always assume full responsibility for the proper use and the safety of all equipment. Damage, breakage, loss of parts, misplacement, or need of reconditioning should be reported immediately to those in charge of the properties involved. When equipment is secured from a stock room or other source, it should be inspected and tested for good condition before the operator accepts it.

The best insurance that equipment will be ready at any time to yield dependable results is *critical and consistent care*. All kinds of apparatus suffer from neglect and misuse. Every operator should develop a respect for, and a pride in, the instruments which serve as his tools.

A very troublesome source of danger to equipment arises from the attempts of careless or ignorant operators to make instruments function in ways for which they were never intended. An experienced operator may be able to modify an instrument or method so that it can safely function outside its ordinary range, but as a rule such diversion of equipment into other uses should be rigidly avoided.

The details of proper care of equipment are often governed by the dictates of good sense and experience. Therefore it is not practicable or necessary here to attempt a complete catalog of general rules on care of apparatus. The following partial lists may be helpful to less experienced operators.

CARE OF FIELD EQUIPMENT

1. With the exception of rough, general-utility articles, all equipment should be provided with specially designed, sturdy field cases or kits, otherwise there is the constant hazard of damage from accidents. Well-built lightweight wooden cases thoroughly painted or otherwise waterproofed both inside and outside are recommended.

2. When an instrument is removed from or returned to its case, it should be handled carefully. In this operation, a little patience and care do much to avoid damage.

3. When on boats, all separate items of equipment should be tied to some fixed object to avoid loss overboard. Equipment lost in deep water is seldom recovered. Submergence in water may damage seriously certain kinds of materials or instruments.

4. Detachable parts of instruments must be fastened securely into place, particularly on apparatus which is to be lowered into water.

5. Small articles of equipment, or parts of equipment, are easily lost in

water, sand, mud, trash along shore, or among vegetation. A special container with subdivided compartments should be provided for such articles so that the absence of any item is detected at once.

6. Protection from rain must be provided for most apparatus. Many instruments are not watertight in construction. Some items of equipment are perishable in rain. Waterproof covers should always be at hand. Ponchos are good for general or emergency protection. A small tent which can be suspended from the limb of a tree affords a quick shelter against a sudden shower.

7. Equipment should never be left unguarded in exposed places.

8. Excessive exposure to direct sunlight is detrimental to certain kinds of equipment, such as silk nets, certain colorimeters, etc.

9. Apparatus composed wholly, or in part, of glass must have special protection in the field. Under the best of care such items are subject to accident and, when possible, extra parts should be available.

10. When it is necessary to transport liquids in the field, they must be carefully stoppered and kept in a specially made kit. Under no circumstance should they be packed into kits containing instruments or other equipment.

11. Extra parts, carried for emergency replacement, must be provided with adequate protection, preferably in a compartment in the same kit which houses the equipment concerned.

12. Lines, not wound on drums, should be coiled and carried in field kits when not in immediate use. They are thus protected against accident, are kept from underfoot, and when so coiled are less likely to tangle when put into use again.

13. Before leaving shore, it should be made certain that one end of the anchor rope is firmly attached to the boat and the other end to the anchor; also that extra anchors are safely fastened in their proper storage places.

14. The inside of field kits should be dried out from time to time. Small holes drilled in the bottoms allow water to drain out but do not provide for complete drying.

15. In so far as it is possible, all apparatus should be cleaned carefully before leaving the field. Means for such cleaning should always be a part of the field equipment.

16. Care in the transportation of equipment cannot be overstressed. Kits should be provided with strong handles and with locking devices. A satisfactory locking mechanism is composed of a common type of hasp and staple held into locked position with a strong harness snap. When transported in trucks or automobiles, equipment must be protected against undue vibration, shock, or chafing. Instruments of delicate construction or adjustment should be carried on a seat or, perhaps preferably, on the operator's lap. When shipped by freight, express, or mail, double and

triple packing in a series of cartons one within the other, each separated by a generous bed of shock-absorbing material, is necessary. When transported on boats, all apparatus should be kept in cases or kits to avoid accidents resulting from rough water. Miscellaneous small accessory items must be kept in canvas carrying bags or other containers.

17. Before leaving the field, a check list of all equipment should be used to insure that everything is accounted for.

CARE OF LABORATORY EQUIPMENT

Rules for the proper care of laboratory apparatus in general apply equally well to limnological equipment and need not be enumerated here. The operation of such procedures rests largely upon the exercise of good judgment.

STORAGE OF EQUIPMENT

When equipment is to be stored for a long time, some items should receive special care. The following suggestions cover some of the ordinary needs:

1. Metal equipment subject to rusting should be covered with a thin coat of oil, grease, or petroleum jelly, care being taken to insure that it extends to joints, hinges, seams, nuts, holes, and other less accessible areas. In some instances, a coat of paint, shellac, or similar protective cover may be advantageous. Wire lines and cables should be dried and then oiled thoroughly; if stored on a reel or drum, the whole coil should be wrapped in oiled cloths and protected against rain and other sources of moisture.
2. Long storage of certain kinds of equipment, such as silk bolting cloth, colored signal cloth, liquid colorimeters, indicator solutions, must be in very subdued light or even darkness.
3. Some kinds of equipment, such as rubber articles, suffer from high temperatures. Storage in attics should be avoided. As a general rule, cool storage is recommended.
4. Metal instruments and chemicals must not be stored in the same containers or compartments.
5. All liquids containing large amounts of water must be protected against freezing.
6. Rubber valves in samplers should be unseated and stored in such a way that they are not under pressure.
7. Wooden staffs, such as meter sticks, sounding poles, and similar articles, must be stored flat on some straight dry support.
8. Nonmetallic lines not on drums should be coiled and hung from hooks in a dry room.
9. Maps may be stored flat in wide cabinets, or rolled, or hung from

hooks if mounted on marginal strips. Storage should be in subdued light or in darkness.

10. Graduated scales on metal or glass instruments may be cleaned by wiping them gently with chamois skin or a soft rag moistened with weak ammonia. They should not be rubbed with cloth, paper, or a rubber eraser since such treatment may damage the graduation markings.

Part I

Hydrographic Mapping and Morphometry

Accurate, detailed hydrographic maps are basic requirements. No body of water can receive more than a casual study in the absence of such a map. In general, only the large, navigable lakes, a limited number of smaller ones and certain streams have been charted in such a way that their maps can be turned to limnological purposes. Federal and state bureaus, engineers, limnologists, and others (see Appendix, p. 356) have charted various lakes and streams in an adequate fashion, but for the vast majority of inland waters, maps suitable for limnological work are still lacking.

The limnologist will search out any already existing maps of the waters with which he is to deal, but he must use the greatest care in choosing any of them as the basis for his work. For him most maps in ordinary geographic sources are worthless. Even if he is fortunate enough to find a hydrographic map constructed on a large scale, he must investigate its history, authorship, method of construction, degree of precision, and other facts salient to a decision as to its dependability. In case of doubt, the cautious worker will remap the waters with which he is to deal. The limnologist is often faced with the necessity of making his own maps. Therefore he must be familiar with dependable mapping methods and be equipped for such work. Knowing the special needs of his work, he can often make maps which are superior for his purposes. The employment of professional surveyors is costly and often impracticable.

Training in the essentials of surveying is much to be desired, but a certain skill in mapping lakes and streams can be secured without going through all of the technicalities of professional surveying. However, no mapping for scientific purposes should be undertaken until there has been sufficient advance experience in (1) construction and operation of instruments, (2) operation of the method in the field, (3) proper checks on field operations, and (4) drafting operations suitable to final charting of field data.

Methods described in this book range from very simple ones usable only on ponds or small streams to more complicated ones suitable for larger waters. In the selection and presentation of these methods the following facts have been kept in mind: (1) different waters vary widely in their mapping requirements, (2) surveying and map-making equipment may be limited in some limnological laboratories, and (3) many students of limnology have had little or no training in formal surveying. Throughout the descriptions the use of technical surveying terms has been avoided when possible. Training in elementary geometry and trigonometry is presupposed.

In the hands of individuals trained in their proper use, or when used under the direct supervision of an experienced person, standard surveying instruments are the most satisfactory equipment for hydrographic mapping. However, some mapping operations do not necessarily demand such elaborate and expensive equipment. This is particularly true of small waters, such as ponds, lagoons, beach pools, and even small lakes having areas of only a few acres. When properly used, simpler devices will sometimes give as accurate results as more elaborate instruments. Certain improvised equipment, described in Chapter 2 (p. 47), may provide a way for work to be pursued, perhaps on a necessarily restricted scale, by those who do not have expensive instruments or whose previous training does not permit the use of complicated standard apparatus. Companies which deal in surveying materials supply simpler instruments, such as the nontelescopic devices. The limitations of all simplified and improvised instruments must be fully understood and all operations guided accordingly.

The user of this book should understand that mapping can be accomplished in many ways other than those described here. Familiarity with the principles and major problems of mapping will go far in enabling him to adapt methods to his particular needs; also to utilize other methods should they seem desirable.

While accuracy must always occupy first place of importance in all mapping procedures, it should be pointed out that, in hydrographic surveying, the high precision required in certain surveying operations on land is not possible. For example, the location of a shore line is often a matter of arbitrary judgment operating through a range of a few feet to perhaps several rods. Also changes in surface elevation of a lake or stream may cause extensive shifts in the position of the water line. Thus the location of a shore-line position to a precision closer than a few feet is frequently of no consequence. Sounding positions should be determined carefully but it seldom happens that any need exists for locating them with more exactitude than within a few feet. Final construction of maps on the usual reduced scales absorbs, in widths of lines representing shore line and submerged contours and in space occupied by numbers representing soundings, much minute detail secured by unnecessary expenditure of time and effort.

SHORE-LINE SURVEYS OF LAKES

From the numerous methods of mapping shore lines, one must be chosen which is suitable for the requirements of the particular lake to be surveyed. Size, physical character of shore line, shape, surrounding vegetation, and other features make some methods more practicable than others. Those outlined here are selected because they meet the needs of smaller inland waters in a satisfactory way without being too complicated and they differ sufficiently to allow for the proper choice of a method adapted to the requirements of different waters.

Surveys of shore lines during periods of ice cover are sometimes easier and less time consuming than in the open season. This is particularly true of surveys which deal with lakes whose margins are boggy, swampy, very rocky, heavily wooded to the water edge, exceedingly precipitous, or are otherwise difficult to survey in the usual ways.

Modifications or combinations of the methods described in this chapter may be made whenever special conditions, convenience, or better results seem to warrant.

Traverse Survey

A series of connected straight lines whose lengths and angles have been determined by appropriate methods and instruments is known as a *traverse.* This method employs a transit and a steel tape as the major items of equipment. It presupposes sufficient understanding of the construction and operation of a transit to permit its use in this kind of survey. If such knowledge is lacking, preliminary study and practice is imperative.

Suitability. A traverse may be made of a lake of any size if certain obstacles are overcome, but on the whole it is most satisfactory for larger lakes whose shore lines are several miles long. For surveying small lakes, ponds, and similar waters other methods are likely to be found more convenient.

Equipment. Transit; reading glass; 100-ft. steel tape; ax; 11 chain pins, each with colored cloth tied into upper end, on carrying ring; 1 or more pieces of 1-in. gas pipe, 2 ft. long; wooden stakes; 2 or 3 range poles; form for records.

corrections and adjustments. Examine transit for proper adjustments (see a standard text or manual on surveying for directions). Test steel tape for general accuracy, extreme precision is not required.

ORGANIZATION OF PARTY. Chief of party; experienced transitman; two tapemen; two flagmen; axman; general assistant.

If the party must be reduced, the chief of the party can act as transitman, and the general assistant may be omitted. The chief of the party must (a) be experienced in all aspects of the survey; (b) have general direction of survey; (c) make preliminary examination of lake in order to anticipate survey requirements; and (d) check operations of transitman when necessary.

PROCEDURE

1. Choose convenient initial transit station (Fig. 1, *B*) on selected position of survey (bank, mid-beach, or water line); drive iron stake at this point and reference it to at least two or three permanent objects (large trees, buildings, road intersections, surveyor's monuments, etc.) by connecting lines whose lengths are measured and recorded.

2. Clockwise along shore from *B*, establish station *C* at the end of as long a straight line connecting *B* and *C* as shore-line configuration will permit without this line's departing in general from selected level of survey. Drive wooden stake at that point leaving 4-in. projection; mark stake

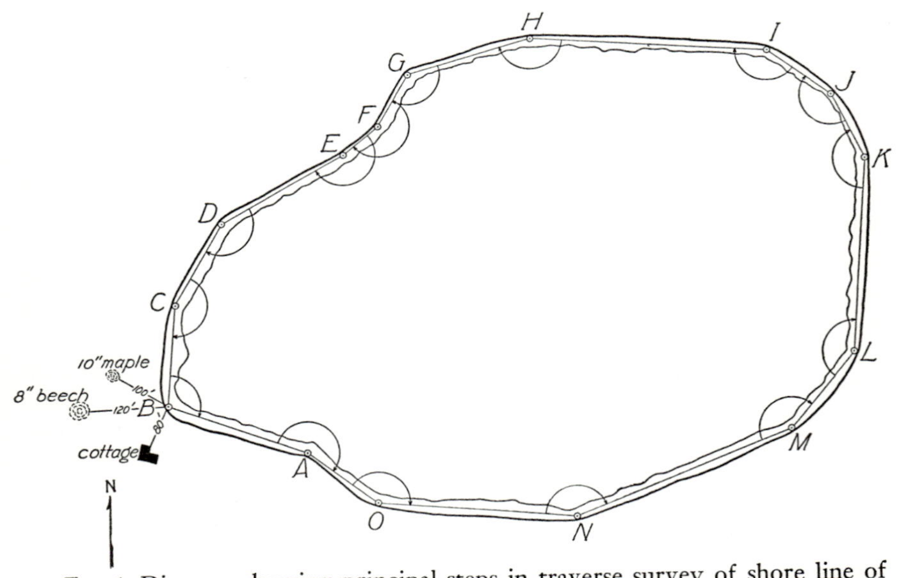

FIG. 1. Diagram showing principal steps in traverse survey of shore line of lake. (*A–O*) Series of transit stations. Curved arrows indicate interior angles. Fixed objects near *B* (cottage, trees) are referencing points for locating *B* by means of distances indicated in feet. Heavy bounding line of figure represents the bank. Lighter wavy line is water edge. For purposes of this diagram, traverse is shown as run on beach slightly below bank.

with station designation *C;* drive nail or tack in center of top of stake. Similarly establish station *A* on other side of *B* and mark it with proper designation.

3. Set up transit over center of *B;* direct telescope on range pole held by advance signalman on tack at *C*. Determine and record magnetic bearing of line *BC* for orientation purposes, using all precautions against local attraction. Then set zero point of lower, movable scale on transit at zero point on vernier. Measure distance *BC*, tapemen being kept in line by transitman if distance exceeds one length of 100-ft. tape. Keep tape level during measurements; check record of total whole-tape lengths with number of chain pins used. Then rotate telescope clockwise and sight on range rod held by back signalman on tack at station *A*. Determine angle *CBA* from transit reading, using *interior angle* as shown in Fig. 1. Check angle by determining *exterior angle* and take sum of both which should equal 360 degrees. Measure distance *BA* as described above; determine and record magnetic bearing of line *BA* for orientation purposes. Turn off compass needle.

4. Move transit and set up over station *C*, centering on tack in stake. Take foresight on range pole set on station *D* which has been selected in same fashion as *C*. Set transit scale on zero vernier mark. Measure distance *CD*. Rotate telescope clockwise and direct on signal set on *B;* determine angle *DCB* from transit reading, using interior angle. Check angle as described above.

5. Determine other angles and distances in same way, as transit is moved successively from *C* to *D*, *D* to *E*, *E* to *F*, and so on to the end of survey, the last foresight being from *A* to *B*.

6. During survey keep series of field notes on such physical features of the shore line between stations as will aid in subsequent mapping procedures.

7. On completion of survey, check accuracy of same by formula, $(N\text{-}2) \times 180^\circ =$ *sum of all interior angles*, in which *N* represents the number of interior angles measured; difference between this value and sum of all measured angles is the *error of closure*. In a traverse in which many angles have been measured, the usual practice is to expect that a closing error should not exceed one half the least count of the vernier times the square root of the number of angles; however, a somewhat larger error may be permissible in lake traverses because of the nature and irregularities of their shores.

8. Provide for mapping check on position of limited number of transit stations by back sights on two or more previously occupied stations so located that angles are not small. For example, in Fig. 1, with transit at *K* determine angle *EKL* and with transit at *L* determine angle *KLE;* with all angles and one side known, distances *KE* and *LE* can be computed and

laid out on map. Choose other significant checks in a similar manner from other transit stations.

FIELD RECORDS. Enter field records in some previously prepared form so constructed that no essential data will be overlooked. Such a form, including a sample record, is presented below (Form 1). A field sketch, prepared as survey progresses and containing all pertinent information, is often very useful.

General Considerations

a. In lakes where beaches are firm but difficult to traverse because of dense marginal vegetation or other obstacles on the bank, the survey can be facilitated if done in autumn when fall in lake-surface elevation may have increased the width of exposed beach. Similarly, such a traverse may be greatly expedited if done on ice cover.

b. In some transits, the horizontal graduated scale is numbered continuously from 0° to 360°; in others there are two sets of figures which increase from 0° to 360° but in opposite directions. Such instruments are convenient for traverse work. However, still other transits have a scale divided into quadrants with two zero points diametrically opposite each other and the graduation numbers increasing each way from 0° up to 90°. In such instruments the transitman must be particularly careful in making his records when the angles extend beyond the limits of one quadrant and into another, both in making the correct summation of the total angle and in reading the vernier.

c. When the survey is completed, all stakes representing transit stations and other positions should be left in place for future reference. To insure that stakes remain in the original positions they should be driven completely into the ground and their locations concealed. The approximate position of such concealed stakes can be determined later by means of the field data and the stakes themselves found by scalping the ground with a broad flat spade.

d. A stadia rod (p. 9) may be substituted for a steel tape in measuring distance if the transit eyepiece is provided with stadia wires and if proper precautions are observed.

e. Ordinarily, measurements of distance between stakes to the nearest 0.5 ft. will be sufficiently accurate.

PLOTTING TRAVERSE SURVEY

1. From estimates, rough sketches, or other sources of information concerning dimensions of lake, determine a *scale* which will yield a map of desired size. Select paper of sufficient size to hold map with extra space on all sides; trim paper so that all sides are straight and at right angles to adjacent sides. Attach paper securely to smooth table by means of pieces of Scotch cellulose tape or other adhesive.

FORM 1

FIELD RECORDS

Traverse Survey of Shore Line

Lake	Transitman	Monuments and Referencing Data
Date	Recorder	
Instruments	Tapemen	
	Signalmen	

Transit Station	*Foresight*		*Backsight*		*Interior Angle*	*Distance*		*Check*	
	Line	*Magnetic Bearing*	*Line*	*Magnetic Bearing*		*Line*	*Length (ft.)*	*Shore Stations*	*Angles*
B	*B–C*	*N* 11° 00′ *E*	*B–A*	*S* 62° 30′ *E*	106° 30′	*B–C*	150		

2. At left side, about 1 in. from edge, draw straight line parallel to left edge and about 6 in. long; attach arrowhead figure to upper end and mark with letter *N* to represent north end of north-south line.

3. Select appropriate position on the paper for first transit station (Fig. 1, *B*) and mark it by pencil dot, surround with small penciled circle, and label it *B*. Through *B* carefully draw line parallel to north-south line mentioned in paragraph 2 above; prolong line to length sufficient for accurate use of protractor.

4. With *B* as center, draw line *AB* in accordance with magnetic bearing taken in field (see p. 5), converting magnetic bearing to true bearing if correction is great enough to be significant (see table or map of magnetic declination in any manual on surveying). Scale distance between *B* and *A* and mark point *A* by circled dot. Then protract (see p. 67 for directions for testing protractors) interior angle *CBA* and draw line *BC;* scale distance *BC*, mark point with circled pencil dot, and label it *C*.

5. Move protractor to *C* and protract interior angle *DCB;* draw line from *C* at proper angle and scale distance *CD*, marking point with circled pencil dot. Label it *D*. Proceed in this fashion with succeeding interior angles until entire shore line is covered. Magnetic bearings of the lines can be used as checks; also in triangles laid out as checks (p. 5), plot, compute, and scale connecting lines described in paragraph 8, p. 5.

Another procedure may be followed; namely, plot approximately one half of traverse clockwise from starting position and the other half counterclockwise, thus providing more protection against accumulation of small errors of plotting.

6. Connect all shore-line positions in correct sequence, as indicated by original designations, by continuous heavy line; draw freehand and with rounding configuration. In drawing shore line between instrument positions, make liberal use of field notes (p. 6) for such information as will help in making more accurate the form of sketched-in portions.

7. When shore line is complete and all doubtful points are checked, erase extraneous lines, marks, and other temporary construction pencilings, leaving only shore line, instrument positions and their designations, and the compass figure.

8. In some appropriate available area outside shore-line figure, enter the information listed on pp. 68–69.

9. Establish, in some suitable space, the permanent north-south compass figure. Make the figure simple in construction, modest in size, and dignified in appearance; avoid ornate patterns.

Stadia Survey

A stadia survey depends upon the measurement of all required distances by means of a stadia rod and a transit or other instrument equipped with stadia wires. Distance is determined by a measure of the vertical space intercepted on a stadia rod between the stadia wires. Commonly stadia wires are so installed in instruments and stadia rods are so graduated that if stadia wires intercept a vertical distance of 1 ft. on the rod, the rod is 100 ft. away (plus the *stadia constant* which is approximately 1 ft. for most transits); if the stadia wires intercept 5 ft. of vertical space on the rod, the rod is 500 ft. distant plus the stadia constant, and so on. Stadia measurements are rapid and very satisfactory if high precision is not demanded and if made by an experienced person. Accuracy in stadia records depends to a large extent upon the instrument and the rod used and upon atmospheric conditions. Distances within 500 ft. can be read with considerable precision, i.e., easily to within a fraction of a foot, and distances approaching 1000 ft. can be read within an accuracy of 10 ft. Since errors of 1 to 5 ft. are often of no consequence in shore-line surveys, readings can be made, satisfactory for this purpose, up to about 700 or 800 ft., and in some instances to 1000 ft. without introducing errors of real significance. In shore-line surveys, the stadia constant can usually be ignored.

For shore-line surveys, stadia measurements may be substituted in the traverse survey for the tape measurements, provided the necessary distance limitations are observed.

Suitability. A stadia shore-line survey of the type described here is best adapted to small lakes whose transverse dimension does not exceed about 1000 ft. and whose long dimension does not require numerous changes in position of the instrument containing the stadia wires.

Equipment. Transit, or similar instrument with stadia wires; stadia rod, with graduations in feet and tenths of foot; ax; 2 iron stakes, 2 ft. long; wooden stakes; forms for records.

Corrections and adjustments. Examine transit or other instrument for proper condition (see surveying manual for directions). Be sure that the stadia wires are horizontal and intercept the vertical distances claimed for them. Test stadia rod for accuracy and uniformity of graduations. Graduations should be bright, have high visibility and have pattern suitable for the task.

Organization of Party. Instrumentman; rodman; assistant.

PROCEDURE

1. Make preliminary reconnaissance of lake in order to anticipate needs of survey. Among other things, make estimates of maximum distances, examine extent of irregularities of shore line, and seek appropriate location or locations for transit station.

2. Choose first transit station (Fig. 2, *A*); drive iron stake flush with ground at this point and reference it to adjacent permanent objects; set up transit over stake.

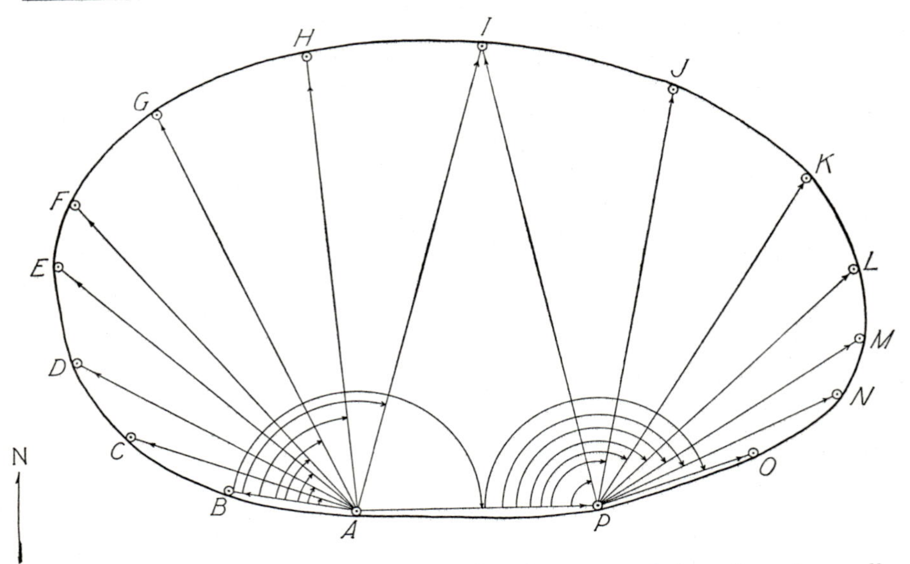

Fig. 2. Diagram illustrating one type of stadia survey of shore line of a small lake. (*B–O*) Different shore-line stations on which stadia rod is erected and to which distances from transit either at *A* or *P* are measured. (*A*) First transit station. (*P*) Second transit station. Curved lines with arrowheads indicate angles measured.

3. Proceeding clockwise along shore from *A*, on chosen level of survey (bank, mid-beach, or water line), select first stadia-rod station *B*, whose distance from *A* depends upon configuration of shore line. (Stadia-rod stations may be much farther apart where shore line is approximately straight or curvature very slight; the greater the curvature the closer must be the stations.) Drive wooden stake at that point, leaving 4-in. projection; mark stake with location number or letter. Erect stadia rod on top of stake, holding it vertical and as steady as possible; direct telescope on *B* and set zero point of lower transit scale at zero point on vernier; take stadia-wire reading on rod to determine distance *AB*. Record with great care magnetic bearing of line *AB* for orientation purposes; guard against local attraction. Rodman must keep stadia rod in position until released by down signal from transitman.

4. Progressing clockwise along shore, select next station *C* appropriate to shore configuration. Drive wooden stake as before and designate; erect rod; direct telescope on rod; make stadia-wire reading to measure distance *AC;* determine angle *BAC;* record magnetic bearing of line *AC* as check.

5. Locate similarly other stations and measure angles and distances until survey from initial transit station *A* is completed. If maximum distances

exceed 800–1000 ft., reset transit as follows: Establish suitable position for new transit station *P;* drive iron stake at that point and erect stadia rod on it; with transit still in initial position *A*, take stadia rod reading for distance *AP;* determine angle *BAP;* record magnetic bearing of line *AP* for check. Move transit to new position at *P;* sight on *A* and set zero point on transit scale at zero on vernier; make stadia reading on *I* and record angle *API* for check; record magnetic bearing of *PA* for confirmation. Resume series of shore-line stations (Fig. 2, *J*) and conclude work of survey as indicated above. If additional set-ups of transit are required, as might occur in a lake of irregular outline, locate new transit positions in same manner as described for location of *P*.

6. Provide data for checking shore-line positions by choice of procedures most appropriate to circumstances of survey. For example, transfer transit to station *D* (Fig. 2) and secure angle and distance data on several of previously used stadia-rod stations; then transfer transit to station *M* and secure similar information for several other stations within working distance.

FIELD RECORDS. Enter field records in some previously prepared form so constructed that no essential data will be overlooked. Such a form, including one sample record, is given below (Form 2).

FORM 2

FIELD RECORDS

Stadia Survey of Shore Line

Lake	Instrumentman	Monuments and Referencing Data
Date	Rodman	
Instruments	Assistant	

Line of Sight	*Magnetic Bearing*	*Distance (ft.)*	*Angle with Zero Line*	*Remarks*
AC	*N* 59° 30′ *W*	236	9° 30′	

General Considerations

a. The rodman should keep field notes descriptive of the details of shore-line configuration between stadia-rod stations. Freehand sketches accompanied by appropriate labels are often preferable to descriptive notes.

b. Ordinarily no attempt should be made to measure distances closer than to the nearest foot.

PLOTTING STADIA SURVEY

1. Adopt scale which will yield map of desired size. Select paper of sufficient dimensions to include map with extra space on all sides; trim paper so that all sides are straight and form right angles to adjacent sides. Attach firmly to table with Scotch cellulose tape or some similar adhesive.

2. At left side, about 1 in. from edge of paper, draw straight line parallel to left edge and about 6 in. long; attach arrowhead figure to upper extremity and mark with letter *N* to represent north end of north-south line.

3. Select appropriate position for first transit station (Fig. 2, *A*) and mark by pencil dot surrounded by small penciled circle; label it *A*. Through *A* carefully draw line parallel to north-south line mentioned in preceding paragraph, prolonging line to length sufficient for effective use of protractor. Using corrected magnetic bearing of *AB* and distance *A* to *B*, draw line *AB* to scale; mark *B* with pencil dot surrounded with small penciled circle and label it outside of developing figure.

4. Put center point of protractor on *A* and pivot until zero line of protractor coincides exactly with line *AB* on paper. Then with angle and distance from field records locate station *C;* mark it with pencil dot surrounded by small penciled circle and label it *C*. In similar fashion locate, mark, and label succeeding stations to conclusion of shore line in case one instrument set-up is sufficient. If, however, another instrument station must be used, proceed as follows: With protractor located as described above, measure angle *BAP;* draw line *AP* and scale distance *AP* thus locating *P*. Move protractor to *P*, center on *P*, and pivot until protractor zero line coincides with *AP;* protract angle *API* and scale distance *PI* for check on previous location of *I* from transit station *A*. Then proceed to protract angle *APJ* and scale distance *PJ*, locating, marking, and labeling as already described; continue in this way with succeeding stations until shore line is concluded. Since all angles are measured from line *AB*, a mistake in its location will be extended to all subsequent lines.

5. Use checks provided by field survey. Then finish map as outlined in paragraphs 6–9 (p. 8) in section on plotting traverse shore-line survey.

Survey by Location of Points from Two Simultaneously Measured Angles

This method of surveying a shore line depends upon the location of properly chosen series of successive points by means of two angles measured simultaneously. Two transits (or other instruments which will adequately measure horizontal angles) are required. A transit is located at each end of a base line whose exact length and magnetic bearing, or other form of orientation, is known. From these two positions the transits, properly set with respect to each other, can be directed simultaneously on one selected point on the shore line and the significant angles measured. From this information the point can be mapped either by direct protraction of the angles or by the use of computed lengths of the sides of the triangle involved. Then other points are similarly located in the proper succession.

Suitability. This method is likely to be found most useful on relatively small lakes with fairly regular shore lines and with margins so difficult (e.g., floating bog margins, swamps, dense vegetation, steep cliffs) that the various shore-line stations must be taken by boat. It also has the advantage of enabling the party to proceed at once with soundings without any change of transit positions.

Equipment. 2 transits, or other instruments measuring horizontal angles; 2 or more 2-ft. iron stakes; wooden stakes; 100-ft. steel tape; signal set, containing 1 white and 1 red flag, each on staff; forms for records; ax; chain pins.

Corrections and Adjustments. Examine transits for proper condition (see surveying manual for directions). Check steel tape for accuracy and good condition.

Organization of Party. Two transitmen; signal-rodman; assistant.

PROCEDURE

1. Make preliminary inspection of lake to determine plan of survey; examine location and form of peninsulas, bays, and other shore-line details; inspect for most favorable position of base line and transit stations.

2. Choose transit stations (Fig. 3, *A* and *K*) so that, if possible, all positions on shore line are visible from each transit; if this is not possible, then choose transit stations favorable for survey of one portion of lake. Choose transit stations in such way that all pairs of lines prolonged simultaneously from transits will form, on intersection, no very small angles. It is preferable to keep all angles greater than 30° and less than 120°. (Note in Fig. 3 the large-angle advantage of stations *A* and *K* as contrasted, for example, to result of using *B* and *S* as transit stations.) Drive iron stakes at *A* and at *K* and reference both.

3. Set up transits on *A* and *K;* carefully center intersection of cross hairs of transit *A* on iron stake at *K* if visible; if not visible, center on vertical axis of transit *K*; set lower horizontal scale of transit so that zero point corresponds to zero point on vernier. Direct transit at *K* on transit at *A*, and set lower scale as just described for *A*.

4. Line connecting *A* and *K* constitutes a base line whose exact length and magnetic bearing must be determined. If base line falls entirely along

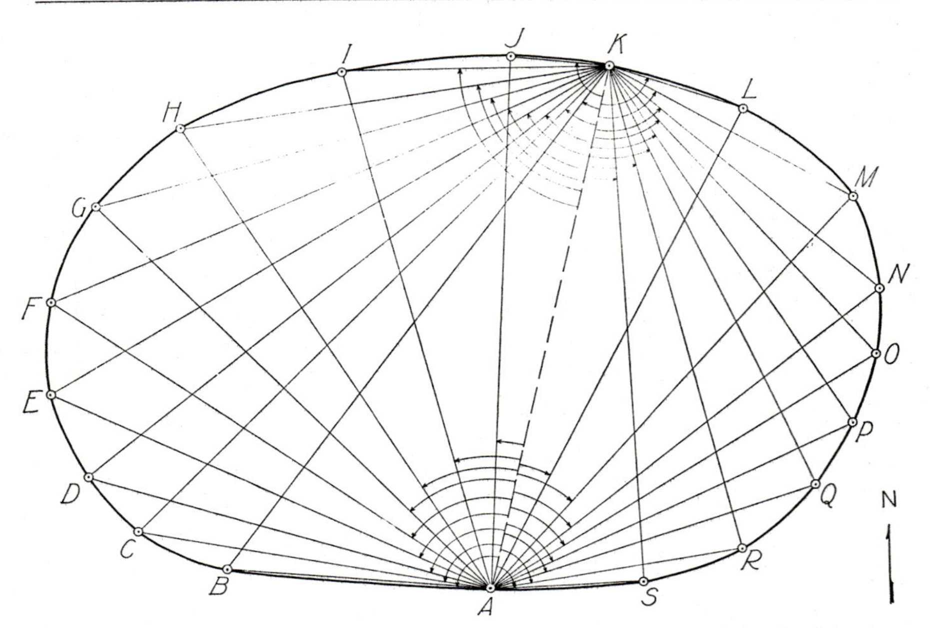

FIG. 3. Diagram illustrating location of positions on shore line of a lake by method of simultaneous measurement of two angles and intersection of lines. (*A*, *K*) Transit stations. (*AK*) Base line of known bearing and length. (*B–S*) Shore-line stations. Curved lines with arrowheads indicate angles measured by transits. Radiating straight lines indicate lines of sight from one transit extended to intersections with corresponding lines of sight from other transit.

shore line, it may be measured directly with tape; if, however, it extends across water it can be measured indirectly as follows: Extend a line from *A* along shore, or at some other position on land, providing a large angle with line *AK;* terminate line at considerable distance from *A* and drive stake. If line can follow shore line, a future signal station may be used, as for example, *B* in Fig. 3. Raise signal on *B;* direct transit *A* on *B* and measure angle *BAK;* also measure carefully distance *AB;* take magnetic bearing of line *AB* for check. Simultaneously, direct transit *K* on *B* and read angle *BKA*. Take magnetic bearing on line *KB* for check. With all angles and one side of triangle *ABK* known, compute length of line *AK* by appropriate trigonometric formula. Use extreme caution in determining length of base line; a mistake in base-line length invalidates the whole map (see paragraph b, p. 15).

5. When preliminaries are completed, survey should proceed as follows: On selected level (bank, mid-beach, or water edge), choose location *B*, whose distance from *A* depends upon configuration of shore line. Drive stake leaving top projecting about 4 in.; mark stake with position

designation (Fig. 3, *B*). Erect *red* flag signal on *B* and direct both transits upon it simultaneously, determining angles *BAK* and *BKA;* take magnetic bearings of lines *AB* and *KB* for checks. Signálman must keep signal erected until released by down signal from both transits. Record angles at each instrument as right (*R*) or left (*L*) depending upon whether angle is to right or left of base line.

6. Progressing clockwise along shore, signalman selects another location *C* appropriate to shore configuration. Drive wooden stake at *C* as before and mark; erect *white* signal; direct both telescopes on *C* and proceed as described above. Locate similarly other positions and determine simultaneously the angles required at each point until survey is completed. Use *red* signal at first, fifth and every fifth position thereafter; use *white* signals for all intervening ones. Take magnetic bearings occasionally as checks. Signalman must keep a tally record of positions as he proceeds; also descriptive notes or descriptive sketch dealing with various significant physical details of shore line at and between positions.

7. Periodically throughout the work, as for example every 10 positions, halt signalman and have transitmen check original orientation (reciprocal bearings) of transits to determine if they have in any way been changed in position. If changes have occurred, restoration to original positions must be made at once and preceding records repeated back to the last check-up.

8. If additional transit set-ups are required to complete survey, such positions must be carefully hung on previous ones by properly measured angles and distances.

General Considerations

a. Transitmen must use great care in reading angles with instruments graduated in quadrants, particularly when angles exceed 90°. Do not attempt to make additions and subtractions in field but enter full information on field records, as for example, 90° + (90 − 46° 45′).

b. On smaller lakes a check on the computed length of the base line may be made by reading distance *AK* on a stadia rod. Such stadia-rod readings should be used only as a check unless the distance involved is only a few hundred feet and the reading has been taken with great care, then checked and rechecked.

c. As a means of avoiding errors in reading and recording angles, transitmen, and recorders if present, should follow the directions outlined on p. xiii.

d. In order to avoid confusion signalman must carry signals with flag ends pointed toward, and near, the ground when going from one position to another; also, when arriving at the new selected position, he should drop to the ground the signal which is not to be used at that position.

PLOTTING SURVEY BY LOCATION OF POINTS FROM TWO SIMULTANEOUSLY MEASURED ANGLES

1. Determine scale which will yield map of desired size; select paper of sufficient dimensions to include map with extra space on all sides; trim paper so that all sides are straight and form right angles with adjacent sides.

2. At left side, about 1 in. from edge of paper, draw straight line parallel to left edge and about 6 in. long; attach arrowhead figure to upper end and mark with letter *N* to represent north end of north-south construction line.

3. Select appropriate position for one transit station, mark with pencil dot surrounded by small penciled circle, and label it. Through transit station thus located, carefully draw line parallel to north-south construction

---

FORM 3

FIELD RECORDS

Survey of Shore Line by Two Simultaneously Measured Angles

Lake	Transitman	Distance Between Monuments
Date	Recorder	Method Used
Instruments	Signalman	Magnetic Bearing on Monument
	Monuments; Referencing Data	

Signal Position	*Angle on Transit* A	*Angle on Transit* K	*Magnetic Bearing*		*Remarks*
			Transit A	*Transit* K	
B (red)					
C					
D					
E					
F (red)					
G					

line, prolonging it to length sufficient for effective use of protractor. Center protractor on *A* and, using corrected magnetic bearing (true bearing) and distance records, draw base line *AK* having proper position and length (exercise extreme care in locating and scaling base line; a mistake invalidates entire map). Mark other end of base line *K* with pencil dot surrounded by small penciled circle and label it.

4. Center protractor on *A* and pivot until zero edge or line coincides exactly with base line *AK* drawn on paper; protract angle *KAB;* indicate boundary of angle with short pencil mark and label it *B*. With protractor in same position, protract, mark and label in proper sequence all other angles measured in the field at *A*.

5. Transfer protractor to position *K* and pivot until zero line coincides exactly with base line *KA;* protract, mark, and label angle *BKA*. With protractor in same position, protract, mark, and label in their proper sequence all other angles measured in the field at *K*.

6. With T square, meter stick, or other straightedge, pivoted on *A* and *K* locate intersection of lines limiting angles *KAB* and *AKB*, which intersection locates position *B*. In same fashion locate other shore-line positions.

7. Test with such checks as field survey provided, using bearings, measured angles, and computed lengths of sides of triangles. Then finish map as outlined in paragraphs 6–9, p. 8.

Plane-table Survey

The plane table, when circumstances permit its use, provides a means of representing directly on paper the angles and distances involved in locating positions. This graphic method has the advantage of actually constructing the map in the field in the presence of the details involved where matters of accuracy can often be determined to best advantage. The method depends upon the prolongation of two lines, one from each end of a base line, which intersect at a single position. These lines are drawn along the lines of sight of an alidade which may or may not be provided with a telescope. The map is essentially complete when the successive positions, so located, are connected by a continuous line. The simplicity of its operation is an outstanding virtue of this method.

Suitability. This kind of shore-line survey is best suited to lakes of smaller size. Obviously, the size of the resulting map can be no greater than the maximum dimensions of the plane-table board.

Equipment. Plane table; 100-ft. steel tape; chain pins; pencils; several No. 5 insect pins; 2 or more pieces of gas pipe, 2 ft. long; several signal poles, 8 ft. long; 2 or more sheets of good grade, heavy white paper cut to cover whole area of plane-table board; white and red cloth; ax; waterproof cover for board; form for records.

CORRECTIONS AND ADJUSTMENTS. Examine alidade for proper condition. Test steel tape for general accuracy; extreme accuracy is not required.

ORGANIZATION OF PARTY. Instrument man; assistant.

PROCEDURE

1. Make preliminary examination of lake to secure general details of planning survey and to discover any special problems which may be involved. Examine for most favorable locations of base line and signal positions along shore line.

2. Lay out base line of considerable length along shore line and drive iron stake at each end. Make base line as long as shore-line configuration will permit and select termini (Fig. 4, *A* and *B*) so that practically all points on shore are visible from each. If more than one instrument set-up is required, then select base line so that all points on one portion of shore line are visible from ends of base line.

3. Select positions along shore line, the number and location to be determined by its configuration (*C–K*); erect at each an 8-ft. pole. Nail *white* cloth flag to top of all poles, except the first, fifth and every fifth

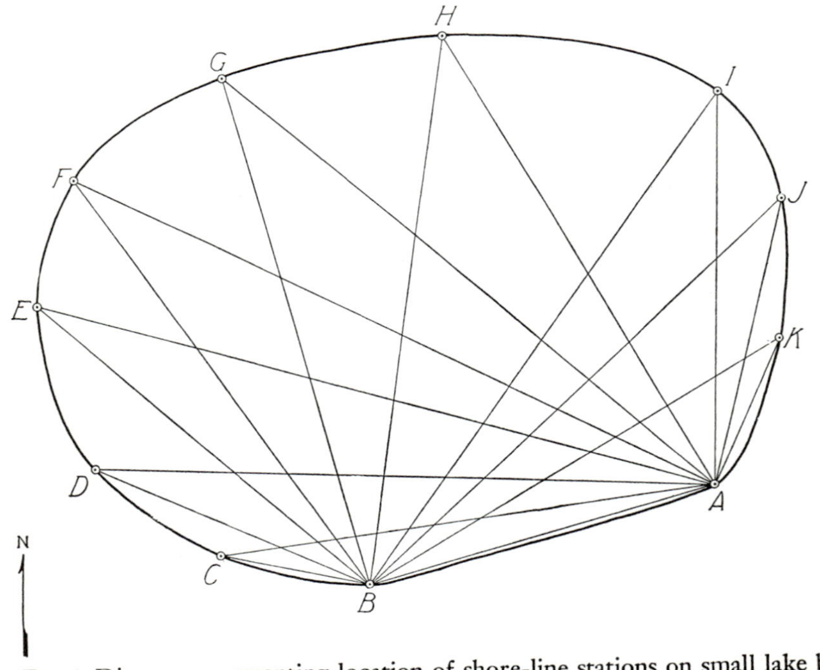

FIG. 4. Diagram representing location of shore-line stations on small lake by means of plane table. (*A, B*) Plane-table stations. (*AB*) Base line. (*C–K*) Shore-line signal stations. Radiating straight lines indicate lines of sight by alidade.

one thereafter, which should be *red*, and give each a serial identification symbol or number.

4. Set plane table over iron stake at one end (*A*) of base line; set legs of tripod firmly in ground. Level board and with aid of compass make one edge of board coincide with north-south direction or, if this is not practicable or desirable, adjust board to any necessary position. Immobilize board and guard rigidly against any subsequent changes of board position.

5. Estimate dimensions of lake and adopt scale of reduction suitable to size of board; estimate space and position requirements of future map on paper. Choose proper position on board for one end of base line *A* and set pin firmly into board at that point.

6. Set alidade with zero point against implanted pin; pivot alidade on pin until line of sight strikes other end of base line (*B*). Draw line along edge of ruler base of alidade; with steel tape, measure actual length of base line *AB;* along line just drawn on paper, lay off to scale the length of *AB* with great care. Designate point so established as other end (*B*) of base line.

7. Rotate alidade, with zero mark pivoted on pin, until line of sight strikes signal on next shore-line position *C;* draw line along ruler edge of alidade and extend it to probable necessary length; label it; again rotate alidade in similar fashion until line of sight strikes next signal *D* and draw line as before; repeat for each of remaining positions (*E–K*).

8. Rotate alidade, sight on end of base line *B*, and check base line drawn on paper. If plane table has been inadvertently changed in position, line of sight will not coincide with base line drawn on paper and error has been introduced somewhere. In case of error, all previous lines must be repeated. If necessary, check on base line more frequently.

9. Transfer plane table to other end *B* of base line; install pin at *B* on map; pivot alidade on pin and make ruler edge correspond to line *BA*. Then rotate plane table board until line of sight of alidade strikes end of base line *A;* check this orientation carefully, then immobilize board.

10. Pivot zero point of alidade on pin until line of sight strikes first shore-line signal *C;* draw line along ruler edge of alidade until it intersects line previously drawn from station *A;* label intersection with same identification symbol as on signal. Pivot alidade clockwise again until line of sight strikes next shore-line signal *D* and draw line to intersection with other line drawn previously from station *A*, and designate. Continue until similar intersections are established at the remaining shore-line signals, designating each in the proper sequence. Rotate alidade to make line of sight strike other end *A* of base line and check position of base line on paper.

11. Provide check on position of some or all signal stations. For example, transfer plane table to *H;* implant pin at *H* already marked on map; pivot zero point of alidade on pin until line of sight exactly coincides with

line *HA* on map. Rotate plane-table board until line of sight passes exactly through station *A;* immobilize plane table; then pivot alidade to coincide with line *HB* on map. If original location of station *H* was correct, line of sight will pass through station *B;* if it fails to do so, error exists in original location of station *H*. Other stations should be tested in similar fashion.

12. With continuous line connect base-line stations and shore-line signal positions to complete map of shore line. When all necessary checks and verifications have been made, the map may be finished as outlined in paragraphs 7–9, page 8.

13. If lake is too large or too irregular in outline to be mapped from one set-up, extend another base line from one end of the original base line, being exceedingly careful that proper orientation, sighting, and scaling is done. Then map next portion of lake as before described, using ends of new base line for plane-table positions.

Simple Methods for Ponds, Lagoons, Beach-pools, and Similar Small Waters

SURVEY BY CONNECTED SERIES OF TRIANGLES

This method depends upon the establishment of a continuous series of triangles the length of whose sides are known and the apices of which occur at the various points on the shore line. Its virtue lies primarily in its simplicity and the small amount of equipment required.

Suitability. This method is suitable only for small bodies of water whose depths are so shallow that direct linear measurements can be made across the open water.

Equipment. Wooden stakes; crayon; steel tape, or graduated rope; compass; ax; field notebook.

Organization of Party. Two operators.

Procedure

1. Select number and position of shore-line stations so that required accuracy of map will be attained; drive stake at each station; designate (number or letter) stakes in regular sequence. Begin at some convenient point (for example, *A* in Fig. 5); measure and record distance *AB;* then successively measure and record distances *BC*, *CD*, *BD*, *DE*, *EB*, *EF*, *FB*, *FG*, *GB*, *GA*, *HA*, *GH*, *AL*, *HL*, *HI*, *IL*, *IJ*, *JL*, *JK* and *KL*.

2. Determine magnetic bearing of *AB* (or some other line) with compass.

3. For checks on field data, make series of measurements along lines not previously used; for example, in Fig. 5, measure distances *AF*, *AI*, and *AJ* and distances *CE*, *CF*, *CG*, *CH*, and *CI*.

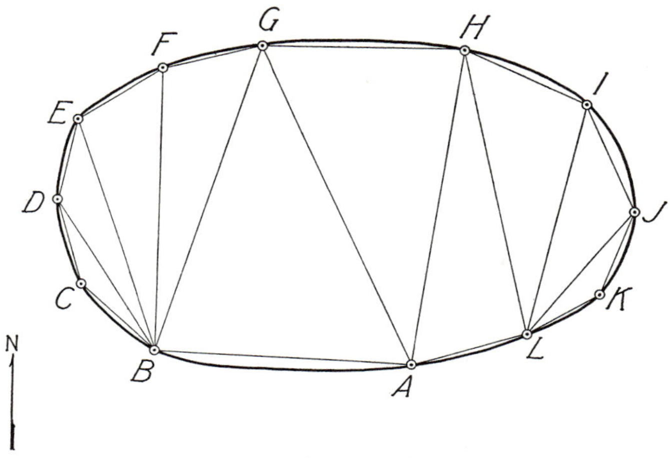

FIG. 5. Diagram illustrating survey of a small shallow pond by connected series of triangles. (*A–L*) Shore-line stations.

Plotting of Field Data

1. Adopt scale which will yield map of desired dimensions.

2. Using field data, draw straight line having corrected magnetic bearing of *AB* in proper position on paper; along this line scale distance *AB* thus locating *B;* label positions *A* and *B*.

3. Scale distance *AG* and with one leg of dividers set on *A* draw arc of circle having radius equal to this scaled distance. Scale distance *BG* and with one leg of dividers set on *B* draw arc of circle having radius equal to this scaled distance. Intersection of these two arcs locates position of *G*.

4. Scale distance *GF* and with one leg of dividers set on *G* draw arc of circle having radius equal to this scaled distance. Scale distance *FB* and with one leg of dividers set on *B* draw arc of circle having this distance as radius. Intersection of these two arcs locates position of *F*.

5. In similar fashion continue with other triangles until all are completed. Connect apices of triangles by continuous line, thus representing shore line.

6. Scale check distances for confirmation.

7. Erect north-south line and add data specified on pp. 68–69.

SURVEY BY TRANSVERSE MEASUREMENTS

This method is based upon a series of measurements transverse to a long base line and at intervals determined by requirements of shore-line configuration.

SUITABILITY. This method is usable only for very small basins whose depths are so shallow that direct linear measurements can be made across the open water.

EQUIPMENT. Wooden stakes; crayon; steel tape, or graduated rope; compass; ax; two line rods; field notebook.

ORGANIZATION OF PARTY. Two operators; assistant.

Procedure

1. Select base line (Fig. 6, *XX′*) along one side, preferably the longest side with no obstructing physical features. Along this base line drive stakes at such intervals (*a–h*) as are required by shore-line configuration; record distances used between stakes; set line rod behind and in contact with first stake at *a*. Attach zero end of tape or graduated rope at first stake *a* and set second line rod some distance back of *a* in such position that the two line rods represent a line at right angle to base line. With the line rods as guide, extend tape to station *A* on shore, and measure distance. Shift tape and line rods to next position *b* on base line, set up as described before, and determine distance *bN* and *bB*. Repeat for each of the remaining positions (*c–h*).

2. With compass, carefully determine bearing of base line *XX′*.

3. For purposes of confirmation, measure distances between significantly situated shore-line stations, as for example, distances *AL* and *AD;* also *HL* and *HD*. Make other similar measurements as circumstances seem to warrant.

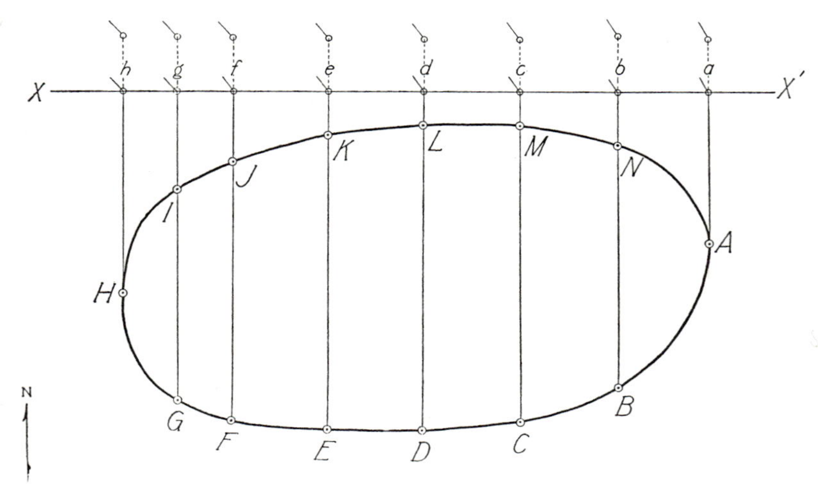

FIG. 6. Diagram illustrating survey of small shallow pond by method of transverse measurements. (*XX′*) Base line. (*A–N*) Shore-line stations. (*a–h*) Positions along base line at which transverse measurements are made.

Plotting of Field Data

1. Adopt scale which will yield map of desired size.

2. Using field data, draw straight line to represent base line *XX′* having proper position on paper, correct magnetic bearing, and accurately scaled length.

3. Select position *a;* draw line through *a* at right angle to *XX′;* along this line scale distance *aA* and locate first shore-line station *A*. Indicate position by pencil dot surrounded by small penciled circle and label.

4. From *a* scale distance *ab* along base line *XX′* to locate *b;* draw line through *b* at right angle to *XX′;* along this line scale distance *bN* and *bB* thus locating *N* and *B*. Indicate position of each by pencil dot surrounded by small penciled circle and label.

5. Proceed in similar way with other transverse lines until all shore-line stations are located. Using measurements taken for confirmation, check location of stations involved.

6. Connect shore-line stations in correct sequence by continuous firm line to represent shore line. Erase all construction lines. Erect compass figure. Add data specified on pp. 68–69.

Cross-section paper may be convenient in plotting of this kind.

SURVEY BY ENCLOSING RECTANGLE

This method depends upon the simple procedure of measuring distances from positions, selected at known intervals along the sides of an enclosing rectangle, to intersections with the shore line. Lines connecting rectangle positions with shore line are erected at right angles to sides of rectangle.

SUITABILITY. This method is most suitable for small basins occurring in open country with low or scant vegetation and having depth or shore-line features which make direct measurement across the open water impracticable.

EQUIPMENT AND ORGANIZATION OF PARTY. Same as in Survey by Transverse Measurements.

Procedure

1. Set up four lines of stakes in form of rectangle (Fig. 7) surrounding lake to be mapped; establish rectangle in most convenient position so that angles at *X*, *X′*, *Y*, and *Y′* are right angles. If the physical features of the shore permit, one or more sides of enclosing rectangle may be made tangent to the shore line. Set stakes *a–v* at measured distances and at such intervals as will bring out essential features of the shape of the shore line. Set line rod behind and in contact with stake at *a* and attach zero end of tape or graduated rope to stake at *a;* set second line rod some distance back

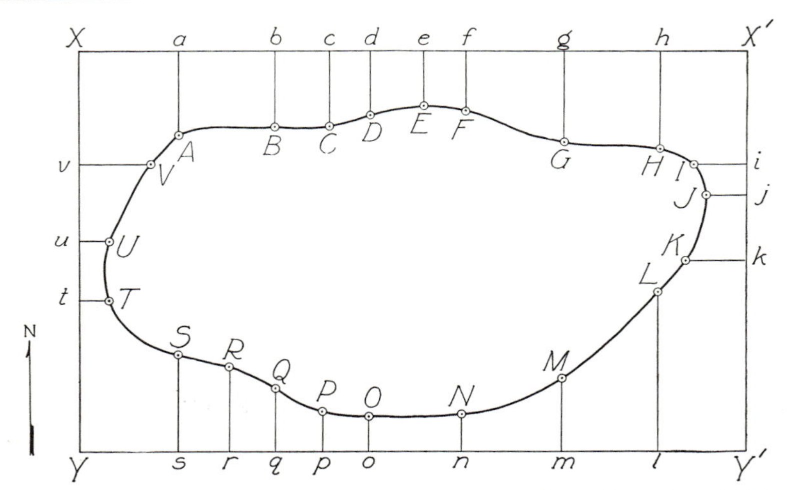

FIG. 7. Diagram illustrating method of mapping shore line of small lake by means of enclosing rectangle. (*XX′Y′Y*) Enclosing rectangle. (*a–v*) Stake positions on enclosing rectangle from which measurements to shore line are made. (*A–V*) Stations on shore line.

of *a* so that the two line rods represent a line at right angles to adjacent side of rectangle. With the line rods as guide, extend tape to station *A* on shore line, drive stake and measure distance. Shift rods and tape to *b*, set up as just described, and measure distance *bB*. Repeat this procedure for positions and distances *cC–vV*.

2. Interpolate measurements between stations of first series if it is later found that additional ones are necessary to the construction of a more accurate map.

3. With compass determine bearings of sides of enclosing rectangle.

Plotting of Field Data

1. Adopt scale which will give map of desired size.

2. Using field data, draw rectangle *XX′Y′Y* in proper position on paper, using bearings and accurately scaled dimensions.

3. Scale distance *Xa* to locate *a;* draw line through *a* at right angle to *XX′*. Along this line scale distance *aA* and mark *A* with pencil dot surrounded by small penciled circle; label.

4. Proceed in same fashion to locate and label other shore-line positions in their proper sequence. Connect all shore-line positions by continuous firm line to represent shore line; erase all construction lines; erect compass figure; and add data specified on pp. 68–69.

Cross-section paper is very convenient in this kind of mapping.

For Selected References, see p. 59.

SOUNDINGS

Sounding positions on lakes or streams may be determined in a number of ways. Certain ones described here are selected because of their general usefulness to the limnologist. Soundings must be located with considerable accuracy if the resulting hydrographic map is to serve its best purposes, although it should be pointed out that, because of the very nature of lake features, high precision is usually not only very difficult but is without justification. This statement holds both for sounding locations and for the soundings themselves (depths) since lakes often have bottoms, or regions of bottom, which are so soft that the precise determination of bottom level is not practicable. In general, soundings should be read to the nearest 1.0 foot or to the nearest 0.3 meter. Closer readings are meaningless, except in very special cases.

Sounding Lines

Since the limnologist deals, for the most part, with waters of relatively shallow depth there are not many demands for sounding machines which use stranded wire or piano wire (p. 339) as a line. If such provision is at hand, it should be used when deeper lakes are sounded, but this method involves expensive equipment and is not adapted to shallow waters. Therefore, most sounding work in inland lakes is done with graduated lines of proper size and construction. Lines suitable for this purpose, their preparation, correction, limitations, and care, are described elsewhere (pp. 337–338). For such lines drums or hoists may be used when occasion warrants, but it is often more convenient and more rapid to handle them by hand.

SOUNDING LEADS

Weights on lines used for sounding by hand are commonly referred to as sounding leads, although they may or may not be composed of that metal. In fact, they may be of any substance having the proper specific gravity, shape, and size. Simple leads are usually elongated, symmetrical, and either cylindrical or tapered slightly at the upper end. Common 6–10 lb. iron window weights may be used. Sounding weights commonly on the market are made of lead, octagonal in transverse section, a 6-lb. weight being about 8 in. long, 1¾ in. in diameter at base and about 1 in. in diameter at top. A 6-lb. lead is usually a good general utility size and is suitable for inland waters when sounding is done from a stationary boat.

However, if sounding is to be done from a moving boat, a 6-lb. lead is suitable only for quiet water not more than 25–30 ft. in depth; for greater depths, and in heavy wave action or strong currents, the weight of the lead should be 15–20 lbs. When sounding work is done with a sounding machine equipped with stranded-wire or with piano-wire line, and operated in such a way that the weight is merely lifted free from the bottom after each depth record, the weight may be very much heavier and streamlined in shape.

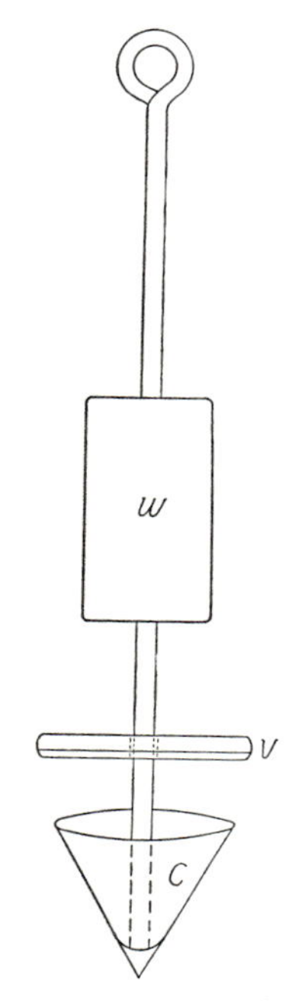

Fig. 8. Combination sounding weight and reconnaissance bottom sampler. (*w*) Weight. (*v*) Sliding cover. (*c*) Conical bottom cup.

It is often convenient to combine sounding with qualitative reconnaissance of the bottom materials. The common type of lead weight described above usually has the bottom of its base excavated to form a conical pit the sides of which carry a series of diagonal ridges to which a bottom sample will usually adhere. Other forms of sounding weights carry at the lower end a cup-shaped receptacle (Fig. 8, *c*) provided with a sliding cover *v* which leaves the receptacle open on the down journey but which drops over the top and is held in place by hydrostatic pressure during the up journey.

In waters, such as many seepage lakes, bog lakes, and others, in which there is much bottom deposit of a loose, semisuspended sort and through which an ordinary sounding lead might sink for many feet, it may be desirable to use a lead which will show distinct retardation in descent on meeting the upper portions of such a deposit. For this purpose, a useful type may be secured by sawing off the upper half of an ordinary 6-lb. sounding lead and attaching to its lower end a circular tin plate of sufficient diameter (4–6 in. or more) to produce the desired retardation. This plate is easily attached by means of an ordinary screw. Such a weight may show a tendency to "sail" if lowered too rapidly.

Sounding Pole

A good sounding pole, of convenient construction and graduation, is almost indispensable for sounding work in shallow water; also for many other purposes. It must be of some light, straight-grained, well-seasoned wood, kept well painted, and stored flat on some straight support when not in use. The lower end may be loaded to facilitate sinking in water. It may also be desirable, for some purposes, to attach to the lower end a flat, circular, metal shoe which will retard the sinking of that end of the rod into the softer bottom deposits. The writer uses a sounding pole made of white pine, 10 ft. long and ¾ × 1¼ in., graduated in 0.1 meter intervals on one side and 0.1 foot intervals on the other. Graduations are marked with black paint and the whole is covered with several coats of shellac.

A sounding pole may be made with a conical cavity in the lower end for securing bottom samples. Adhesion of bottom materials is facilitated by a coating of grease or tallow on the surface of this end pit.

If a boat hook of the usual type is carried on board the boat as a regular part of its equipment, permanent graduations may be marked on the boat-hook pole thus providing a satisfactory substitute for a sounding pole.

Methods of Sounding

LOCATION OF SOUNDING POSITIONS BY TWO ANGLES MEASURED SIMULTANEOUSLY ON SHORE

This method is essentially the same as that described on pp. 12–17 for locating shore-line stations by two angles measured simultaneously, and the principle involved need not be restated here.

Suitability. This method can be used on any lake, but under most circumstances it is probably most useful for lakes other than the very small ones. It is one of the most common methods, is convenient and accurate when well done, and, in experienced hands, proceeds with considerable speed.

Equipment. 2 transits, or other instruments measuring horizontal angles; 100-ft. steel tape; signal set, containing 1 white and 1 red flag, each on staff; graduated sounding line; sounding lead; graduated sounding pole; forms for records; boat of appropriate size and equipment.

corrections and adjustments. Test transits for good condition. Examine sounding line and sounding pole for general accuracy.

Organization of Party. Two transitmen; soundingman; signalman-recorder; boatman; assistant.

Procedure

1. Examine map of shore line, if available, or examine lake itself and determine whether one or more transit set-ups will be required. Select transit stations so that all positions on lake surface are visible from both instruments, or, if more than one transit set-up is necessary, select transit stations so that all points on surface area of portion to be covered by first set-up are visible. Also select transit stations well apart from each other and in such a way that angles at lines-of-sight intersections will be kept large (larger than 30° and smaller than 120°), as for example, in Fig. 3 with transits set on *A* and *K*. Use already established positions and stakes if available, or establish new ones if necessary, drive iron stakes at transit

positions, and carefully reference them to convenient permanent objects near by.

2. Set up transits on selected stations; direct transits on each other, sighting either on top of iron stake or on axis of opposite transit. Carefully set scale of each transit so that, when telescope is properly directed on opposite transit, zero mark on lower scale corresponds exactly with zero on vernier; be sure that no mistake is made in this set-up.

3. Determine exact distance between transits, either by direct measurement or by indirect means such as establishment of triangle and computation of length of desired side (see p. 14 for directions). This step is not necessary if distance between transits is already known from shoreline survey.

4. With each transit determine magnetic bearing of base line (line connecting transits). Note that bearing read on one transit should be reciprocal of that on other transit; guard against local attraction.

5. Organize boat party; locate soundingman and signalman-recorder in close proximity in boat; review with whole party prearranged signals of all sorts. Supply transitmen and signalman-recorder with forms for records on which sounding positions are indicated in vertical column by consecutive numbers, beginning with 1, and on which 1, 5, 10, and every fifth number thereafter are written in red or circled with red. Supply oarsman, engineman, or helmsman with small sketch map of lake on which is indicated the pattern of courses that the boat should follow in order to cover the lake surface adequately. These courses may be steered blindly, or they may be steered with guidance of preselected, conspicuous objects on shore.

6. All preliminary arrangements having been completed, sounding operations may now begin. Sounding boat chooses position No. 1 (not in or near the line connecting transits) and comes to halt; soundingman makes sounding with line and lead, or with sounding pole if in shallow water, and signalman-recorder elevates red flag, keeping flag elevated for time interval previously agreed upon. Soundingman announces sounding to recorder, recorder makes record and repeats entry aloud, and soundingman reads line or pole the second time and repeats announcement or announces correction if error was made. Simultaneously each transitman directs his telescope on boat signal when elevated, reads and records angle shown by his instrument between base line and line of sight from transit to sounding position, and rereads instrument for check.

7. Boat moves along course to new position using approximate distance interval previously agreed upon; the next sounding is made in the same manner as just described, as are all subsequent soundings. At sounding No. 5 and every fifth sounding thereafter, the red flag is elevated on boat as

a check on all simultaneous operations; the white flag is used at all other positions. If records at instruments and on boat fail to agree on red-flag positions, some error has occurred and corrections must be made at once.

8. Periodically (for example, every 20 soundings), halt sounding boat and have transitmen check their original position readings to determine if instrument positions or set-ups have been altered. If alterations have occurred, restoration of the original instrument orientations must be made at once and the preceding group of records regarded as erroneous.

General Considerations

a. The sketch supplied to the boatman designating courses to be followed may be in any form which will provide that soundings are spread over the lake with some approach to uniformity. A series of courses across the smaller dimension of the lake followed by a lengthwise series provides a satisfactory pattern.

b. It is inevitable that, when plotted, soundings will fall short of uniform distribution and that certain spaces on the map will lack them. If desirable, these spaces may be filled to the necessary extent by noting on the map the position of such spaces; then, on returning to the field, instruments are set up as before and the sounding boat is directed into and across the unsounded areas. Data so secured can then be plotted on the map.

c. It is usually desirable to agree in advance that the instrumentmen will sight on the signal, particularly when large lakes are being sounded. On small lakes it is possible to sight on the soundingman if that seems desirable.

d. Since in sounding work "down signals" by instrumentmen to the signalman cannot be used, instrumentmen must follow the course of the boat between soundings by sighting over the top of the telescope and must take position records *when the signal goes up*. The signalman should keep the signal in upright position for an interval not to exceed 10 seconds —an interval which can be conveniently determined by counting 10. At other times, particular care should be taken to keep all signals flat on the bottom of the boat.

e. If a lake of considerable size is being sounded, the leader of the party will find a pair of field glasses useful in following the various operations.

f. The sunshades regularly provided for the telescopes should be used on bright days.

g. If circumstances demand that sounding positions be confirmed, check readings may be provided as the work progresses by the use of a third transit on shore or a sextant on board the boat if all of the conditions demanded for the use of a sextant are met (see pp. 38–47).

Plotting of Field Data

1. On a map of the shore line previously made, put 14-inch protractors on instrument positions representing those used in making soundings in the field. Protract the angles measured simultaneously in same fashion as described on p. 17. Intersection of lines of sight made simultaneously represents the sounding position; enter the number indicating sounding at that point. Continue similarly for other pairs of lines of sight until all soundings are located on map. (See pp. 67–75 for additional instructions for construction of hydrographic maps.)

2. If checks on sounding positions were provided in field work by use of some other method, as for example the sextant, plot confirmation records as described for method used (see pp. 46–47).

LOCATION OF SOUNDING POSITIONS BY TRANSIT AND STADIA ROD

This method depends upon the location of positions on water by means of a transit, located at some convenient, known shore position, which determines angles and reads distance directly on a stadia rod elevated on the boat at sounding positions. When employed under proper conditions, it is dependable and rapid but its limitations must be fully understood. General facts concerning stadia surveys are stated on pp. 8–12.

SUITABILITY. This method can be used only on small lakes and on calm water. Since stadia-reading distance is limited to not more than 1000 ft., preferably less, only restricted portions of larger lakes (bays, connecting channels, local areas) can be sounded in this way. Furthermore, since there must be no vertical motion of the stadia rod at the time of reading, even the slight movements of the boat make the records uncertain. Calm water is a necessity.

EQUIPMENT. Transit equipped with stadia cross hairs; signal set, containing one white and one red flag, both on staffs; graduated sounding line; sounding lead; sounding pole; stadia rod; forms for records; boat of appropriate size and equipment.

CORRECTIONS AND ADJUSTMENTS. Test transit, sounding line, sounding pole, and stadia rod for general accuracy and good condition.

ORGANIZATION OF PARTY. Transitman; soundingman; rodman; boatman; recorder.

Procedure

1. Examine map of shore line or examine lake itself to determine whether one or more transit set-ups will be required. Select as first transit station a station already used in previous shore-line survey and marked on map; set up transit on it.

2. Swing telescope to left and direct line of sight on shore-line station

whose location is determined by known distance from transit station and which gives a line virtually parallel to the shore line from the transit station. (Position *B* in Fig. 4 represents such a location; such a position may be one of the shore-line survey stations or it may be established especially for transit orientation.) Set zero point of transit scale on zero of vernier. Set in this way, all angles can be measured clockwise and their magnitude read directly. If such a form of transit orientation is undesirable, some other line can be used as the line of reference from which the angles are measured, provided that the position of such a reference line is fully described and can be located with complete certainty in subsequent mapping activities. Sometimes it may be convenient to orient the telescope in the magnetic meridian and use the latter as the line of reference.

3. Organize boat party, locating soundingman and rodman in close proximity in boat; review with entire party all operating signals. Supply recorder and rodman with forms for records on which future sounding positions are indicated in a vertical column by consecutive numbers beginning with 1 and on which 1, 5, 10, and every fifth number thereafter are written in red or circled with red. Supply boatman with a small sketch map indicating the pattern of courses to be followed by boat. Assign to recorder the duty of recording transit readings.

4. Boat takes sounding position No. 1; boat comes to full stop, rodman elevates stadia rod, keeping it vertical, as immobile as possible, and faced toward transit. Transitman directs line of sight on stadia rod and reads distance from transit to rod, also angle with reference line. Simultaneously on boat, soundingman makes sounding with rod or line and announces depth to his recorder (rodman) who records it, repeating announcement aloud, and soundingman checks his reading making second announcement. On shore, transitman announces distance and angle to his recorder; recorder enters record, announcing it aloud; and transitman checks readings and makes second announcement for confirmation. Boat proceeds to next sounding position and the second record is made in same fashion, and so on with subsequent positions. Rodman elevates red signal at every fifth position as a check. At each position rodman must keep rod elevated until waved down by transitman, and boatman must keep boat in same position as nearly as possible.

General Considerations

a. If the line of reference is the magnetic north-south line, boat positions may be read and recorded in bearings, as for example, a record may be $N\ 26°25'W$. Bearings are convenient when a transit is used which has its scale divided into quadrants instead of a continuous graduation from 0° to 360°.

b. In very shallow water the stadia rod may be held on the bottom while the distance reading is being taken. In depths up to about 10 feet the stadia rod may be attached to the upper part of a weighted pole and the readings taken with the lower end of the pole on the bottom.

c. If more than one transit station is required, the new positions may be established as already described (p. 11).

d. If the line of reference is some line other than the magnetic meridian, bearings on boat positions may be made as checks.

Plotting of Field Data

1. On a previously constructed map of the shore line, center protractor on position representing transit station used when field work was done; rotate protractor until zero line coincides with zero reference line used by transit. In that position angles are protracted in same direction as measured in field.

2. Protract first angle, scale distance from transit to boat along line of sight, and enter sounding value at that point. Repeat this procedure for all subsequent soundings.

3. If more than one transit position was used in field work, successively center protractor on other transit stations on map and protract angles, scale distances, and enter soundings made from that station.

4. Complete map as described on pp. 68–69.

LOCATION OF SOUNDING POSITIONS ON ICE COVER

If the difficulties of low temperatures, snow cover on ice, and the labor of cutting holes through the ice are not too great, sounding through the ice cover has certain obvious advantages. This method depends upon the location of positions by angles and distances and the use of the sounding line through a hole in the ice at each selected location. Sounding positions can be located with great precision, if such precision is desired, and the distribution of soundings over the lake surface can be accomplished with considerable uniformity. There are various methods of making such soundings, and the following are both simple and accurate.

A. Transit and Stadia Rod

This method depends upon the use of a transit for (a) measurement of angles from a line of reference, (b) control of the lines of sounding holes, and (c) reading of distances directly on the stadia rod.

Suitability. This method is suitable for waters whose size and shape are such as to make it possible to keep stadia rod readings within the limits of acceptable accuracy.

Equipment. Transit, equipped with stadia cross hairs; stadia rod; sounding line; sounding lead; ice spud, or other tool for cutting holes through ice; forms for records.

CORRECTIONS AND ADJUSTMENTS. Examine transit, sounding line, and stadia rod for good condition.

ORGANIZATION OF PARTY. Transitman; soundingman; rodman; assistant.

Procedure

1. Examine map of shore line already made, or the lake itself, to determine number of necessary transit set-ups; choose as suitable initial set-up one of the permanently marked shore-line positions used in constructing map of shore line.

2. Select reference line (magnetic meridian, or any other line which can be accurately related to map of shore line); swing telescope into this reference line and set at zero.

3. Choose sounding positions on ice according to any suitable plan (in straight lines radiating from transit station; in lines in other directions; or spotted irregularly). Angles with reference line and distances on stadia rod are read by transitman; holes are cut through ice by assistant and soundings are recorded for each position. Choose sounding positions so that reasonable spread of soundings over whole lake is assured. Check vocally or by signal every fifth location number to insure simultaneous transit and sounding readings; continue until lake is covered.

General Considerations

a. If soundings are made along lines radiating from the transit, the transitman must keep sounding crew in locations on lines of sight.

b. If the transit must be moved to new locations in order to cover the whole lake, care must be taken to orient new locations properly with respect to previous transit stations.

c. Boring machines, designed for sounding through ice, greatly minimize the labor of cutting holes and also speed up the work (see Appendix).

d. Ice spuds, sounding leads, and any other instruments used through or about the sounding hole must be tied to the operator or to some adjacent large object to avoid loss.

Plotting of Field Data

Proceed essentially as described in Location of Soundings by Transit and Stadia (p. 32).

B. Transit and Tape

The method described above (pp. 32–33) can be operated by substituting a steel tape for a stadia rod if facilities for stadia readings are not available. Measurements by tape are accurate but slower than stadia readings. When a tape is used, it will probably be most convenient to select sounding holes along straight lines radiating from the transit station.

C. Plane Table with Stadia Rod or Graduated Tape

Soundings through the ice can be mapped directly by means of a plane table and some suitable means of measuring distances. If the plane table is equipped with a telescopic alidade it will be possible to use a stadia rod as described above (pp. 32–33); if the alidade is nontelescopic it will be necessary to determine distances by a tape or graduated line. Orient and operate plane table as described (pp. 17–19). If distances, wind, or other circumstances prevent accurate reporting of soundings vocally, the instrumentman will merely locate positions of sounding holes on his map, numbering them serially, and enter soundings at a later time from the records of the soundingman. As usual, every fifth position should be checked to make certain that soundings and plane-table records are kept simultaneous.

Adverse weather conditions may make plane-table work difficult or impossible.

To finish map, either enter soundings on map during field work or mark positions and enter soundings at a later time. Remove construction lines and add information mentioned on pp. 68–69.

D. Ranges and Tape

On lakes whose shape and size are such as to make it possible and practicable, a method may be used which reduces to a minimum the equipment required. The method depends upon the establishment of a series of parallel ranges (straight lines extending across a lake from one shore to the other) of known distance apart and along which the sounding holes through the ice are made at known distances. The essential features of such a survey are shown in Fig. 9. Ordinarily such lines are equally spaced, but the interval between them may be varied if any useful purpose is accomplished by so doing. The number of ranges will be determined by the degree of accuracy desired in the map.

EQUIPMENT. 100-ft. steel tape; sounding line and lead; compass; ice spud, or other tool for making holes through ice; ax; forms for records.

ORGANIZATION OF PARTY. This method can be operated by two men if necessary, although another helper will facilitate the work.

Procedure

1. Establish initial end of first range line at some convenient point (Fig. 9, *A*) at one side or end of lake, mark permanently and reference it, or use some already established point on shore whose position with reference to map of shore line is already definitely known. Erect pole at *A* with flag at top; with compass prolong line from *A* to point *A′* on opposite shore along known magnetic bearing; erect signal pole at *A′*; record magnetic

bearing of line *AA′*. Establish range *BB′* at selected distance from and parallel to range *AA′* and erect signal pole at each end; establish other ranges in like fashion until lake surface is covered.

2. **Return to *A* and** measure off distance *A–1* along range AA′, sounding hole being kept on range line by person standing behind signal pole at *A* and sighting to range pole at *A′*. Cut hole through ice, make sounding and record. Move along range, measure distance *1–2* to next sounding position; and proceed as before. Continue operation until *A′* is reached; transfer activities to range *B′B* and continue in same way until whole lake is covered.

The time and labor involved in setting up numerous ranges may be saved by erecting fewer parallel ranges and then after the soundings along

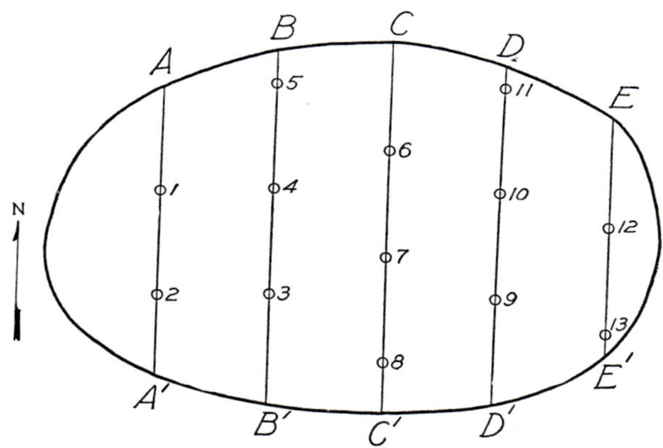

Fig. 9. Diagram illustrating location of sounding positions on permanent ice cover by means of simple method of ranges and tape. Lines (*AA′*, *BB′*, *CC′*, *DD′*, *EE′*) represent selected ranges along which sounding positions (*1–13*) are distributed.

them have been completed use range connecting the signal of one range with that of another on the opposite side of the lake, as for example, in Fig. 9, set up the ranges *AB′*, *B′C*, *CD′* and *D′E;* also *E′D*, *DC′*, *C′B* and *BA′*.

Checks on location of sounding positions may be provided by beginning at some selected fixed point on shore and measuring distances to selected positions in triangle form; for example, distances *A–5*, *5–4*, and *4–A;* then set up chain of triangles by measuring distances, *4–6*, *6–7*, and *7–4; 7–10*, *10–9*, and *9–7;* and so on.

Plotting of Field Data

1. On a map of the shore line previously made, draw lines representing ranges used in field work, using same scale employed in mapping of shore

line and making certain that range lines are properly located. Scale distance *A–1* along the first range line; enter the sounding on the map. Scale distance from sounding position *1* to sounding position *2* and enter the sounding; pass to the next range line and proceed in the same way. Continue this procedure until all soundings are entered on map.

2. Plot check records for confirmation.

3. Remove construction lines; draw contours and enter descriptive data as described on pp. 68–69.

LOCATION OF SOUNDINGS BY PLANE TABLE AND RANGES

This method depends upon the use of the plane table from a known position on shore and upon boat courses (ranges) so planned and executed that the lines of sight from the plane table intersect the ranges at sounding points to form angles of as large a size as is practicable. When properly used, it is rapid, accurate, convenient, and requires a small party and few items of equipment. Furthermore, the labors of subsequent mapping, required by some of the other methods, are greatly reduced since the map is constructed while the field work is in progress. The instrumentman can watch the distribution of the soundings over the area of the lake, detect the unsounded regions and if they are large enough to be of consequence such gaps can be filled by locating new ranges before the party leaves the field.

SUITABILITY. This method is best suited for work on lakes of small sizes. It requires conditions of reasonably calm water, unless a sizable powerboat which can maintain a straight course on the range in rough water is used for soundings. This method should use as its basis the map of the shore line made by the plane-table method just in advance of the sounding work (see pp. 17–20, Fig. 4). As mentioned before, the size of the resulting map can be no greater than the maximum dimensions of the plane-table board.

EQUIPMENT. Same equipment as listed on p. 17, with the addition of white and red signal flags on staffs.

ORGANIZATION OF PARTY. Instrumentman; boatman; signalman-recorder; soundingman; assistant. In small lakes where soundings and sequence numbers can be reported vocally from the boat to the instrumentman as they are made, a signalman-recorder is not needed.

Procedure

1. If the plane table is located at *B* (Fig. 10), swing the line of sight (alidade) on *A* as a check on position of instrument; if the line of sight on *A* fails to coincide with line *BA* on map, make the necessary adjustment of board to effect such coincidences.

2. Select range, as for example, range *KG* with boat starting near *K;*

on the map on plane-table board draw a straight line connecting ends of selected range (*KG*). Select other ranges which will serve to distribute soundings over lake; draw lines on map representing these ranges and indicate order and direction of use. Make a sketch map of lake showing ranges and their sequence for guidance of the boatman.

3. If rowboat is employed, locate assistant at easily visible range pole *a* several feet behind signal pole *K* and assign to him the task of keeping boat on straight course between *K* and *G* by use of hand or flag signals made to oarsman. In this position assistant sights over *K* to *G*, thus defining the range, and any swing of boat to right or left is easily detected. If powerboat is used, the same method of keeping on range may be employed, or assistant may erect an easily visible range signal *b* at some distance behind *G*, thus supplying the wheelman on boat with means of steering on course by keeping *G* and *b* in line.

4. Boat takes position on range, comes to stop, sounding is taken, recorded, and designated as No. 1; simultaneously instrumentman rotates

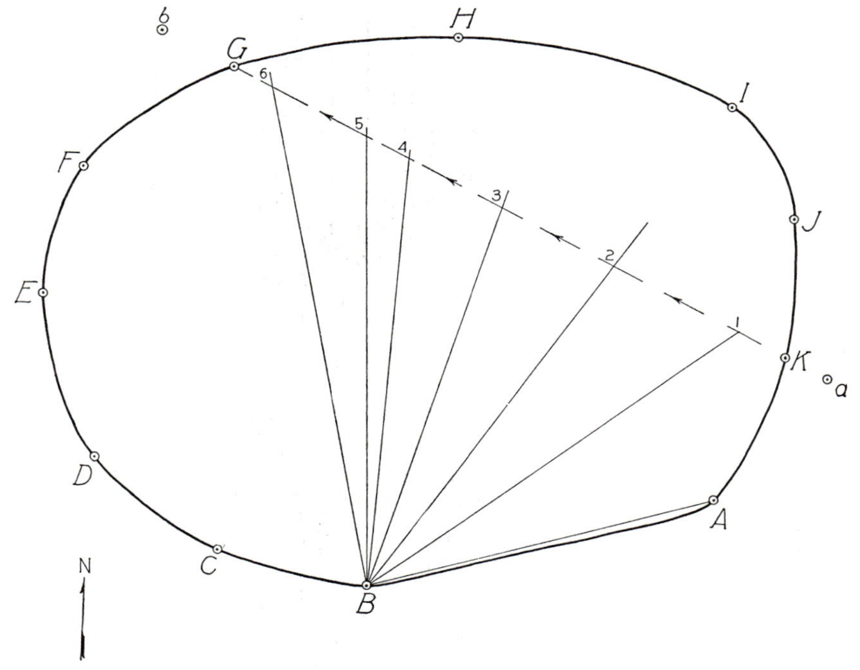

Fig. 10. Diagram illustrating method of locating sounding positions by plane table and ranges. (*A–K*) Fixed stations on shore line, the positions of which are indicated on shore-line map. (*KG*) Selected range over which sounding boat passes. (*a*, *b*) Range poles set some distance back of *K* and *G*, respectively, serving as aids for keeping boat on course. (*1–6*) Positions at which soundings are made, located by lines of sight from alidade intersecting range.

alidade on zero point, sights alidade on soundingman (or on signal if lake is large and distances considerable), draws line along edge of ruler base of alidade until it intersects range line, this intersection being the sounding position; if lake is small and the day calm, sounder may announce vocally the sounding and instrumentman can enter sounding at once on map at intersection. If vocal announcing is used, soundingman must call information in loud clear voice, instrumentman must call it back to soundingman, and finally soundingman must announce it again for confirmation. Record of sounding should be kept on boat for future reference. If lake is large and distances great at times, vocal announcing must not be attempted, but instead, signalman-recorder on boat should keep record of soundings and their sequence by number (1, 2, 3, 4, etc.), while instrumentman designates the intersections in same sequence, entering soundings on map after field work is completed. When sounding No. 1 is made on large lakes, signalman-recorder should elevate red flag.

5. Boat now moves along range for desired distance, and next sounding is taken and treated as described above. At the first and fifth sounding position and every fifth one thereafter, red flag should be raised; use white flag on all other positions.

6. When end of range is reached, change course of boat to next range in sequential order, and proceed as already described, continuing in this way until all ranges are completed. If untouched areas of lake seem too large, establish new ranges to cover them and proceed with additions.

Plotting of Field Data

1. If lake is small and air conditions quiet so that vocal announcements of soundings can be made directly and dependably from boat to instrumentman, the soundings can be entered on the map as made and the map virtually completed before the party leaves the field. Remove construction lines and finish map by adding data listed on pp. 68–69.

2. In instances of large lakes on which instrumentman cannot know soundings as made, soundings must be entered on map at a later time, care being taken that the proper sequence is followed from field records. Remove construction lines and finish map by adding data listed on pp. 68–69.

LOCATION OF SOUNDINGS WITH SEXTANT

This method provides for the location of positions on water by means of a sextant operated on board the boat at the time soundings are made. It requires the establishment of a series of objects or signals along or near the shore the position of each of which is definitely known on the shore-line map of the body of water being sounded. The sextant, an instrument operated in the hands, measures the horizontal angle at the boat between two signals. If the horizontal angle between the second of the pair of signals already sighted upon and a third signal is also read immediately then the two angles so read on the sextant will locate the point at which the sounding was made.

SUITABILITY. This method can be used on any lake whose size is such that signals on shore can be seen from the boat and definitely identified. This qualification depends upon the kind of sextant used. If the sextant is equipped with a telescopic sight tube, a larger lake can be mapped than is possible with a sextant having a nontelescopic sight tube. However, it should be noted that the size of a lake which can be mapped with a sextant having a nontelescopic sight tube will depend somewhat upon how distinct the shore signals are. In general, determination of position by sextant-read angles taken on board the sounding boat is one of the most satisfactory and commonly used methods. It has the advantage of being rapid, is adapted to work on a boat in motion, and may be operated with as few as two persons in the sounding crew. In fact, under conditions of quiet water, sounding work can be done by one person if the party must be that reduced and if speed is no consideration.

EQUIPMENT. One sextant (two sextants if available and if the party can include another instrumentman); sounding lead; graduated sounding line; graduated sounding pole; forms for records; boat of appropriate size and general equipment. If the sextant has a nontelescopic sight tube, a pair of field glasses may prove useful at times in establishing the exact identity of shore signals.

Since few limnologists seem to be familiar with the sextant its essential features are described here for their convenience. More extended accounts will be found in general treatises on surveying and navigation. In this discussion attention is restricted to the sextant as it may be used in work on ordinary lakes and streams.

The construction of a sextant, even in its most precise form, is relatively simple. Figs. 11 and 12 show the construction of a simple type and of a standard precision type. The sight tube and the horizon glass (mirror) are fixed in position; the index mirror is attached to the base of the index arm and moves with it. One-half of the horizon glass is silvered on the back so that it functions as a mirror while the other half is clear. In some instruments the upper half of the frame holding the horizon glass is not filled with glass at all. At its free end the index arm carries a vernier which swings over a graduated scale fixed on the frame. Most sextants have an index arm about 7 in. in length. The graduated scale is divided into degrees and fractions of degrees. It usually consists of an arc of 60° but on the instrument the scale indicates this 60° span as 120°. From the principle of the instrument this is necessary since the angles formed between the various positions of the index arm and the zero point are *one-half* the size of the true angle between the two distant objects on which the sextant is

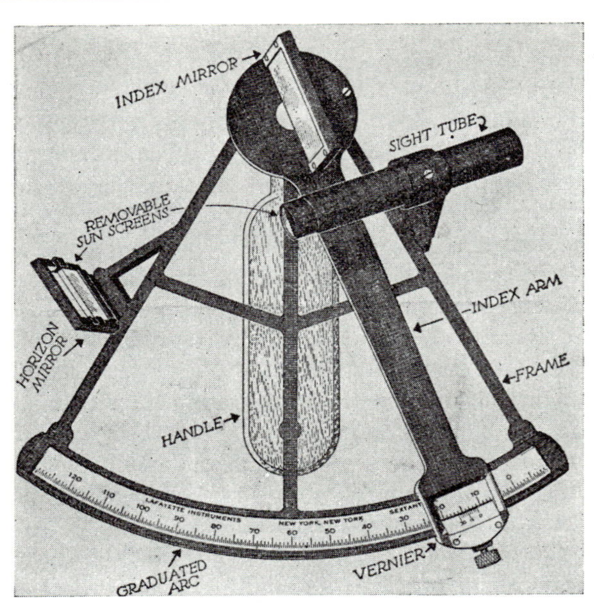

FIG. 11. A simple type of sextant. (Courtesy, Yoder Instruments, East Palestine, Ohio.)

being sighted. Mathematical proof of this relation will be found in almost any standard work on surveying. The smallest reading which may be made with the vernier (*least count*) will differ in different kinds of sextants. For hydrographic work sextants need only to read to minutes; in fact for much limnological work a least count of 3′ is adequate.

CORRECTIONS AND ADJUSTMENTS. The sextant should be checked by the following procedure: Set index arm exactly at zero; hold sextant horizontally; look through sight tube and through upper clear portion of horizon glass at some *distant* object, as for example, a telephone pole. Determine if reflected image of same object in lower portion of horizon glass lines up exactly with the object seen through upper portion; if direct and reflected images line up exactly, then the *index error* = 0 and the instrument is in proper adjustment. If the direct and reflected images do not coincide, move index arm, one way or the other, until images do line up; then clamp arm. The value on the scale is the *index error* which must be subtracted from, or added to, each reading depending upon whether the initial index arm position was on the scale (between 0° and 120°) or off the scale (beyond 0° in the other direction).

An index error may be reduced to 0 by adjusting the horizon glass but this should be attempted only by an experienced operator; usually it will be preferable to determine the index error and correct all readings.

FIG. 12. A standard type of sextant. (Courtesy, Leupold & Stevens Instruments, Portland, Ore.)

ORGANIZATION OF PARTY. One sextantman (two if two sextants are available); soundingman; recorder; boatman; assistant. The size of the party may be less than the five indicated above if necessity demands.

Procedure

1. Examine shore-line map of lake, or the lake itself if preferred; select carefully a series of natural objects, or erect artificial signals along or near the shore line the exact location of each of which can be accurately indicated on the shore-line map. These objects are to serve as *signal stations.* Each signal station must be distinctive and positively identifiable from sounding boat; they must be far enough apart so that no angle between any two of them is very small, but on the other hand they must not be more than about 120° apart, and must be so distributed about the lake margin that from all lake-surface areas at least *three adjacent signal* stations are clearly visible. For suggestions concerning artificial signals, see pp. 45–46.

2. Organize the boat party and review with the whole party all the procedures to be followed in the field. Supply oarsman, engineman, or helmsman with a small sketch map showing courses to be followed by the boat to cover the lake area adequately—courses to be steered blindly or with guidance of shore objects.

3. Sounding operations may now begin. Sounding boat chooses position No. 1 and comes to halt; soundingman makes sounding with line and lead, or with sounding pole if in shallow water, and announces sounding

to recorder, recorder makes record and repeats entry aloud. Soundingman reads line or pole second time and repeats announcement or announces correction if error was made. While sounding is being made, sextantman chooses a group of 3 easily visible shore signals (for example, 8, 9, and 10, if numbered signals are used), and with index arm of instrument set at zero, or at correction position, directs sight tube on shore signal (No. 8), then moves index arm until adjacent shore signal (No. 9) seen in lower part of horizon glass coincides with first signal seen through upper part of horizon glass. Index arm is then clamped and the vernier read, the angle between the two signals is read aloud to recorder who enters reading on record form and repeats the entry aloud. Sextantman

FORM 4

FIELD RECORDS

Location of Sounding Positions by Sextant

Lake	Sextant No.	Kinds of Signals Used
Date	Sextantman	Sounding Apparatus
Wind	Recorder	Corrections
Sky	Boatman	Sextant Index Error Sounding Line Errors
	Soundingman	

Position No.	*1st Angle*		*2nd Angle*		*3rd Angle*		*Sounding (m.)*	*Remarks*
	Signals Used	*Angle*	*Signals Used*	*Angle*	*Signals Used*	*Angle*		

reads vernier a second time and repeats announcement or announces correction if error has occurred. Then sextantman resets sextant at zero point, or at correction point, and directs sight tube at middle signal of selected trio (No. 9); moves index arm until third (No. 10) signal coincides with first signal (No. 9). Index arm is clamped and the angle between the second and third signals is read and recorded as described above. The two angles so determined locate the sounding position.

4. Sounding boat proceeds to another position and the operation described in paragraph 3 is repeated with the exception that sextantman may choose same shore signals as before if satisfactory for his purposes, or he may choose another trio. Trios of shore signals are chosen in such combinations and in such positions as best facilitate the location of sounding positions.

5. There are some advantages in distributing sounding positions along ranges, although this procedure is sometimes ignored.

General Considerations

a. Sextant should be checked frequently for index error.

b. When shore signals or objects are definite and readily identifiable by the unaided eye, a sextant without telescopic equipment is no disadvantage; in fact under some circumstances there may be an advantage.

c. When used on small boats, the sextant should be equipped with a strap or cord by means of which it can be attached to the observer as a safety measure.

d. Considerable practice is necessary in order to become expert in the use of a sextant. Inexperienced persons should have drill, preferably under direction, before undertaking to locate sounding positions in the field.

e. If soundings are made from a boat in continuous forward motion, two sextants and two sextantmen are needed, one measuring one angle and the other the second angle simultaneously. The soundingman must be experienced in line casting, if sounding is done by hand, in order that the sounding value may be secured when the line is vertical. If the sounding is done by means of a sounding machine, the angle of inclination of the line must be measured and the proper correction made.

f. In measurement of horizontal angles with the sextant, it is important that the distances between sextant and shore signals be considerable. The construction of the instrument is such that when close-up signals are used errors are introduced.

g. Ordinarily, in the measurement of angles between two objects on shore, the sight tube should be directed at the fainter one if there is any appreciable difference. Since the index mirror is to the right of the sight tube it will be necessary to hold the sextant upside down if the fainter object is on the right. However, the operator will be able to judge whether sighting on the fainter object is necessary.

h. If difficulty is experienced in picking up certain signals with the sextant, the work will be facilitated if the sextant can be set at roughly the angle to be measured. This approximate setting of the instrument may be accomplished as follows: Hold the hand at full arm's length in front of the eyes and spread the thumb and fingers to their widest position; then determine the angle subtended between the ends of the thumb and the little finger. With this rough value determined it is possible, with the hand in this position, to step off an approximate angle between two objects on shore.

i. Whenever there is more than one choice of signals in determining any position on the water, those which yield the strongest position–that position easiest to observe and plot with certainty–should be selected. Strong positions occur: (1) When sounding boat is within the triangle formed by the three shore signals. (2) When the three signals are nearly in line, or the middle signal is nearer the sounding boat than the others, and all angles have a magnitude of 30° or more. (3) When two signals, remote from each other, are in range and the angle between the second and third signal is not less than 30°. Under these circumstances the boat position is determinable from the one angle (between middle signal and the one not in line) but it may be desirable to measure an angle between the signal not in range and a fourth one just for a check.

j. Further guidance in the choice of shore stations may be secured from the following directions: (1) Avoid small angles as a general procedure. (2) Avoid the use of fixed shore objects as signals if they have considerable difference in elevation and one of the pair is near the observer. (3) Avoid the use of signals when the distance between two of them is small compared with the distance from the observer. (4) Avoid using signals which are very close at hand; instead, choose signals on opposite side of lake or on some other remote situation, provided the distance is not so great as to detract from the accuracy of the work. (5) Avoid the selection of any three shore signals whose positions happen to fall on the circumference of the same circle drawn through them. If such a selection must be used then it is imperative that the sounding boat positions be kept remote from the periphery of such a circle since such positions are not determinable. Only by the use of a third angle can such a position be located.

k. Shore objects or signals which are to serve as fixed positions for sextant sounding work must be in visible positions; must be individually and certainly identifiable from the sounding boat; must be situated in carefully chosen and strategic places; and must be *accurately located* on the shore-line map. If natural physiographic features on shore or artificial permanent constructions on shore meet the requirements just mentioned they may be used. However, occasions are likely to be rare when such objects are sufficiently distinctive and adequately distributed. Consequently, specially made signals must be provided. Satisfactory and relatively simple signals may be made in the following ways: (1) Boards, at least 8 in. wide and 4 ft. long, nailed together in the form of a cross, painted white, and fastened substantially to the top of an 8- or 10-ft. post. Such crosses may be spiked to trunks of trees along shore if such mounting does not interfere with their visibility. Combinations of white and red painting of such crosses will facilitate individual identification if such aid is needed. (2) Narrow boards or barrel staves may be arranged in the form of Roman numerals and nailed across the top of a tall pole, thus indicating the identity of the signal. Visibility is improved if these boards or staves are painted white. (3) Two boards, 3 in. wide and 2 ft. long, may be nailed crosswise to a pole or tall post, one at the top and the other at a distance below it about equal to its length, forming two horizontal arms. Sheets of cloth—red, white, or both—are stretched across the space between the arms and tacked firmly to them. (4) If one or more signals must be set in shallow water, a satisfactory construction is made as follows: three pieces of iron pipe, each about 12 ft. long and of suitable diameter, are securely tied together at one end; the untied ends are spread well apart in the form of a tripod and sunk firmly into the bottom materials; flags of white, red, or both white and red cloth, about 1.5 ft. square, nailed to staffs the lower ends of which will just fit snugly inside the pipes, are installed at the top of this tripod. Sometimes an addition to this type of signal consists of two strips of cloth wrapped around the tripod about midway between the water surface and the top.

l. On occasion it may be desirable, or even necessary when other well-located shore objects or signals are lacking, to measure an angle on a well-defined tangent to the end of a peninsula, the end of an island, or some other permanent physiographic feature. Such procedure may give acceptable results if carefully done and if the features used are not too remote from the sounding boat.

m. If signals are to have any permanence, as might be desirable in many instances, the following suggestions should be considered: (1) Construct principal parts of substantial and weather-resistant materials; (2) install shore signals above high-water and ice-action line; (3) set posts deep into the ground; and (4) paint both wood and metal parts.

n. Artificial shore signals should be located where good judgment indicates. Avoid using a larger number of such signals than is strictly necessary. The exact position of each should be marked by an iron stake, two feet long, driven completely into the ground and its position referenced in relation to adjacent permanent objects. Such procedure will make it possible to restore the signal if it is destroyed.

o. Artificial shore signals may, if desirable, be located at some distance landward from the shore line provided they meet all other qualifications of good signals.

p. At least some of the artificial shore signals should, if possible, be located at traverse stations or at local triangulation stations used in mapping the shore line. Such stations presumably are accurately located on the map and will facilitate the location of other signals.

Plotting of Field Data

1. This method presupposes the availability of a shore-line map of the water sounded. Such a shore-line map must be of such a scale that soundings can be located on it with reasonable ease, accuracy, and without crowding. Very small maps are not usable. If necessary, enlarge the shore-line map by any one of the approved methods (p. 77). Shore-line maps which are just included within a 36 $\times$ 36-in. sheet will usually be adequate for most inland lakes.

2. On the shore-line map locate with all accuracy possible the positions of the natural objects or the artificial shore signals used in the field work. If the original stations used in the shore-line survey have been retained on the map and if some of such stations have been used for signal positions for the sounding work, the plotting will be greatly facilitated as will also the location of the positions of other signals used. Unfortunately, such original survey stations are too often omitted from finished hydrographic maps. If the shore-line map has been made under the same authorship as that of the sounding work, shore-line signal positions will be properly located and available; but if a shore-line map of another authorship is employed, the shore signals must be located by whatever means best suits the circumstances. Then recourse must be had to objects definitely located and identified on the shore-line map, such as adjacent road intersections, large permanent buildings, lighthouses, large permanent piers, causeways, mouths of inflowing streams, beginnings of outlets, old sawmill sites; angular peninsulas, and other usable features.

3. Sounding positions which have already been located in the field by the three-point method can be plotted on the map most conveniently and rapidly by the use of a three-arm protractor. The high-class, metal three-arm protractors are superior instruments but because of cost may not be available. Somewhat less expensive celluloid three-arm protractors may be secured. If no three-arm protractor is available, a usable substitute may be provided as follows: For each sounding record, draw three lines on a piece of tracing paper or other translucent material so that the two measured angles are included between the lines. Shift the pattern of lines so prepared

over the shore-line map in the general vicinity of the shore signal positions used for that particular record until the three lines pass exactly through the three points; in this position, the point of union of the three lines (the apex) represents the position of the sounding position and may be pricked through into the underlying map with a needle. Three-arm protractors are used in a similar fashion, the sounding position being determined when the proper edges of the accurately set protractor arms cut through the middle of the three points. The plotting of positions is likely to be facilitated if some uniform method of placing the protractor or the three-line pattern on the map is followed, such as that of placing the middle arm or middle line on the middle shore position with the other arms or lines about equidistant from the corresponding shore positions and then pushing the center of the instrument or pattern toward the shore signal positions, shrinking the distances about equally on the two sides until the proper edges of the protractor or the lines of the pencil pattern exactly bisect the station points.

4. When a sounding position is located on the map, the sounding value should be entered on the map as described on p. 68. These procedures are repeated in like fashion until all soundings are recorded.

OTHER METHODS

COUNTING OAR STROKES. If instruments are lacking, or if the usually expected accuracy is not required, ranges may be established somewhat as illustrated in Fig. 9 and soundings may be made along such ranges by using a selected number of oar strokes to determine distance between soundings. If the average distance covered by one oar stroke by oarsman is determined, and the positions, directions, and lengths of ranges are known, it is possible to chart soundings made in this way. Obviously, such a method can give only gross results. Calm water conditions are a necessity.

Plot field data by laying off selected ranges on shore-line map; scale distances along each range line and locate soundings as determined by field data. Conclude map in the usual way.

LOCATION BY TIME INTERVALS. Under conditions somewhat similar to those mentioned in the section above, distances between soundings may be determined by the time interval involved. Ranges are laid out and the boat follows them in a definite order. If the approximate speed of the boat is known, the distance between soundings taken at stated intervals can be determined roughly. The limitations of this method are obvious. Plot data on map as described above and finish with data listed on pp. 68–69.

IMPROVISED INSTRUMENTS

In the absence of standard surveying instruments, satisfactory mapping can sometimes be done with improvised equipment if proper care is used, if the principles of the methods are understood, and if only small areas and distances are involved. The limitations of such home-made apparatus must be thoroughly understood and not exceeded. The following suggestions will indicate something as to the possibilities in this direction.

PLANE TABLE. By means of sturdy attachments, mount an ordinary drawing board, about 20 × 20 in. in dimensions, on a heavy tripod of convenient height and provide means of rotating and immobilizing the board. Select a good grade, flat, wooden ruler, 12–14 in. in length, graduated accurately to 1/16 in. or finer on one edge and in centimeters and millimeters on the other edge; near each end of the ruler in mid-line position drive a small brad in a vertical position, leaving about ½ inch of each brad projecting above surface of the ruler. A ruler, so modified, constitutes a rough alidade, the brads serving as sights. Add to this equipment a small pocket compass, range rods, tapes or graduated ropes, and chain pins, all of which can be improvised if necessary, and a workable plane-table outfit is provided. Very accurate maps of small areas may be made with such rough equipment. In such an outfit all distances must be measured by taping.

INSTRUMENT FOR MEASURING ANGLES. Mount, by sturdy attachment, a small drawing board about 16 × 16 in. in size on a heavy tripod. Attach firmly to the upper surface of this board a full-circle, 12-in. diameter, protractor, graduated to ¼ degree and with the whole degrees numbered by two sets of figures which increase from 0 to 360 in opposite directions. Prepare a slender wooden or brass pointer, slightly longer than the diameter of the protractor; attach to each end of the pointer, on mid-line, a slender pin to serve as a sight. Near one end of the pointer cut a slender, oval window about 1 in. long and seal a very fine wire or thread lengthwise on the top surface of the window to coincide exactly with the mid-line connecting the two sights; install a vertical axis or pivot in the middle of the pointer and set firmly at center of the protractor. Construct the pivot so that the pointer can be rotated, not loosely, over the protractor in very close proximity to but not in actual contact with it. In this position the pointer window should extend across the graduated scale on the protractor in all positions and the hair or thread across the window serves as a means of reading the angles. This window must be wide enough to enable the observer to read the graduated scale accurately.

Such a rough instrument may, on occasion, be used as a substitute for a transit and with proper care and understanding good results may be obtained in simple mapping of small areas.

AERIAL PHOTOGRAPHIC MAPPING

Rarely if ever is it possible for the limnologist to make aerial photographic maps. He must depend upon those prepared by certain branches of the government or perhaps by commercial companies which specialize in that kind of work. In certain parts of the United States, large areas have already been mapped from the air and excellent maps may be secured from the various sources, although cost is still a consideration. Since such maps are not usually made specifically to serve the needs of the limnologist, careful selection is necessary. When made on a scale large enough to be serviceable, or when enlarged to usable size from the originals, these maps may yield valuable information otherwise difficult to get. They serve as excellent checks on ground surveys, particularly in such features as form of shore line, fixed points on shore, and position and extent of shallowly submerged shoals. Because of their nature, the accurate interpretation of aerial photographic maps requires experience on the part of the observer, hence conclusions of a beginner should be carefully checked by someone familiar with the problems of air-map reading. It must be understood that, for limnological purposes, no aerial photographic map is a substitute for a standard hydrographic map since it does not supply information on many salient features such as detailed soundings, form of basin, and character of bottom.

Additions to Lake Surveys

Limnological needs in lake surveys often require information in addition to that supplied by the conventional hydrographic map. Bottom materials and deposits in the various areas, character of the banks, degree of shading of banks, aquatic vegetation in the various areas, and special features of inflowing water, flood levels, artificial features, and any other significant characters require field study for their incorporation on the maps. (See pp. 71–75 for suggested symbols, descriptive designations and other characters useful in such mapping.) Information basic to the charting of these features should, if possible, be secured immediately after the shore line and basin have been surveyed, while all stakes and other established points are still in position. Surveys of the various bottom areas, vegetation areas, and similar features may be made by modifications of methods already described. Transects, spread in significant patterns and definitely located with respect to fixed points on shore, constitute one of the best methods. Stations along transects or scattered on the open water may be located by any of the methods described for location of sounding positions.

For Selected References, see p. 59.

STREAM SURVEYS

Streams, like lakes, may be mapped by various methods. Those described here are useful for streams of smaller size (creeks and brooks). Larger units of the running water series (rivers) may be mapped by the proper adaptation of certain methods described in this chapter and in the preceding ones.

Deflection-angle Traverse

In a deflection-angle traverse the direction of each successive axis line is established by measuring the angle that the last line makes with the forward prolongation of the preceding one. The same result is obtained if, instead of a deflection angle, the angles which are formed by the two lines of a course without prolongation are used, but deflection angles are mostly smaller in magnitude, easier to plot, and on certain kinds of instruments are easier to read.

Suitability. This method, as outlined here, is especially suitable for mapping small streams whose widths and depths are such that the traverse can be made in the channel, or for streams whose shore conditions are such that, if the traverse cannot be in the channel, it can be run on either bank or along either water edge.

Equipment. Transit; stadia rod; range rod; ax; wooden stakes; chain pins on carrying ring, with colored cloth tied in upper ends; crayon for marking stakes; 100-ft. steel tape, or 100-ft. graduated rope; sounding pole; megaphone; forms for records.

corrections and adjustments. Examine transit, stadia rod, and tape or rope for accuracy and good condition.

Organization of Party. Transitman; stadia-rodman; range-rodman; two tapemen; assistant.

PROCEDURE

1. Make preliminary examination of stream or portion of stream to be surveyed in order to anticipate any special problems peculiar to the task. It is necessary to determine in advance whether it is practicable to run the traverse in the stream bed, whether it must be run on the bank, or whether a combination of channel and bank course must be used. The following description will assume a traverse run in the channel.

2. Select initial transit station (Fig. 13, *A*); drive iron stake at this

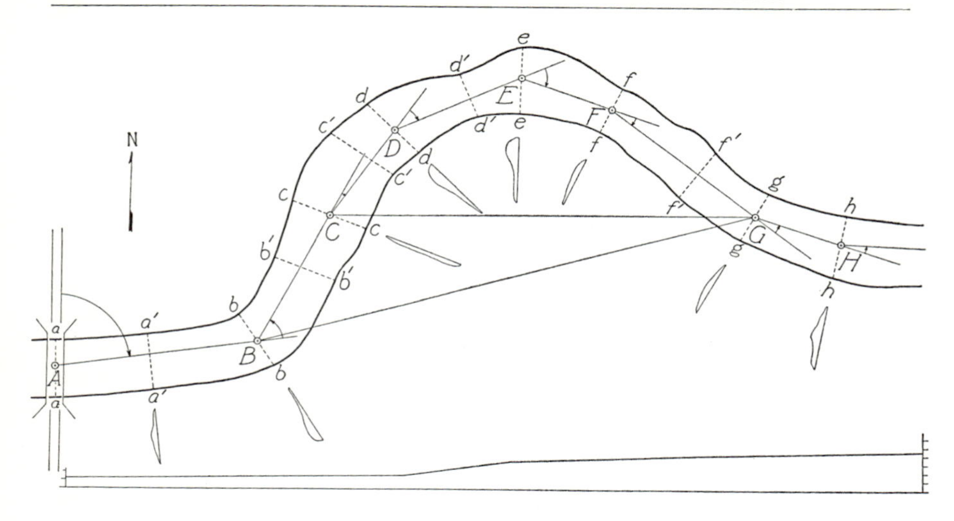

FIG. 13. Diagram illustrating deflection-angle traverse of a portion of small stream. Traverse is run in stream channel. Curved arrows represent deflection angles. (*A–H*) Transit stations. (*aa–hh*) Positions of measurements of channel widths at transit stations. (*a′a′–f′f′*) Positions, intermediate between transit stations, at which measurements of channel widths were taken. (*CG, BG*) Checks on position of transit station *G*. Survey is represented as beginning at bridge where road crosses stream.

Diagrams opposite *a′*, *b*, *c*, *d*, *e*, *f*, *g*, and *h* represent transverse *depth profiles* of channel at those positions. In these profiles, scale for width is same as that of map of stream; but scale for depth has been exaggerated by multiplying all depth data by some stated value, a procedure made necessary by very shallow depths of water. Different scales may be necessary in profile mapping, depending upon conditions of stream involved. However, scales of profiles must always be definitely stated in detail.

Profile at bottom of figure represents fall in level of stream made according to some convenient scale. Value of such scale must always be clearly stated.

point and reference it to adjacent, permanent objects. It may be convenient to begin survey at road intersections, bridges, or other easily identifiable positions.

3. Set up transit on initial station *A;* choose next station in direction in which survey is to proceed and as far in advance as contour of stream will permit line of sight to remain well within borders of channel (unless distance is unusually long, then intermediate stations must be established). Mark and drive stake *B;* direct transit on signal at *B* and carefully take magnetic bearing; also, if starting at intersection of some already established survey line or road, determine angle which line *AB* makes with such line or road.

4. Erect stadia rod on *B* and determine distance *AB*.

5. At *A* and at significant intervals *a′a′* between *A* and *B* make transverse measurements of stream, water edge to water edge, or bank to bank, or both as desired. Make such measurements so that distances, right and left, from line of sight *AB* to each channel margin are recorded, also determine distance of each transverse measurement from *A*. Drive stakes at all such positions and mark on each with crayon its identification symbol or number.

6. At all positions of transverse measurements, make water-depth measurements as follows: in middle of stream and at least at mid-distance between middle and each bank (three measurements). Distances of depth measurements from each other and from line of sight *AB* must be known and recorded. Use more than three depth measurements in very wide streams.

7. Transfer transit to *B;* direct telescope on range rod set on *A* and set transit scale at zero; either reverse telescope on its horizontal axis ("plunge telescope") or reverse telescope on its vertical axis ("reverse in azimuth") exactly 180° thus prolonging line *AB*. Select position next in advance *C*, mark and drive stake, label and erect signal; direct telescope on *C* and determine angle of line *BC* with prolongation of *AB* (deflection angle). Record deflection angle as *Right* or *Left* depending upon how the new line of sight relates itself to the prolongation of the previous one; erect stadia rod on *C* and determine distance *BC*.

8. Make transverse-channel and water-depth measurements at *B* and between *B* and *C* as described in paragraphs 5 and 6 and keep all necessary records.

9. Proceed with survey, successively establishing positions *D–H* and determining all distances, deflection angles, transverse measurements, and depth measurements as already described. Record magnetic bearings of some or all lines of sight for checks.

10. In surveying tortuous streams in open, unwooded areas, check sections of traverse by setting up triangles from place to place, as for example in Fig. 13 establish *G* at beginning of survey of that section of stream and mark with signal. Determine angle *GBC* when transit is at *B*, angle *BCG* when transit is at *C;* with these angles and the length of line *BC* known the position of *G* can be computed or plotted. Or, if distances and visibility will permit, defer check until *G* is reached in traverse, determine distances *GB* and *GC* by tape or stadia rod, then lengths of all sides are known, also one angle, and the position of *G* can be determined by plotting or computation.

11. For additions to the survey, especially for biological purposes, see pp. 71–75.

General Considerations

a. When the traverse must be run on the bank, or at the water's edge, most of the necessary modifications of the method as described above are obvious and need not be enumerated here. A bank traverse often requires considerable clearing of vegetation from lines of sight.

b. If necessary or desirable, a combination of channel and bank traverse can be employed by merely transferring with known angles and distances, the general course of the traverse from channel to bank or from bank to channel.

c. If a stadia rod is not a part of the equipment, all distances can be measured by means of steel tapes, chains, or graduated ropes.

FORM 5

FIELD RECORDS

Deflection-angle Traverse

Stream | Transitman

Section Surveyed | Recorder

Date | Rodmen

Transit No. | Tapemen

Tape No.

Station	*Deflection Angle*	*Distance*	*Bearing*		*Remarks*
			Magnetic	*True*	

d. The number of transverse-channel and water-depth measurements must depend upon the variation in the physical features of the stream and upon the precision desired in the survey. For ordinary purposes and in sections of the stream where the channel is quite uniform, the number of transverse measurements can be reduced to a minimum.

e. In a channel traverse, great care must be used in setting up the transit to insure against changes in position due to water flow, walking about the tripod, and other causes. Soft bottoms are particularly troublesome. In case of doubt revert to backsight and reread the angles.

f. It is possible, and sometimes desirable, to combine with the survey the determination of stream channel elevation so that a level profile can be constructed. However, this should not be attempted unless the transitman or leader of the party is familiar with the principles and practices of measuring levels.

g. A traverse similar to the kind described in this section can be made, with less precision, by using a surveyor's field compass instead of a transit. In such procedure, angles must be read in the compass box and depend upon magnetic bearings; also all distances must be measured by lines or tapes.

PLOTTING OF FIELD DATA

1. Determine scale suitable for size of map desired.

2. Locate initial station (Fig. 13, *A*) of transit so that map will occupy paper in desirable position; make pencil dot, surround dot with small penciled circle, and label it *A*.

3. Establish north-south line through *A*.

4. Draw transverse line *aa* through *A* and scale channel width at that level; mark positions.

5. With protractor centered at *A*, draw line *AB* according to magnetic bearing recorded in field data; check with any angular measurements provided in field notes.

6. Scale distance *AB* and locate *B* with pencil dot surrounded by small penciled circle.

7. Scale distance *A* to intersection of *a′a′*; draw transverse line representing *a′a′* at right angle to line *AB;* scale distance *a′a′* and mark positions *a′* and *a′*. Proceed similarly with other transverse measurements of channel between *A* and *B* if more than one was made in field.

8. Transfer protractor to *B;* protract deflection angle from a prolongation of *AB* and draw line *BC;* scale distance *BC* and locate *C* by pencil dot surrounded by small penciled circle.

9. Bisect angle *ABC* and draw transverse line *bb;* scale distance *bb* and locate *b* and *b* as determined by field data. Mark positions *b* and *b*.

10. Scale distance *B* to intersection of *b′b′;* at right angle to *BC* draw

line *b′b′* and scale distance *b′b′* as determined by field data; mark positions *b′* and *b′*. Proceed similarly with other transverse measurements of channel between *B* and *A* if others were made.

11. Connect with firm pencil line positions *a*, *a′*, *b*, *b′* and any intermediate ones to outline one bank of channel and similarly connect positions *a*, *a′*, *b*, *b′* and any intermediate ones on other side to form other bank of channel.

12. Proceed in similar way with other lines and deflection angles until map is completed. Apply all check information taken in field operations to test accuracy of map.

13. On one side of map of channel when completed draw cross-section depth profiles, using same horizontal scale as employed in mapping channel (or if desirable use an enlarged scale), at each position representing levels at which depth measurements were made (Fig. 13). In very shallow streams, a much larger *depth scale* will be required to make profiles show sufficient depth to be useful. Be sure that the various scales employed are indicated on the map.

14. When channel map and profiles are completed, remove all construction lines, and add usual map information as specified on pp. 68–69.

15. If channel-water levels have been measured in connection with field work, then construct at the bottom of the map an elevation profile, using some convenient height scale, which will show channel slope.

16. If desired, add to the map such words, symbols, and abbreviations as will indicate the character and distribution of the various stream characters which have limnological significance, as for example, bottom materials, character of banks and aquatic vegetation (pp. 71–75). To avoid congestion of the map it may be necessary to make more than one copy in order to represent properly all such features. Colored shading may be used effectively to incorporate areas of different kinds of bottom materials and similar data. For specific indication or naming of details, whole words may be written on the map at appropriate places; in other instances, properly chosen symbols or abbreviations (pp. 69–75) may be preferable.

Plane-table Traverse

In stream mapping the plane table if available can often be used to considerable advantage, particularly if a telescopic alidade with stadia cross wires is employed. This method has the well-known virtue of constructing the map on the plane-table board as the survey progresses thus making possible the retaking of readings and the adding of extra measurements when necessary. It has the disadvantage, when compared with the transit traverse method, of requiring the use of a scale small enough to bring the map within the dimensions of the plane-table board and in stream surveying this scale is often far too reduced. If large-scale maps are necessary they can be made directly by plane table only by using numerous sheets of plane-table board size, each sheet representing such section of the stream as can be included, and later assembling these sheets in proper sequence. To make such assembly of sheets, care must be taken to secure in the field full information necessary to construct this assembly in correct form. Some of the features of plane-table operation are described on p. 17.

SUITABILITY. In general, same as described for the deflection-angle traverse (p. 51).

EQUIPMENT. Plane table; stadia rod; range rod; 100-ft. steel tape; chain pins on carrying ring, and with pieces of colored cloth tied into upper ends; 2–3 iron stakes; wooden stakes; ax; crayon for marking stakes; several No. 5 insect pins; waterproof cover for plane-table board; compass; forms for records; pencils; several sheets of good grade, heavy white paper cut to cover whole area of plane-table board.

CORRECTIONS AND ADJUSTMENTS. Examine plane table, stadia rod and steel tape for good condition.

ORGANIZATION OF PARTY. Instrumentman; stadia-rodman; range-rodman; two tapemen; assistant.

PROCEDURE

1. Make preliminary examination of stream in order that survey can be properly planned. Among other things, determine whether traverse can be run in the channel or whether it must be run on bank.

2. Select first instrument position (Fig. 13, *A*); drive iron stake and reference it to adjacent, permanent objects. If convenient and desirable, begin survey at road intersection, bridge, or some other previous survey location.

3. Set up plane table over *A;* orient board in most desirable position and lock. Establish and mark north-south line on paper; establish next position *B*, drive stake, mark with sequence designation, and elevate stadia rod on it.

4. Select scale to be used for mapping on plane-table board and record it on one side of paper. Estimate space and position requirements of future map on paper; choose proper position of *A* and set pin firmly at that point.

5. Set alidade with zero point against implanted pin; pivot alidade on pin and direct telescope on stadia rod at *B;* draw line along ruler edge of alidade. Read distance *AB* on stadia rod and lay off same to scale along the line just drawn; mark point and designate *B*.

6. Pivot alidade to position at right angle to line *AB;* measure distance to water's edge (or bank) on stadia rod, and draw line and scale distance. Mark point on map; drive stake at shore position *a* and mark.

7. Place stadia rod at position halfway between *a* and *A* and measure

water depth at that point. (Distance and water depth can both be read on the stadia rod if water is not too deep.) Record depth in field records; measure depth at *A* and record.

8. Pivot alidade 180° and measure distance to opposite edge (or bank), scale and mark; drive stake at shore position and mark. Place stadia rod halfway between *A* and *a;* determine and record water depth; measure depth of water at *A* and record.

9. Choose position of next transverse measurement *a′a′* and drive stake at both positions; without moving plane table, pivot alidade and sight on stadia rod at *a′;* read distance, scale, and mark on map. Pivot alidade to stadia rod on other side of stream at opposite end of *a′a′;* read distance, scale, and mark. Make water-depth readings along *a′a′* at middle and halfway between middle and each water edge (or bank) with tape or graduated rope; record depths.

10. Move plane table to *B;* set up carefully and rotate board until line *AB* on plane-table board coincides exactly in direction with line of sight *BA*, when alidade is rotated in working position about pin at *B*.

11. Establish next position *C* in advance and proceed with the survey as described above, including transverse-width and water-depth measurements at *bb* and *b′b′* and any others intervening between *B* and *C*. Continue thus until survey is complete.

12. In open, unwooded areas, check survey by reading backward on selected positions, as for example, in Fig. 13, with plane table at *G*, read direction and distance on stadia rod at *B;* scale distance and scaled point should fall on *B* on map.

13. Remove all construction lines; add information designated on pp. 68–69.

General Considerations

a. Items *a*, *b*, *c*, *d*, and *e* on pp. 54–55 relating to deflection-angle traverse apply in general to the plane-table survey.

b. A nontelescopic alidade can be used in this method, but if so, all distances must be measured with tapes or graduated lines.

c. If desired, complete map by inclusion of the descriptive features of the stream as mentioned and indicated on pp. 69–75.

Soundings in Streams

Methods for making soundings described on pp. 25–37 may be readily adapted for use on streams and need not be considered here. Sounding requirements on very wide streams approximate those for lakes. For narrow but deep streams, a method of making soundings along a series of parallel, transverse ranges will probably be found satisfactory and reasonably rapid. Directions for the recording and mapping of sounding records as given on p. 68 should be followed also in work with streams.

Aerial Photographic Maps of Streams

As in the instance of lakes, aerial photographic maps (p. 48) when available and made on suitable scale, yield information of much value. This is particularly true when the exact form and length of a stream, or part of a stream, is desired. Also facts concerning the drainage basin may be secured. However, for the purposes of detailed stream survey or stream investigations, there is at present no satisfactory substitute for the type of ground mapping described on pp. 51–58.

Additions to Stream Surveys

Certain biological needs in stream surveys require information other than stream configuration, channel widths, and channel or water depths. These additions should be made if possible during the progress of the survey or immediately afterward while all of the stakes are still in position. For each area definitely marked off by stations and stakes, make field records describing the features concerned. Prominent among these features are the following: bottom materials; character of banks; nature of flow of water; shading from bank; aquatic vegetation; character of inflowing water; high-water and flood levels; and artificial features. (See pp. 69–75 for list of suggested symbols, descriptive designations, and other characters useful in mapping.)

One of the best ways of making necessary field notes on the items mentioned above is the use of freehand sketches of the stream on which the various features are indicated and labeled. Such sketches may be made one section at a time and then pasted end to end at the conclusion of field work. It will usually be found convenient to use more than one such freehand sketch to avoid congestion of labels and sketched areas.

Selected References

(For Chapters 1–3)

Bouchard, H.: "Surveying," 2d ed., 625 pp., 187 fig. Scranton, Pa., International Textbook Co., 1940.

Brown, C. J. D. and O. H. Clark: Winter lake mapping, *Michigan Conservation*, 8: 10–11, 1939. (Reprint of same paper issued to authors in enlarged form.)

Embody, D. R., Goodrum, C. A., and S. D. Edmond: "A Statistical Analysis of Width Measurements on a New Hampshire Stream," Biol. Surv. Merrimack Watershed, Survey Rept. No. 3, New Hampshire Fish and Game Dept., 1938, pp. 198–200.

Hawley, J. H.: "Hydrographic Manual." 170 pp. Spec. Pub. No. 143, U.S. Coast and Geodetic Survey, 1928.

Lea, S. H.: "Hydrographic Surveying." 172 pp. New York, Eng. News Publ. Co., 1905.

"Surveying Tables." 392 pp. Technical Manual, War Dept., Washington, D.C., 1940.

Swainson, O. W.: "Topographic Manual." 121 pp. Spec. Pub. No. 144, U.S. Coast and Geodetic Survey, 1928.

Wainwright, D. B.: "Plane Table Manual." 97 pp. Spec. Pub. No. 85, U.S. Coast and Geodetic Survey, 1922.

WATER LEVELS

Since water-surface elevations fluctuate, sometimes markedly, and since some measure of such variations is vital to an understanding of certain limnological phenomena, it is often necessary to provide means of securing dependable records of water levels and their changes.

Water levels may be expressed in two ways: (1) with reference to an arbitrarily selected level; or (2) in actual elevation above sea level. In the first instance, total change of water level over a period of time is secured and for some purposes this information may be adequate. Often, no other alternative is possible since a *bench mark*—a permanent point of known elevation—may not be available. For the determination of actual elevations in terms of sea level an authoritative bench mark is required. The U.S. Coast and Geodetic Survey and the U.S. Geological Survey have established many bench marks throughout the United States, the locations of which can be secured from the offices of these bureaus. These bench marks consist of circular bronze tablets, about 4 inches in diameter, cemented into rock or masonry structures or into the top of concrete posts. On the bronze tablets are inscribed certain data, usually including the elevation at that point. Sometimes other dependable elevation bench marks have been established by railroads, cities, or scientific organizations. The extension of a level datum from a bench mark to a position on a water-level recording instrument is a task to be performed only by an experienced surveyor or by someone thoroughly familiar with the running of levels. Also only instruments suitable for such work can be used. If the only available bench mark is at a considerable distance and the country is rugged, it may not be feasible to attempt to extend such a level unless the project of measuring water levels is to be one of more than usual importance and duration. Local bench marks, once established, should be fully protected against damage or alteration; also the level datum which they represent should be extended to other adjacent permanent objects and so marked that the elevation datum can be recovered with a minimum of effort in case it is lost.

Graphic Water-level Recorders

A complete, continuous record of water-level changes may be secured by the use of a graphic water-level recorder of which there are many models on the market. These instruments are expensive and require, for their best operation, an elaborate installation, but when used properly they have many points of superiority. The essential features of such an instrument are as follows: a float which rides on the surface of the water; a cylinder revolved at some uniform speed by clockwork mechanism; a recording pencil or pen so adjusted that it traces the rise and fall of the float; and paper charts of suitable pattern attached to the revolving drum on which the recording pencil or pen inscribes a continuous graph. This graph is both a *time* and a *water-level change record* for the period over which the instrument was in continuous operation.

Since these recorders require complete protection and since the most satisfactory record is secured when the float operates in a *still well,* they are usually installed in some kind of gage house which combines a still well with a superimposed platform on which the recording mechanism of the instrument is permanently supported. A still well is a device for minimizing the effect of waves. It consists essentially of some kind of a well, whose depth extends below all possible surface levels for the lake or stream to be measured and which is connected by a pipe of small diameter to the open water. Plans for a gage house may be secured from the office of the U.S. Geological Survey.

Hook Gage

The essential feature of all types of hook gages is a recurved metal hook so set in one end of a staff that its free pointed tip is directed toward the opposite end. When used, the hook-bearing end of the staff is submerged in the water; then the staff is carefully raised until the free end of the hook just pierces the water surface. The distance from the tip of the hook to some superimposed bench mark is then read on the staff. The virtue of the recurved hook lies in the fact that under ordinary circumstances it is easy for the observer, operating the gage from above the water surface, to detect the level at which the ascending end of the hook just breaks the surface film.

For precise work there are hook gages on the market, equipped with verniers and slow-motion screws. Such a gage may be fastened permanently to some support located in an operating position. With such instruments, readings to 0.001 foot may be made, assuming that the conditions under which the instrument is used are favorable enough to warrant a record of that precision. A still well or some similar device for minimizing surface disturbance is necessary for very precise measurements.

For many limnological purposes a satisfactory hook gage may be constructed as follows: select a staff composed of some light, well-seasoned wood, about ⅜ × 1.0 in. in transverse dimensions, and with length suitable to the task. Construct a U-shaped hook from ⅛-in. metal rod and fasten one end firmly into the lower end of staff. Round off free end of hook to form blunt point; make free end of hook correspond exactly in level with lower end of staff. With zero point at lower end, graduate staff carefully by marking position of each foot of distance and further subdivide each interval into divisions of 0.1 foot. Cover with several coats of shellac. Choose favorable and convenient place for water-level measurements; provide some sturdy, immovable support; fasten to support and in vertical position a guide made of two 0.5 $\times$ 2-in. pieces of wood, straight, planed, and about 2 to 3 ft. long, nailed together at edges to form L-shaped trough. From adjacent bench mark extend some known level to wooden guide; establish this level on guide by permanent transverse mark and label with elevation value at that point. When ready for use, place hook staff in inner angle of vertical guide; slide staff down until hook on staff is completely submerged; raise staff until free end of hook just pierces surface of water. Read distance from water surface to bench mark, using accurate, finely graduated hand ruler if necessary to secure more precise reading. Record distance as taken if bench mark has only an assumed value; if bench mark represents actual elevation above sea level, subtract water surface-to-bench mark distance from bench-mark value and the difference will be actual elevation of water surface at that time.

The hook gage just described operates more accurately in a still well or under some other circumstance in which wave action is reduced to a minimum. If it must be used in the presence of wave action, the reading should be taken by lowering the hook until its free end is as nearly midway between the trough and the crest of the waves as can be estimated. Obviously, such readings are less exact than those made on still water. On still water and with care, it is possible to measure water level with the kind of gage just described to a precision of 0.05 inch.

The supporting device must be protected against disturbance after it has been installed and the bench mark has been evaluated. Such a bench mark should be checked from time to time in order to discover if changes have occurred.

For some purposes a Stevens anchor gage may be convenient. It is essentially a hook with two balanced recurved points mounted opposite each other in such a way that the whole device resembles an anchor. It is usually operated on a steel tape wound on a small hand lift. When used in a still well, levels can be read with rather high precision by means of an overhead light which is reflected from the water surface on which the dimples made by the two points are readily seen.

Staff Gage

For ordinary and less precise purposes, the height of the water surface may be measured by a *staff gage*. Such a device consists of a graduated board established in a vertical position at a suitable, convenient place on the side of a still well, retaining wall, dock, boat slip, breakwater, pile, stake, or similar structure. It is so placed that the lower end is always below the minimum water level. Such a staff gage may be constructed in any form and size suitable to the circumstances under which it is to be used. For large lakes and streams such a gage should be made of some well-selected board, about 6 in. in width, 1 in. thick, and with a length which exceeds the known variation in water level. It should be painted white with graduations marked with black paint at intervals of 1 ft. and tenths of a foot. The 1-ft. intervals should be numbered in some convenient sequence. Metal staff gages, covered with porcelain enamel and graduated in various systems, are now manufactured by certain supply companies, and for many purposes are superior.

Water-streak Gage

For running waters, surface level may be determined as follows: in the lower end of a 6–10-lb. sounding lead, install a peg about the size of an ordinary lead pencil and about 4 in. long; taper the free end of the peg to a rounded blunt point; attach the lead to a graduated rope or tape; suspend the sounding lead from some fixed overhanging support, on which is established a bench mark or a reference mark, and lower until the free end of the peg just touches the moving water, making a water streak. The distance from the bench mark to the free end of the peg subtracted from the bench-mark value gives actual surface elevation. If the reference mark is an arbitrary one, the reading provides a relative value. This method lacks the precision of certain others but for many purposes it is useful.

Other Methods

If reduced light conditions, distance of water surface from eye of observer, or any other circumstance makes it impossible to see the dimples made by a hook gage or by an anchor gage, the water level may be determined by the use of an inelastic tape or line long enough to reach from the reference or bench mark to the water surface. Attach a weight at the lower end of the line and smear the lower two feet of the line with carpenter's blue chalk. Lower the line until the weight and part of the lower chalked section of the line is submerged. Mark line at top to correspond with reference mark and raise line. The water line will show on the blue chalk. Measure distance between wet mark on chalk and the upper reference mark and make any further computations necessary to determine the water level.

Certain pastes are available on the market which, when smeared as a coating on a rod or tape, markedly change color when brought into contact with water, producing a sharp line at the water surface.

Selected References
(For Chapter 4)

Lee, L.: "Equipment for River Measurements." 3 pp. and 9 sheets of illustrations and specifications. Water Resources Branch, Geol. Survey, U.S. Dept. of Interior, 1934.
Stevens, J. C.: "Data Book," 4th ed., 144 pp. Leupold & Stevens Instruments, Portland, Ore., 1945.

MAP CONSTRUCTION

Equipment and Materials

The following easily obtainable items of equipment and materials should be provided:

1. Drawing Table. Substantial, smooth-surface table so located that the operator can work all around it. Size not less than 3 × 3 ft., preferably larger.

2. Meter Sticks. Two or more good quality, wooden meter sticks. They must be *straight* in all positions; should be used only for map construction work; and should be stored, when not in use, on a flat, straight surface and kept dry.

3. Rulers. Good quality 1-ft. rulers; straight.

4. Triangles. Two or three sizes; celluloid type is preferable.

5. Parallel-line Rulers. Not absolutely necessary but very convenient.

6. Paper. Some good grade, light-colored paper having a surface which will withstand reasonable erasures without serious damage; heavy enough to rest flat on board.

7. Protractors. Large circular or semicircular protractors, 12–14 in. in diameter and graduated to ¼ degree; paper, celluloid, or metal. More elaborate protractors can be secured but for purposes described herein are seldom necessary. Smaller protractors should be avoided. Celluloid protractors have the advantage of being translucent.

CHECKING ACCURACY OF PROTRACTORS. Protractors may be in error either because of imperfect construction or lack of adjustment, the latter being a possible fault in three-arm protractors. Some types of protractors can not be readjusted but must be tested for significant error. Metal test plates, made especially for this purpose, afford a convenient and quick means for checking. If such a test plate is not available, a test *pattern* can be constructed as follows: Across the center of a large sheet of heavy paper draw two straight lines *exactly perpendicular* to each other. These two lines must be set up with the greatest care possible. Extend lines to edges of paper. Bisect the right angle in each quadrant and extend straight line through bisection point and the central intersection of all lines. Superimpose protractor over this figure, make the central point of both protractor and test figure coincide; rotate protractor in a series of positions and determine the amount of agreement in each instance.

8. Other Equipment. Erasers of several kinds including block of art gum; Scotch cellulose tape; good T square. For further information and directions relating to drawing tools and materials, consult some standard work on cartography.

Construction of Hydrographic Map

Irrespective of the method used in making soundings in the field, the map should be constructed as follows:

1. Enter all numbers representing soundings in the normal reading position, i.e., so that all numerals are in the same position and readable when bottom of map is nearest reader.
2. Use simple, clean-cut, open style of numeral to represent soundings; lightface italic style is recommended.
3. Use numeral of easily readable size; numerals whose maximum height is 2–3 mm. are recommended.
4. Place numeral on map so that it is centered on the sounding position.
5. Draw submerged contours at desired intervals. Construct such contours by drawing a continuous line connecting all points having same depth; pass contour line through all numbers having exactly the value of the chosen contour; where the exact chosen value is absent pass contour line between those soundings which are less and greater than the chosen value, estimating position of contour line between the two values.
6. Make all contour lines of uniform thickness; use line distinctly lighter than shore line.
7. Remove all construction lines not necessary to future interpretation of map.
8. On every hydrographic map, when finished, there should appear certain essential information and explanations. Maps will differ somewhat in these requirements depending upon their nature and detail. However, most or all of the following items should appear on every map:

   a. General title, if map is one of a series
   b. Series number, if map is one of a series
   c. Official name of lake or lakes, stream or streams
   d. Geographic location (state, county, township, section)
   e. Scale, or scales
   f. Compass figure
   g. Low-water datum and its use
   h. Date of field work
   i. Date of map construction
   j. Names of field party
   k. Names of draftsmen
   l. Abbreviations used and their meanings
   m. Key to symbols used

n. Units used in soundings (feet; meters; fathoms)
o. Methods used in field work
p. Distance between guide lines, if guide lines are drawn
q. Value of submerged contours

Still other information may be included if circumstances warrant.

9. Since water-surface elevations fluctuate, causing depths to have a different value in the same place at different times, soundings should be referred to some known elevation. It is a standard practice to reduce soundings to *low-water datum* for that body of water. This low-water datum may be the lowest known surface elevation for that body of water, or it may be an assumed one, preferably the former if a sufficient number of records are available. To reduce all soundings to low-water datum, it is necessary to know the surface elevation at the time the soundings are made; then the difference between the low-water datum and the surface elevation is subtracted from the sounding as made in the field, and the value so reduced is entered upon the map. It must then be remembered that on any hydrographic map in which all soundings are reduced to low-water datum, depths on the map are minimal depths to which must be added some value in order to determine the true depth at a certain position on any particular date.

Map Colors

Colors are often useful in portraying on a map certain kinds of physical features of lakes or streams. The following ones are commonly used.

1. For depths of water:
   a. Blue, deep shade—very shallow water
   b. Blue, light shade—deeper shallow water
   c. White—deep water

2. For finely divided bottom materials:
   Yellow—sand
   Blue—fibrous peat
   Tan—pulpy peat
   Brown—muck
   Dark gray—marl

Application of Colors on Maps

It is often desirable to represent certain limnological features on maps by the use of colors. There are various ways of applying colors to maps, some of which are both simple and effective.

Colors from Wax Pencils. These colors may be applied permanently by the following simple method: Secure a supply of a good grade of colored wax pencils, such as "Venus" or "Dixon" brands, representing the desired colors; with rapid pencil strokes fill in roughly the area to be covered; then by means of a blunt-pointed paper pencil (composed of closely and spirally wound paper) dipped in benzine, redistribute the color into a uniform tint by rubbing it evenly. Under this treatment the color can be spread with satisfactory results. When the benzine evaporates the color is permanent; it does not rub off or smudge. Results are usually acceptable when the coloring is done on any paper of fair quality and appropriate color, provided the application has been done with reasonable care. If such coloring is done on waterproof paper the product is likely to have some superiority.

Somewhat similar, but perhaps less satisfactory, results may be secured by the use of any ordinary good grade, colored crayon or pencil followed by smoothing accomplished by rubbing the penciled surface with a bit of cloth soaked in good grade gasoline. For finer work the cloth should be wrapped around the end of a toothpick or the blunt end of a wooden penholder; for work on larger areas the cloth can be bunched into a soft mass. However, much depends upon the kind of crayon or pencil used and the operator must make a series of preliminary tests in order to arrive at a satisfactory selection.

An even tint may be brushed onto maps by means of ordinary water colors applied with a fine sable brush. Water colors adhere well to paper and do not spread. Powdered aniline dyes when dissolved and applied with a brush are likewise serviceable for producing tints on maps. Also, some use may be made of waterproof colored inks applied with a brush. However, the satisfactory application of water colors, aniline dyes, or colored inks requires more time, patience and skill than does the wax pencil-benzine method described above and often produces no better, or even inferior, results. These colors must be painted onto the surface of the paper, and if an even tint is to be secured, speed and accuracy are required. In painting on colors, the back of the map should be elevated somewhat so that the paper is tilted downward toward the operator. If a considerable area of surface is to be colored it should be subdivided into portions of convenient size. Each subarea should be painted separately and a white line left between adjacent subareas. This white line is filled in later with a half-dry brush, a procedure which gives a much better result than overlapping the painted edges. Painting a surface should proceed *downward* and the advancing front of the colored materials must not be allowed to dry during the progress of the work. Since it is very difficult to improve faulty or uneven painting, every care must be taken to apply the color in the best way possible. Areas too heavily colored may be lightened somewhat by the judicious use of an eraser, and those which are too light can be darkened with a half-dry brush, but there are limitations to these remedial measures.

Descriptive Terms and Their Abbreviations

No set of descriptive terms has been adopted as standard for designating those various physical features often important in the analysis and mapping of lakes and streams. The following lists are offered as a suggestion of how such terms and their abbreviations may be assembled. Abbreviations may be entered upon a map in such a way that location and extent of the features concerned are indicated in a useful pictorial fashion. These lists may be modified, altered, and extended to meet special needs. Suggestions given on pp. 72–75 for the proper use of symbols on maps also apply to the use of descriptive abbreviations.

BOTTOM MATERIALS

bedrock	— *br*	miscellaneous debris	— *md*
boulders	— *b*	mud	— *m*
coarse gravel	— *cg*	muck	— *mk*
false bottom	— *fb*	peat	— *p*
fine gravel	— *fg*	rubble	— *r*
hardpan	— *hp*	sand	— *s*
intermixtures	— *im*	silt	— *st*
marl	— *ml*		

CHARACTERS OF BANK

boggy	— *bg*	low	— *l*
cliffs	— *c*	medium height	— *mh*
cultivated	— *cv*	marshy	— *ms*
depositing	— *dp*	rock outcrop	— *ro*
eroding	— *er*	shrubs	— *sh*
flood level	— *fl*	swampy	— *sw*
high abrupt	— *ha*	topsoil	— *ts*
high sloping	— *hs*	wooded	— *w*

Also other descriptive terms and abbreviations from list on bottom materials.

INFLOWING WATER

boggy	— *by*	surface runoff	— *sr*
clear	— *cl*	small volume	— *sv*
large volume	— *lv*	turbid	— *tr*
springs	— *sp*		

NATURE OF FLOWAGE OF STREAMS

eddies	— *ed*	rapid	— *rp*
falls	— *f*	smooth flow	— *sm*
pool	— *p*	turbulent	— *tb*
riffle	— *rf*		

SHADING

dense	— *ds*	open	— *o*
partly shaded	— *ps*		

Symbols

Much detailed information can be incorporated on maps by means of symbols. Various systems may be devised for special purposes provided that the design of none of the characters used duplicates that of other symbols. No standard code of symbols devised for limnological purposes has been adopted. Among the suggested ones which follow are a few standard topographic symbols used by certain branches of the Federal government; others have been selected from various sources; and many of the biological ones are proposed by the writer.

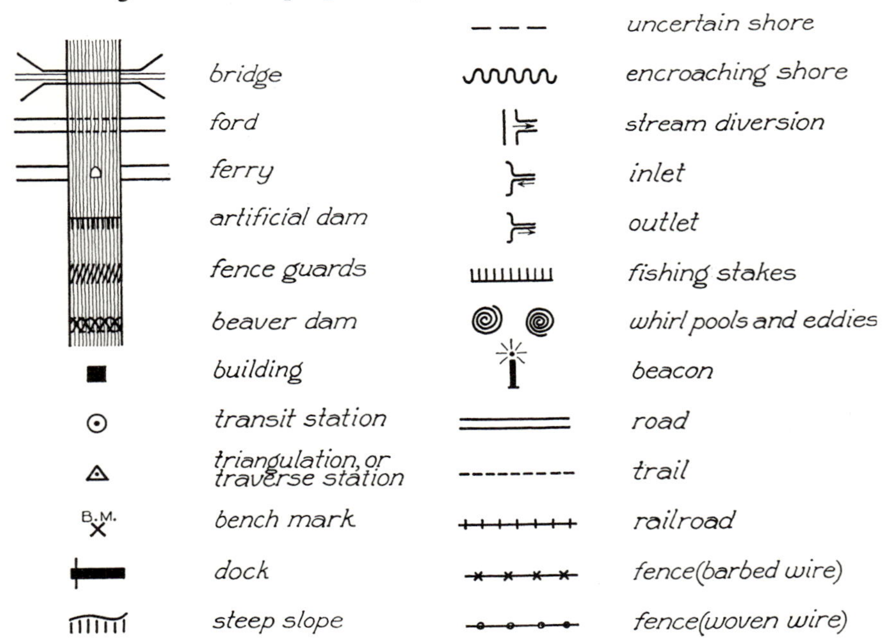

FIG. 14. Symbols for physical features.

In the use of symbols on maps, the following suggestions should be considered:

a. Use symbols no more generously than is necessary. Undue use causes congestion in that area of the map and tends to obscure its major features.

b. Make symbols no larger than clear legibility demands. Their size should be in keeping with the size of the map as a whole.

c. Symbols should be so constructed on the map that they are background rather than foreground characters.

d. When uniformity of an area is to be indicated by repeating its symbol over the space concerned, keep the distribution of the repeated symbol as thin as possible without sacrificing the result intended.

e. In so far as is practicable, put all symbols on the map in the vertical position. Maps with symbols turned upside-down and at various angles are difficult to use.

f. Symbols should not be entered upon a map until all of its major features have been completed.

g. The design of a symbol should be as simple and as distinctive as possible. Avoid ornate patterns.

h. Union of symbols to represent a combination of two or more meanings should be done with great discretion; ordinarily it should be avoided.

i. Incorporation of many different kinds of symbols into the same map or area should be avoided since it may render the map more or less unintelligible. It is much preferable to divide the various sets of symbols among two or more duplicate outlines of the same map.

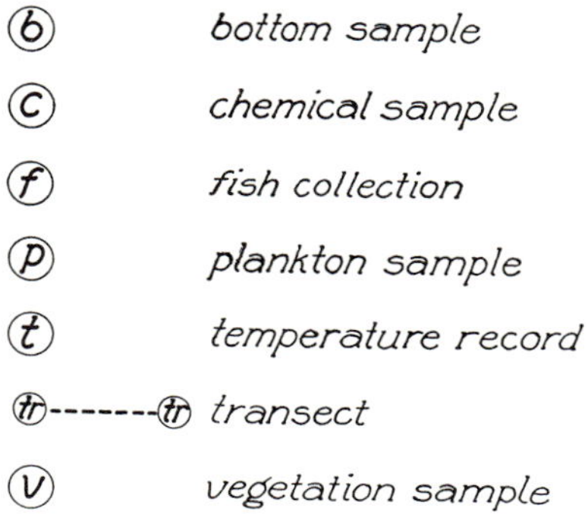

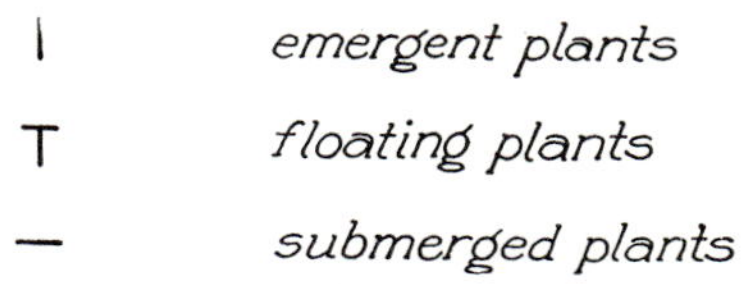

ⓑ	bottom sample		
ⓒ	chemical sample		
ⓕ	fish collection		
ⓟ	plankton sample		
ⓣ	temperature record	I	emergent plants
(tr)-------(tr)	transect	T	floating plants
ⓥ	vegetation sample	—	submerged plants

Fig. 15. (*Left*) Symbols representing sampling stations.

Fig. 16. (*Right*) General symbols representing vegetation. Used when rough division of aquatic vegetation into *emergent*, *floating*, and *submerged* classes is adequate.

DUPLICATING SYMBOLS ON MAPS

If symbols are to appear scantily on a map they may be entered upon the areas involved by rapid freehand sketching with pen. However, many limnological maps require numerous repetitions of various symbols in order to produce the necessary distribution record. Under these circumstances freehand sketching is too slow and tedious to be practicable. A relatively simple but effective solution to the problem of duplicating symbols can be found in the use of rubber stamps constructed as follows: To the end of a cylindrical handle of convenient size and length seal a block of rubber having a density about that of an ordinary lead-pencil eraser. On the free end of the rubber block, sketch with pencil, in the form and size desired, the symbol to be used taking particular care to make the margins of the figure very definite and regular. Turn into a vertical position and fasten the handle in a vise, with the rubber block uppermost. Install over the rubber block and in working position a large-diameter magnifying glass. Using a very sharp, fine-pointed scalpel, excavate to a lower level all rubber surface not included in figure of symbol, taking particular care to keep the margins of the figure smooth and regular. Remove pencil marks from the figure of the symbol which now stands in relief. Press stamp, so made, on an ink pad and make trial impressions on paper to determine whether further work on the stamp is necessary to produce an acceptable result. Judicious smoothing of cut edges may improve the impression. When in satisfactory form, a stamp of this type makes it possible to repeat the symbol many times on a map when such repetition is required. Since duplicator and ink-pad inks usually fade, care should be taken to use a pad saturated with an ink which is permanent and which provides the desired color. Various modifications of this simple method of making

Alisma
Carex
Cyperus
Decodon
Eleocharis
Equisetum
Hippuris
Iris
Juncus
Megalodonta
Phragmites
Polygonum
Pontederia
Ranunculus
Sagittaria
Scirpus
Sparganium
Typha
Zizania

FIG. 17. Symbols representing *emergent* aquatic plants. Assembled from various sources.

stamps can be devised. If preferable, the sketches of symbols may be sent to a manufacturer of rubber stamps. Such commercially made stamps are likely to yield impressions superior to those of the homemade variety.

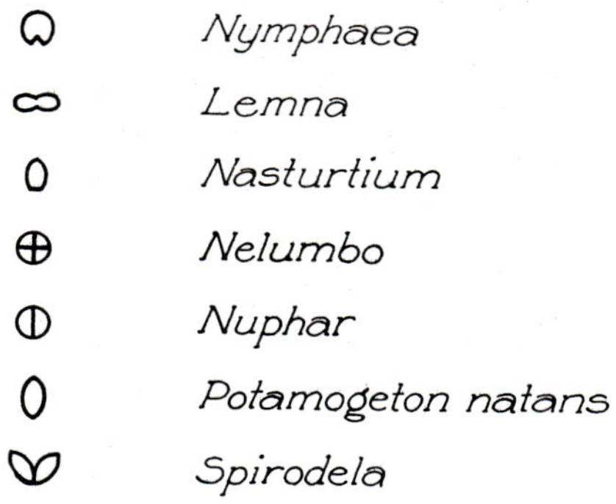

Nymphaea
Lemna
Nasturtium
Nelumbo
Nuphar
Potamogeton natans
Spirodela

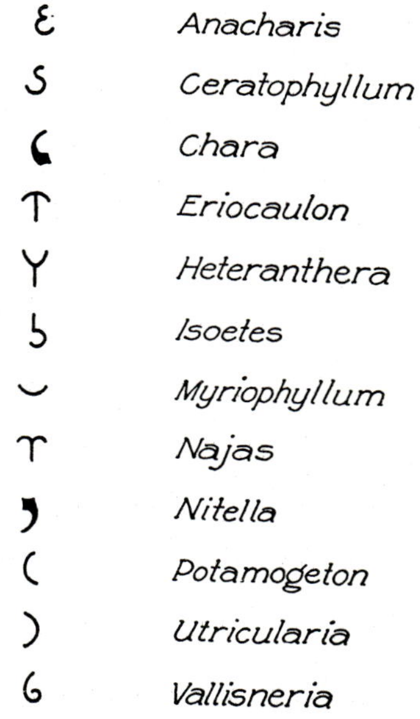

Anacharis
Ceratophyllum
Chara
Eriocaulon
Heteranthera
Isoetes
Myriophyllum
Najas
Nitella
Potamogeton
Utricularia
Vallisneria

FIG. 18. (*Left*) Symbols representing *floating* aquatic plants. A selected series of common genera. Assembled from various sources.

FIG. 19. (*Right*) Symbols representing *submerged* aquatic plants. A selected series of common genera. Assembled from various sources.

SELECTED REFERENCES

(For Chapter 5)

Brown, C. J. D.: "Coloring Lake Maps." 1 p. Spec. Pub. No. 12, Limn. Soc. Am., 1943.

Deetz, C. H.: "Cartography," 2d ed., 84 pp., 30 fig. Spec. Pub. No. 205, U.S. Coast and Geodetic Survey, 1936.

"Forest Service Map Standards." 31 pp. Forest Service, U.S. Dept. Agr., 1936.

Raisz, E.: "General Cartography." 370 pp. New York, McGraw-Hill Book Co., 1938.

Ridgway, J. L.: "Scientific Illustration." 173 pp., 22 pl. Stanford Univ. Press, Stanford University, Calif., 1938.

"Standard Symbols Adopted by the Board of Surveys and Maps, United States of America." 1 sheet. U.S. Geol. Survey, 1925.

"Suggested Symbols for Plans, Maps, and Charts." 12 pp., 21 pl. Nat. Resources Comm., Washington, D.C., 1938.

Swainson, O. W.: "Topographical Manual." 121 pp. Spec. Pub. No. 144, U.S. Coast and Geodetic Survey, 1928.

MORPHOMETRY

That branch of limnology which deals with the measurement of morphological features of the basin of a lake or stream and its included water mass is known as *morphometry*. Certain fundamental conditions of biological productivity arise directly from the structural relations of inland waters. Therefore it is necessary for the limnologist to make various measurements of morphological features of lakes and streams in order to determine the role which they play in limnological phenomena.

In order that morphometric data may be of value, measurements must be based upon reliable hydrographic maps of appropriate size and construction. Other things being the same, the larger the map the more accurate the results. It is futile to try to use diminutive maps. Small maps, if otherwise suitable, may be enlarged (1) directly by some photographic process; (2) traced from a lantern slide projected on a screen; or (3) by means of a pantograph. Other forms of enlargement may also be used if preferred. However, care must be taken that certain limitations inherent in small maps are not accentuated in enlargements.

Lakes

MAXIMUM LENGTH

Length of line connecting two most remote extremities of lake.

The form and position of this line must be such that it represents as correctly as possible the true open-water length. This line may be straight, as in lakes of regular, ovoid form, but often must be curved in form, as in instances of ox-bow lakes or other lakes with irregular shape. Such a line should not cross any land other than islands.

Measurements of maximum length of irregularly shaped lakes based upon a straight line connecting the most remote extremities and crossing portions of the mainland have little or no limnological value and should be avoided. Some lakes are of such shape that it is difficult to select one position for a maximum length measurement; also in others, as for example, lakes of a stellate form, there may not be a definite, single, maximum-length axis. In all such instances, the method and position of measurements should be definitely specified in descriptions.

MAXIMUM EFFECTIVE LENGTH

Length of *straight* line connecting most remote extremities of lake along which wind and wave action occur without any kind of land interruption.

Maximum total length and maximum effective length may be, and often are, identical. However, when islands or groups of islands are so located that they effectively interrupt the continuity of wave action and virtually divide the lake into two or more portions, maximum total length and maximum effective length are different.

MAXIMUM WIDTH

Length of straight line connecting most remote transverse extremities and crossing no land other than islands; line approximately at right angles to maximum-length axis.

MAXIMUM EFFECTIVE WIDTH

Length of straight line connecting most remote transverse extremities of lake along which wind and wave action occur without any kind of land interruption.

MEAN WIDTH

Area of lake divided by its maximum length.

MAXIMUM DEPTH

Maximum depth known for lake expressed in feet, meters, or fathoms.

MEAN DEPTH

Volume of lake divided by its surface area.

MEAN DEPTH-MAXIMUM DEPTH RELATION

Mean depth divided by maximum depth. Expressed as a decimal value, it serves as an index figure which indicates in general the character of the approach of basin shape to conical forms.

MAXIMUM DEPTH-SURFACE RELATION

Maximum depth divided by the square root of the surface. Expressed as a decimal value, it is an indication of the relation of depth to horizontal extent.

DIRECTION OF MAJOR AXES

Expressed in general compass directions, as for example, NE–SW, or NNE–SSW.

AREA

Only in occasional instances is it feasible to determine the area of a body of water by direct measurements in the field. Consequently, methods for measuring area usually require a map of suitable scale and accuracy.

Planimeter Method

This method depends upon the use of a well-known instrument called a planimeter. It is specially designed for measuring the area of a plane surface regardless of the shape of its outline. It is the most rapid and, when carefully operated, the most accurate means of determining areas. Planimeters are made in many forms, some of which are costly. The simple polar planimeters (Fig. 20) are modest in price and meet many of the needs of the limnologist. Manufacturers usually supply detailed instructions for the operation of their instruments. The following description relates to the use of one of the simple polar planimeters.

EQUIPMENT. Polar planimeter; reading glass or hand lens; map of suitable character and scale; Scotch cellulose tape, or some similar adhesive.

CORRECTIONS AND ADJUSTMENTS. If the planimeter has an adjustable tracer arm, the instrument must be set properly for the desired unit of measurement. In planimeters having fixed arms, the value of the unit of measure is determined by the manufacturer and cannot be changed.

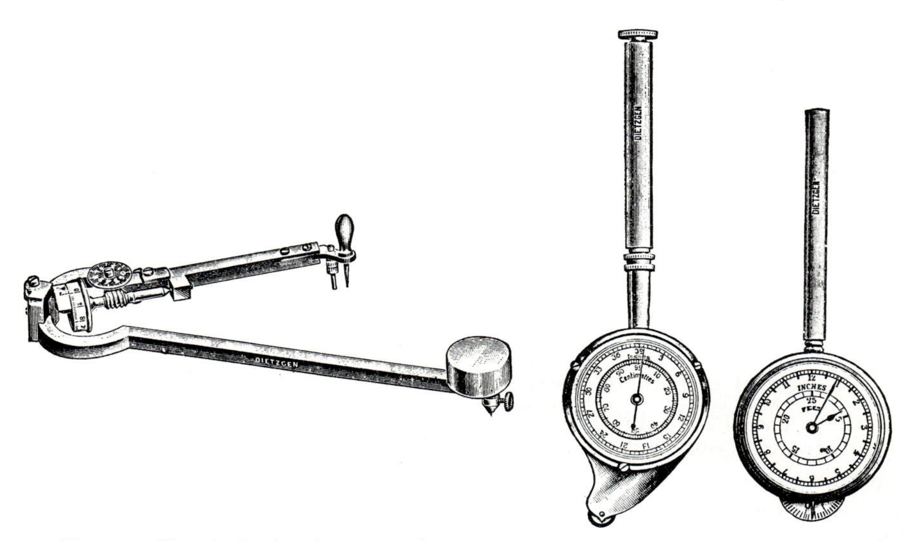

FIG. 20. (*Left*) A simple type of polar planimeter. (*Right*) Two types of map measurer, one at right with graduated tracing wheel. (Courtesy, Eugene Dietzgen Co., Chicago, Ill.)

Test the accuracy of the planimeter by one of the following methods:

1. The simplest and most reliable method of testing the setting and the accuracy of a planimeter is by the use of a *trial disk* sometimes furnished by the makers. It is a round metal disk whose grooved rim surrounds a known area and around which the tracer point can travel in a groove with exactitude. Remove glider if present; set planimeter in the proper way; trace the area on the trial disk; compare reading on planimeter with known area of the disk. Or if preferred, make several continuous nonstop tracings around the plate and divide the total reading on instrument by the number of circuits around the plate; then compare average reading on planimeter with known value of plate. If discrepancies occur they must be due to (a) wrong adjustment of planimeter arm, (b) to bad condition of planimeter, or (c) to error in reading planimeter.

2. Another mechanical method of checking a planimeter is by the use of a planimeter *testing rule* which consists of a graduated metal strip of convenient dimensions anchored at one end but free to rotate about a pivot. At known distances from the center of the pivot there are small holes. Remove glider if present; set planimeter in proper way, making sure that the free end of the testing rule can rotate completely without making contact with any part of the planimeter. Set tracer needle of planimeter in hole in free end of testing ruler; mark with pin point the exact starting position of free end of rule; set planimeter scales in initial reading positions; swing tracing needle throughout full circle determined by testing rule, returning exactly to starting point (or make several circuits and compute average). Using distance from pivot to tracing-point hole as radius of circle described by tracing point, compute area of circle; check computed value with area read on planimeter.

3. With pair of dividers draw on suitable surface a circle of known diameter; make circle as large as possible within operating distance of one setting of planimeter; compute area of circle so constructed. Measure area of circle three times with planimeter, use average of three readings and convert reading to same units as computed value; compare computed and measured values of area.

Procedure

1. Spread map to be measured as smoothly as possible on firm, flat horizontal support, preferably a drawing table; fasten map securely to table with Scotch cellulose tape, gummed tape, or some other suitable means.

2. If map is too large to be measured by one set-up of planimeter, rule on map a series of equally spaced horizontal and vertical lines thus dividing area of map into sections; make such subdivisions as large as can be conveniently covered by planimeter.

3. Select an apparently suitable position *outside* of area to be measured and set needle point (fixed point, or fulcrum) into paper; make trial swing-around over area with tracer point to determine if all parts of periphery can be reached; if not, another set-up must be sought.

4. With suitable set-up position determined and instrument properly balanced, place tracer point on outline of area; make slight indentation or mark on paper to indicate starting point and set tracer point on it; with needle point and tracer point carefully kept in position, lift wheel from paper and set all graduated scales at zero; lower wheel to rest on paper, check setting of instrument, and make further adjustments if necessary.

5. Make tracer point follow, gently and as accurately as possible, outline of figure to be measured, proceeding in a *clockwise direction* and returning to starting point. When tracer point reaches starting position, read and record planimeter values; reread for check.

6. Repeat tracing of outline at least twice, resetting instrument each time at initial starting point, and divide sum of all readings by number of tracings of figure. Or if preferred, make at least three continuous, nonstop tracings of figure and divide total reading of instrument by number of circuits around figure.

7. If size of map requires that it be measured in sections, repeat operations described above for each section and add average readings of all sections.

8. By use of scale on which map is constructed, draw on paper a square which will contain exactly one unit of area (for example, 1 sq. mi.) in which final results are to be expressed. Carefully measure this square with planimeter and determine number of planimeter units in square representing one actual unit of area of lake. Compute area of square by use of dimensions and use value as check on planimeter reading. Divide total number of planimeter units in whole lake by number of planimeter units in square and result will be area of lake in terms of unit used (square mile, hectare, square rod).

General Considerations

a. While potentially the planimeter method is the most exact of the various methods for determining area, the personal error in instrument operation may be considerable if the operator is inexperienced. Preliminary practice is always desirable and for the beginner it is imperative if results are to be dependable.

b. It is possible to operate a planimeter with the needle point (fixed point; fulcrum) *inside* of the figure to be measured and the detailed instructions supplied by dealers usually describe the necessary procedure. However, the inside position for the needle point is usually not recommended since the method is complicated and the results obtained are generally not as accurate as those obtained by the outside position.

Hatchet-planimeter Method

A hatchet planimeter, sometimes called a "jack-knife planimeter" or a "Prytz planimeter," is an exceedingly simple device which when properly constructed and operated will measure the area of a plane figure, either regular or irregular in outline, with an accuracy close to that of an ordinary polar planimeter.

CONSTRUCTION. A hatchet planimeter (Fig. 21) consists of a metal rod bent at right angles at the ends, forming a long arm *a* and two short legs. One leg is tapered to a smooth tracing point *tp;* the free end of the other leg is ground or flattened into a sharp blade *b*. Any good grade metal rod, about 4 mm. in diameter, will serve. Some rods can be bent in a vise without preliminary treatment; others, particularly steel, should be heated to a dull red just prior to bending. The main arm *a* may be any convenient length but there will probably be few occasions for constructing an in-

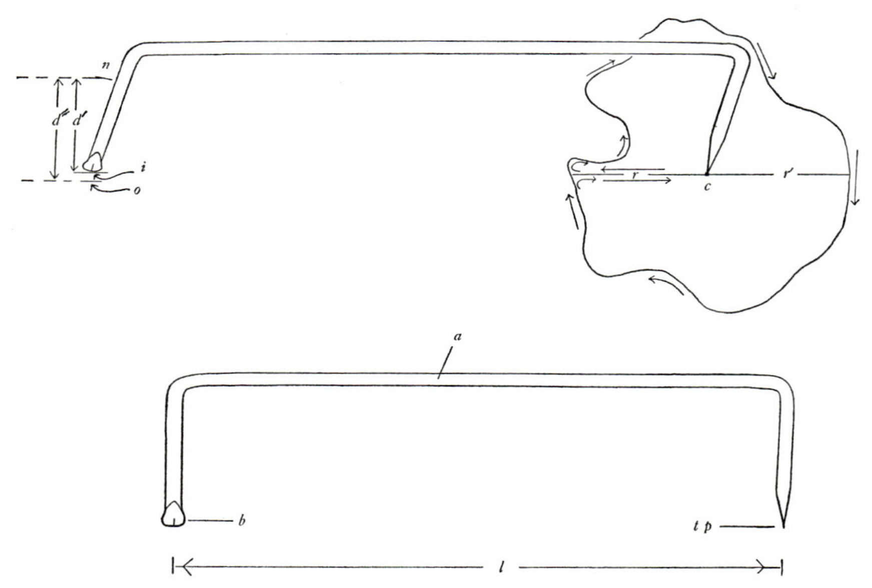

FIG. 21. Hatchet planimeter. Lower figure shows construction of instrument. Upper figure shows method of use. (*a*) Arm. (*b*) Blade. (*c*) Center of gravity of figure. (*d'*) Distance *i* to *n*. (*d''*) Distance *o* to *n*. (*i*) Original blade position. (*l*) Distance from center of tip of tracer point to end of mid-axis of blade leg. (*n*) Terminal position of blade at end of first measurement. (*o*) Terminal position of blade at end of second measurement. (*r*) Line from center of gravity of figure to perimeter. (*r'*) Line from center of gravity to perimeter, opposite line *r*. (*tp*) Tracing point.

strument less than 25 cm. long. Longer ones are useful in the measurement of large surfaces. The legs should be about 7 cm. long. Certain critical points of construction must be observed, namely: (1) The rod must be straight, rigid and heavy enough to insure that the blade will maintain continuous contact with the paper when the tracer point *tp* is in motion. (2) The long axis of the blade, when extended, must pass through the tip of the tracer point *tp*. (3) The edge of the blade must be kept razor sharp. (4) Each end of the edge of the blade must be smoothly rounded upward to insure free passage over the paper.

Procedure

1. Cover smooth flat table with large sheet of *unglazed* paper; attach edges to supporting surface with Scotch cellulose tape or other similar adhesive; place map or figure to be measured on top of unglazed paper in a convenient position and immobilize it with adhesive.

2. Locate approximately the center of gravity of map or figure (Fig. 21, *c*); erect line *r* from center *c* to any selected point on periphery of figure.

3. Place tracer point *tp* on center of gravity *c*, holding planimeter so that legs are vertical and blade is at some convenient position; mark initial position of blade *i* by pressing edge slightly into paper, producing an indentation; mark indentation and label it.

4. Hold tracer leg of instrument lightly but steadily between thumb and first two or three fingers, using tip of little finger as glider support; hold instrument so that legs are vertical; do not allow instrument to tilt at any time during a measurement.

5. Make tracer point traverse line *r* from center of gravity *c* to perimeter of figure; then trace perimeter clockwise as shown by arrows in Fig. 21, returning to center of gravity *c* along line *r;* mark new position of blade by pressing edge slightly into paper and label it *n*.

6. Measure carefully distance *d* between original blade position *i* and terminal one *n;* check this distance by one or both of following methods: (1) With blade exactly at *n* and tracer point at *c*, retrace figure as outlined in paragraph 5 above but in counterclockwise direction; if blade returns exactly to *i*, which it rarely will do, tracing has been duplicated exactly and presumably distance *i* to *n* is as correct as conditions of test will permit; if new blade position occurs near, but not on, initial position *i*, make impression in paper by pressing blade into it, mark and label *o*, then distance sought *d* is derived from formula $\frac{d' + d''}{2}$ in which d' is distance *i* to *n*, and d'' is distance *o* to *n*. Repeat operations in case of doubt. (2) Repeat independently, three or more times, the operations

indicated in paragraphs 3–5 given above, each time shifting figure being measured so that blade of planimeter will make its tracings on new areas of paper; use average of all measurements of distance i to n.

7. Measure carefully distance from center of tip of tracer point to that point of blade which coincides with end of mid-axis of blade leg (Fig. 21, l).

8. Calculation.

Area of figure traced is computed by formula, $a = d \times l$, in which a is area desired; d is distance between initial i and final n or average position of blade; and l is distance between tip of tracer leg and middle of blade, expressed in same linear units as d.

INHERENT ERRORS. When distance d in the formula given above is small as compared with the length l and when the planimeter has been operated under the best of conditions, the instrument measures area directly without significant error; if, on the other hand, the distance d is large, serious error appears. This error becomes greater as the size of d increases. If the greatest dimension of the area to be measured is less than one-half the length (l) of the instrument, the inherent error is negligible, but if d exceeds one-half of l, the figure must be subdivided and the parts measured separately. Consequently, the longer the planimeter the larger the area which can be measured within one tracing without requiring the application of correction values or subdivision of the figure.

If this instrument is used to measure the area of undivided figures the dimensions of which exceed the limit indicated above, then it will be necessary to construct a table of corrections as follows: Either draw a series of conventional figures (triangles, squares, parallelograms, circles, the areas of which are easily computed mathematically) in graded sizes from small to large, or use cross-section paper. Following procedures outlined in paragraphs 3–8, carefully make measurements of a graded series of areas; compare computed values with corresponding ones secured with the planimeter and determine corrections necessary to make the planimeter readings accurate.

ERRORS OF OPERATION. The operator should practice on regular figures of known areas until some skill in use of the instrument is acquired. Important precautions against errors of operation are: (1) In tracing a figure the tracer point must be so handled that the blade describes its various courses without any interference. The tracer leg must turn freely between the thumb and fingers. (2) In all operations the blade must be kept on *unglazed* paper. (3) The blade must operate over a uniform, uninterrupted surface; imperfections in the paper, edges of paper, and other deviations from uniformity lead to error. (4) At all times during a tracing, the tracer leg must be kept in a vertical position. (5) The sharper the blade the better the results; a dull blade must be avoided. (6) The rod must be of such weight that the blade is held continuously on the paper; however, overweight should be avoided.

General Considerations

a. The successive terminal positions of the blade (*i*, *n*, *o*,) fall on the arc of a circle the radius of which is the length *l*, and distance *d* (*i* to *n*; *n* to *o*, etc.) should be measured along this arc. However, measurement along the chord is much simpler and if distance *d* is relatively small the error is negligible, but as *d* increases the error becomes greater and corrections are required.

b. In regular figures (circle, square, triangle, etc.) the center of gravity *c* is easily determined by well-known methods, but in irregular ones the location of this point is difficult and often must be arrived at by approximation. Fortunately the precise location of this point is not necessary and small departures from it do not result in significant errors. Only when this starting point is taken some distance from the true center of gravity does the error become consequential.

c. If it is not possible to locate with certainty the center of gravity, a test may be made as follows: Measure the area of the figure with the planimeter arm extending to the left using *r* as the going and returning line; then without changing the position of the figure, repeat the measurement with planimeter arm extending to the *right*, using *r′* as the new going and returning line; compare results obtained. Essentially the same result is obtained by turning the paper bearing the figure through 180° and repeating the tracing with the planimeter still on the same side.

Weight Method

The weight method is based upon the fact that if a carefully cut-out paper model of a map is weighed on a balance of suitable sensitivity and if a unit area (e.g., 1 sq. mi.) is cut from the same kind of paper according to the scale of the map and then weighed, the area of the body of water is obtained when the weight of the whole model is divided by the weight of the unit area. If the proper conditions and safeguards are observed this method yields satisfactory results.

EQUIPMENT. Map of suitable character and scale; balance having sensitivity commensurate with quantity and thickness of paper used; ordinary laboratory scalpel with sharp point and rounded edge; and paper of suitable weight and uniformity on which to copy map.

CORRECTIONS AND ADJUSTMENTS. Many grades of paper used in map making, map printing, or map copying are sufficiently uniform in thickness and weight to make it certain that the differences in weight of various areas within a sample of the paper fall within the other errors and limitations of the method. However, in case of doubt, or when unusual materials are considered for this purpose, test for weight uniformity as follows: fold a large sheet several times in such a way that by one operation several pieces of *equal size* can be cut out simultaneously from various areas on the sheet; make the cut pieces as large as circumstances will permit; weigh each piece separately on a balance sensitive enough to make weight comparisons of the individual pieces significant. Repeat on at least two other large sheets from same stock. If variations occur which seem to exceed the general error of the method, discard the proposed material and try another.

Procedure

1. If duplicate copies of map are available and if map is printed on paper suitable for weight method, cut model directly from printed copy. However, if model cannot be cut from printed map, transfer outline to another sheet of acceptable paper by any one of several convenient ways, such as carbon copy or direct copying by transmitted light.
2. Fasten map, or copy of map, on large piece of smooth beaverboard or some other similar material against which cutting may be done; with sharp scalpel held vertically between thumb and first finger, make a continuous series of close, vertical thrusts through the paper, cutting through the middle of the line representing shore line; continue until entire shore line is thus treated.
3. Remove model from surrounding paper; with sharp scissors, trim edge of model carefully, removing all tag ends and irregularities due to cutting, especially those between scalpel thrusts. Weigh model of map.
4. In similar manner, cut from same sheet of paper one unit area (1 sq. mi., or some other desired unit area) as determined by scale of which map is drawn and weigh it.
5. Divide weight of model of lake by weight of model of unit area and quotient will be area expressed in unit used in unit area.

General Considerations

a. Islands, if present, must be cut from the model before it is weighed.

b. If this method is employed in instructional work, the average weight of all of the models made by a class divided by the average weight of all of the unit areas will probably yield a more accurate value.

c. Both model and unit area should be preserved with the other original records of the work.

d. A paper made in very uniform weight and designed especially for this purpose is sometimes obtainable.

e. Materials other than paper, e.g., wax plates, may be used, provided they meet the necessary requirements of convenience and uniformity.

Cross-section Paper Method

When the outline of a lake is superimposed upon cross-section paper (squared paper) its area can be measured by dividing the total number of squares included by the number of similar squares included within an area representing, according to the scale of the map, that unit in which it is desired to express the area.

EQUIPMENT. Map of proper character and scale; sheets of suitable cross-section paper.

CORRECTIONS AND ADJUSTMENTS. Cross-section paper should be tested for uniformity of ruling before it is used. Check in both vertical and horizontal directions; also in several different areas on same sheet.

Procedure

1. By means of tracing table, carbon paper, pencil carboning on back side of map, or any other convenient method, transfer outline of shore line of map to cross-section paper. Retrace until outline is distinct at all places.

2. Count all squares *completely* within shore line of map; then *count as whole squares* those squares about the periphery of the map whose areas are *one-half or more* within the shore line, but ignore those that are less than one-half within the figure. Assemble all counts into one total.

3. By means of scale on map, draw on same kind of cross-section paper a closed figure representing one unit of area, e.g., square mile. If drawn as a circle, count squares within this figure as indicated in the foregoing paragraph, expressing total in same size squares as that in total for whole lake area; if laid off in form of square, count all complete squares and estimate *all* incomplete squares. Divide total within entire map by total within unit figure and quotient will represent desired area expressed in units chosen.

General Considerations

a. Some system of marking squares as they are counted is necessary in order to avoid mistakes.

b. Within practical limits, the smaller the squares on cross-section paper the smaller the error of the method.

c. Cross-section papers ruled in large squares which in turn are subdivided into smaller squares have a counting advantage over those ruled only in one size.

d. The incomplete squares about the periphery may all be estimated as to the portion included within the figure, as for example, 0.5, 0.1, 0.9, etc., and the total of all incomplete squares thus computed. Errors of estimation in the incomplete squares are likely to be compensating to some extent since in a large number of estimates there tend to be about as many *underestimations* as there are *overestimations*. However, estimating *all* incomplete squares is slow and laborious, and the end result is not likely to show any improvement over the method of counting as whole squares all those which are one-half or more within the figure and ignoring those less than one-half within the figure.

e. Counts of squares should be repeated for verification both on the map and on the unit area.

f. With proper materials and with the necessary personal care, good results are obtainable with this method.

Method of Geometrical Figures

This method consists of the division of a map, whose area is sought, into a series of geometrical figures (triangles, rectangles, and trapezoids) the area of which can be computed from the scale on which the map is drawn by the use of well-known mathematical formulas.

EQUIPMENT. Map of suitable character and scale—within convenient limits, the larger the map the better; protractor; rulers with fine, correct graduation; dividers; triangles.

Procedure

1. On map to be measured, erect within shore line the largest rectangle, triangle or trapezoid which it will contain and compute its area.
2. Divide remaining portions of area not so included into triangles and compute their areas. Continue until all of map has been covered.
3. Add areas of all geometrical figures. Using scale of map, determine area representing one actual field unit (i.e., square mile). Divide total area of lake by that of unit area and quotient will be area of lake in terms of unit area selected.

General Considerations

a. This method is practicable only for bodies of water which have a very regular outline.

b. When the lengths of three sides of a triangle, a, b, and c, are known in terms of the scale of the map, its area may be determined by use of the formula

$$Area = \sqrt{s\ (s\text{-}a)\ (s\text{-}b)\ (s\text{-}c)}$$

where $s = \dfrac{a + b + c}{2}$

Method of Ordinates by Simpson's Rule

EQUIPMENT. Map of suitable character and scale; rulers; dividers; triangle.

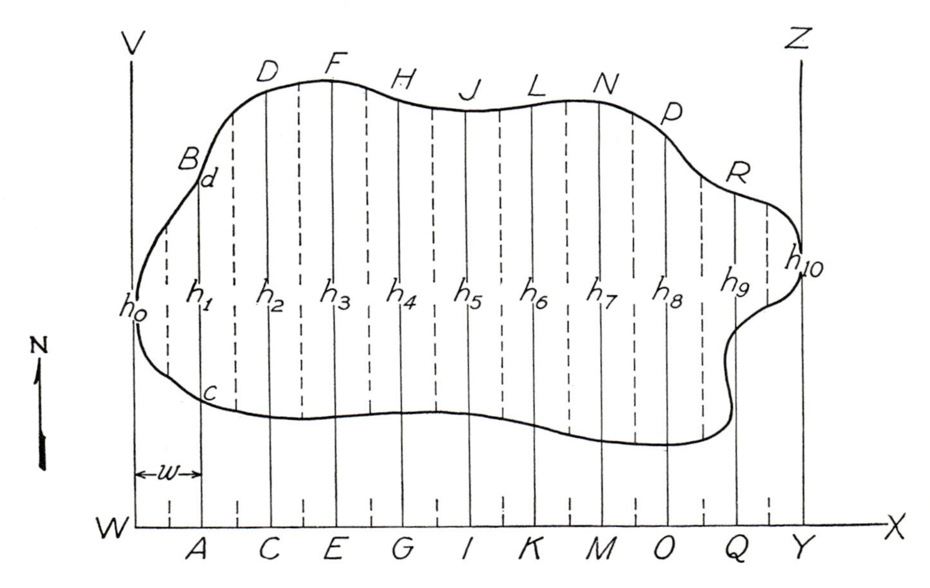

FIG. 22. Diagram illustrating measurement of area by method of ordinates by Simpson's rule; also measurement of area by method of average ordinate rule.

Procedure

1. As in Fig. 22, draw line *WX* below map to be measured and in horizontal direction; draw vertical line *WV* tangent to left end of figure; draw vertical line *YZ* parallel to *WV* and tangent to right end of figure.

2. Divide distance *WY* into an *even* number of *equal* parts and draw ordinates *AB*, *CD*, *EF*, and others. Label left tangent ordinate h_0 and the others successively to right as h_1, h_2, h_3, etc., ending with right tangent.

3. Measure length of ordinate h_1 (c to d) within periphery of map and record it; measure and record all other ordinates in same fashion. Substitute values in formula

$$Area = \frac{1}{3} w\,(h_0 + 4h_1 + 2h_2 + 4h_3 + 2h_4 + 4h_5 + 2h_6 + 4h_7 + 2h_8 + 4h_9 + h_{10})$$

where w = distance between ordinates. Note that h_0 and h_{10} in the figure are each zero in value.

4. Determine in same units the size of one unit area (e.g., 1 sq. mi.) by use of scale of map; divide area within lake outline by area within unit area (1 sq. mi.) and quotient will be area of lake in square miles.

General Considerations

a. In general, the greater the number of ordinates used the greater the accuracy of results.

b. It must be understood that this method gives only approximate values and seldom will yield the accuracy of methods described on preceding pages.

c. Results are better with lakes having very regular shore lines.

Method by Use of Average Ordinate Rule

EQUIPMENT. Map of suitable character and scale; rulers; dividers; triangle.

Procedure

1. As in Fig. 22, draw line *WX* and erect tangents *WV* and *YZ*. Divide line *WY* into *any* number of *equal* parts.
2. Bisect lines *WA*, *AC*, *CE*, etc., and at middle of each draw ordinates (broken lines in Fig. 22).
3. Measure lengths of all ordinates within map. From sum of lengths of all ordinates, determine *average length* of these ordinates; multiply by length of line *WY* and result will be area within figure.
4. Determine in same units the size of unit area (e.g., 1 sq. mi.) by use of scale of map. Divide area within lake outline by area within unit area (1 sq. mi.) and quotient will be actual area of lake in square miles.

General Considerations

a. In general the greater the number of ordinates used the greater the accuracy of results.

b. This method gives only approximate results.

c. It is sometimes convenient and timesaving to determine the lengths of ordinates by means of dividers. Lay off successively on a long sheet of paper the ordinate lengths in one connected straight line, and then measure total length of this composite line to get sum of lengths of ordinates.

LENGTH OF SHORE LINE

Sometimes the length of a shore line may be determined directly in connection with the field survey. Often, however, it is necessary to get this information from maps and the following methods are presented for that purpose.

Map Measurer Method

A map measurer (Fig. 20, *right*), sometimes called a map measure or a chartometer, is an instrument designed to measure the length of lines by means of a tracing wheel whose revolutions are transmitted to a revolving hand operating over a graduated dial. Graduations on the dial are so constructed that they represent units of distance traversed by the tracing wheel, usually inches and fractions of inches, or centimeters and fractions of centimeters. Map measurers are made in various forms and with various scales. In some models the tracing wheel is also graduated and an index pointer mounted on the frame and superimposed over the exposed part of the wheel facilitates the reading of the wheel position.

EQUIPMENT. Map measurer; map of suitable character and scale.

CORRECTIONS AND ADJUSTMENTS. Test map measurer for accuracy as follows: Choose meter stick which has been carefully checked for straightness of edge and for accuracy of graduations; lay meter stick flat on smooth table surface; set hand of map measurer at starting point of scale. Choose some graduation mark at one end of meter stick as starting point; set map measurer against edge of meter stick with axle of tracing wheel exactly at starting graduation. Holding map measurer steadily against edge of meter stick and tracing wheel constantly against table surface, trace length of meter stick, stopping instrument at some graduation near opposite end. See that axle of wheel coincides with end graduation; compare distance covered on meter stick with distance indicated on dial of instrument. Make at least three trials. If meter stick and its graduations are correct within expected limitations and if map measurer fails to agree with meter stick, it will be necessary to prepare a correction table for the instrument.

Procedure

1. Mark with pencil dot some starting point on shore line of map; set map measurer hand at zero point on dial; carefully set axle of tracing wheel directly over starting dot on shore line; or if tracing wheel is graduated, set zero mark of wheel over starting dot.

2. With handle of map measurer held vertically, run tracing wheel along shore line in direction which causes dial hand to move continuously clockwise. Continue tracing until wheel stops exactly at desired destination. Dial will then register distance over which tracing wheel passed. If necessary, apply corrections to results.

3. In similar way and with same precautions, run map measurer over unit on scale of map (1 mi.); if necessary, apply corrections to results; divide length of shore line on map by length of unit (1 mi.) on scale of map and quotient will be actual length of shore line in miles.

General Considerations

a. Accuracy of results depends largely on the care exercised in using the instrument. Since at first the personal error may be large, preliminary practice with the instrument is highly desirable.

b. At least three tracings should be made in order to check results.

c. In using the instrument, keep a moderate pressure on the handle to insure that the tracing wheel maintains close and continuous contact with the surface of the map.

Pins and Thread Method

If map measurers are not available and if the map used is not too small, good results may be secured by a simple method in which a thread is made to lie on the shore line and conform closely to its configuration. The length of such a thread is converted into units of shore-line length by means of the scale on which the map is drawn.

EQUIPMENT. Map of suitable size and character, preferably larger than $8\frac{1}{2} \times 11$ in.; generous supply of No. 5 standard-length insect pins, or their equivalent; linen, silk, or other kind of *inelastic* thread of convenient size; beaverboard or some soft smooth board against which the map can be fastened securely and into which pins can be set with ease; meter stick or rule of convenient type.

Procedure

1. Set pins vertically along edge of shore line in form of palisade, number of pins per unit of length depending upon irregularity of shore line; along convex (toward the water) portions of shore line, set pins on outer edge of line; along concave (toward the water) portions of shore line, set pins on inner edge of line; set pins thick enough to insure that thread drawn against their bases will follow contour of shore line closely.

2. Mark with pen or pencil dot some point on shore line as starting point; tie knot in one end of thread, thrust pin through knot, and set on starting point; lay thread along outside of bases of concave rows of pins and against inside of bases on convex rows; draw thread to remove all slack and to cause it to lie upon and coincide smoothly with shore line; when free end of thread reaches pinned end, mark former with ink or pencil mark.

3. Remove thread and measure length between its two marked points; measure length of unit (1 mi.) on scale of map; divide length of shore line on map by length of unit (1 mi.) of scale on map and quotient will be actual length of shore line in miles.

General Considerations

a. The accuracy attained by this method depends much upon the use of a map of adequate size, and upon personal care and patience in operating the method. When carefully employed quite satisfactory results are obtainable.

b. Make three trials with thread about pins to check results.

Dividers Method

Under suitable conditions, shore line may be measured by stepping off small, uniform intervals of known length on the line with a pair of dividers, counting the total number of such intervals, and dividing that total by the number of similar intervals in the length of some unit of the scale on which the map is constructed.

EQUIPMENT. Map of suitable size and character, preferably larger than $8\frac{1}{2} \times 11$ in.; pair of good dividers.

Procedure

1. Mark starting point on shore line with pen or pencil dot; set divider points at some convenient interval and lock in that position, making interval as small as practicability will permit.
2. With one point of dividers on starting dot, step off a continuous series of uniform divider intervals around the entire shore line; mark all extremities of divider intervals with mark or puncture; count total number of such intervals required to cover entire shore line.
3. With dividers set for same interval, determine number of intervals necessary to cover one unit (e.g., mile; meter) on scale on which map is built; divide this number of intervals into total number of intervals for entire shore line and result will be shore-line length in terms of unit provided in scale of map.

The accuracy of this method depends upon the same conditions mentioned in the previous method. With proper care and a large map satisfactory results are obtainable. Obviously, the smaller the divider interval and the larger the map, the greater the accuracy.

SHORE DEVELOPMENT

Degree of regularity or irregularity of shore line is expressible in the form of an *index figure*. The term *shore development* refers to the ratio of the actual length of shore line of a lake to the length of the circumference of a circle the area of which is equal to that of the lake. If a lake had a shore line in the form of a circle—a very rare circumstance—the shore development would be 1. As the value of this ratio increases beyond 1, increasing irregularity is indicated. This method of indicating form of shore line is very convenient and quite generally used. The following simplified formula may be employed:

$$\textit{Shore development} = \frac{s}{2\sqrt{a\pi}}$$

in which s is length of shore line; a is area of lake.

In the past, occasional use has been made of another ratio known as *absolute shore development* which is computed by dividing the length of the shore line by the square root of the area.

HYPSOGRAPHIC CURVE

A hypsographic curve for subsurface horizontal areas is constructed by plotting depth along the vertical axis (ordinate) and area along the horizontal axis (abscissa). Such a curve not only represents certain elements in the form of the lake basin but it also provides a means whereby areas at any depth level may be determined.

Procedure

1. On well-constructed hydrographic map, measure area circumscribed by each submerged contour. For best results such contours usually should be drawn at depth intervals not exceeding 2 meters. By means of scale on map convert measured areas into actual areas in lake.

2. Examine cross-section paper to be used; determine appropriate and convenient scale for values to be plotted along each axis. Place zero point for both sets of values at upper left intersection of first ordinate and first abscissa. Plot depth along vertical axis and areas along horizontal.

3. If actual areas within contours are plotted, the curve may be called a *direct hypsographic curve*. If, on the other hand, the areas within submerged contours are expressed in terms of per cent of surface area of lake, then the curve is referred to as a *percentage hypsographic curve*.

PROFILES

A profile is a pictorial representation of basin configuration along a selected line. This line is usually straight, but profiles may be drawn along one of any shape if occasion demands. A profile also provides one of the best methods of portraying basin slope along a selected transect. Profiles may be constructed in several ways one of which is as follows:

1. On a hydrographic map of suitable scale and character, draw a selected line of profile. Then on another sheet of paper, preferably a sheet of cross-section paper, draw a line the same length as the profile line on the map; this is to be the base of the profile.

2. Choose a suitable vertical scale. Lay paper bearing the base of the profile upon the hydrographic map in such a way that selected line of profile and the base-of-profile line are edge to edge with the ends coinciding.

3. Using the chosen vertical scale, mark, at as frequent intervals as the map provides, the depths perpendicular to profile base line. Connect all points so located by a continuous line; this line will be the profile.

General Considerations

a. In certain instances some vertical exaggeration may be necessary, particularly when the lake is shallow or when the profile is drawn on a very small scale. The amount of such exaggeration will depend upon circumstances and the purposes of the profile but should not be more than strict necessity demands. Even then, care must be taken that exaggeration does not give false impressions. In all instances the amount of exaggeration must be clearly stated.

b. It must be remembered that a profile gives information which is applicable only at the particular transect involved. Additional information must be secured by making other profiles along transects drawn elsewhere on the map.

c. The position of a profile must always be definitely indicated.

VOLUME

The volume of a lake can be determined by computing the volume of each horizontal stratum as limited by the several submerged contours on the hydrographic map and taking the sum of the volumes of all such strata. Several formulas for computing volumes of these strata are available but it appears that many of them give essentially the same results, the differences being well within the other errors of the method.

Procedure

1. Determine surface area of lake (see p. 79); also determine area circumscribed by each submerged contour.

2. Compute volume of first (uppermost) horizontal stratum bounded by the surface and the plane circumscribed by first submerged contour, using the formula

$$Volume = \frac{h}{3}(a_1 + a_2 + \sqrt{a_1 a_2})$$

in which h is vertical depth of each stratum, a_1 is area of upper surface and a_2 is area of lower surface of stratum whose volume is to be determined.

3. Compute in like fashion the volumes of succeeding strata. The sum of volumes of the several strata is total volume of lake.

VOLUME DEVELOPMENT

In order to express the form of the water mass or the form of the basin, an index figure known as *volume development* is used. This expression represents the ratio of the total volume of a lake to the volume of a cone whose area of base is equal to the surface area of the lake and whose height is equal to the maximum depth of the lake. It may also be computed by the use of the simplified formula

$$\textit{Volume development} = 3 \frac{(md)}{mxd}$$

in which *md* is the mean depth of lake, and *mxd* is maximum depth of lake.

When the value of volume development approaches unity, the lake basin is close to the form of a cone whose height is the maximum depth of the lake and whose base is equal to the surface area of lake; when the lake basin walls are essentially convex toward the water the index figure may be less than unity; when the basin walls are concave toward the water the index figure is likely to be greater than unity.

VOLUME CURVES

Volume curves not only exhibit certain structural relations of the lake basin but also make it possible to derive information from them by interpolation. A convenient volume curve is one constructed by plotting depths along the ordinate and the superimposed volume above each submerged contour along the abscissa.

Procedure

1. From data already obtained in determining total volume of lake, compute that volume of lake superimposed above level of each submerged contour.

2. Examine cross-section paper to be used and determine an appropriate and convenient scale for plotting along each axis. Place zero point for both sets of values at upper left corner. Plot depth along vertical axis and superimposed volume along horizontal.

3. If actual values of superimposed volume are plotted, then the curve is called a *direct volume curve;* if the data on superimposed volumes are computed in terms of per cent of total lake volume and then plotted, the curve is called a *percentage volume curve.*

SLOPE OF BASIN

The slope of the basin of a lake is usually expressed in per cent, although formulas are available for expressing the same result in degrees. In limnological practice a need for one or both of two approaches to bottom slope measurement may arise, namely, the determination of the slope between two submerged contours or isobaths, and the determination of the mean slope for an entire lake having several submerged contours.

Slope between two adjacent contours may be calculated by means of the formula

$$S = \frac{C_1 + C_2}{2} \cdot \frac{I}{A}$$

in which S is the slope in per cent between the two contours, C_1 and C_2 are the lengths of the two contours involved, I is the vertical distance or interval between the two contours, and A is the area of the bottom included between the two contours.

The mean slope of the entire lake may be calculated from the formula

$$S = \left(\frac{\frac{1}{2}\,C_0 + C_1 + C_2 + C_3 + \text{------} + C_{n-1} + \frac{1}{2}\,C_n}{n}\right)\frac{D}{A}$$

in which S is mean slope in per cent; C_0, C_1, C_2, C_3, are the lengths of the contours; n is the number of contours; D is maximum depth of the lake; and A is surface area of the lake.

Streams

LENGTH

The length of streams may be determined in two principal ways: (a) direct measurements in the field; and (b) measurements from maps of suitable character, scale, and accuracy. Direct measurements may be available from the original field survey. In smaller streams results of reasonable accuracy may be secured by rapid chaining along the bank or in the channel, or approximate results by pacing and by pedometer. Measurements from maps may be done satisfactorily by means of any one of the methods described for the measurement of lake shore lines (p. 90). The most convenient method is that of the map measurer.

WIDTH

Stream widths are usually measured in the field. The problem of measuring stream widths directly presents difficulties owing to the fact that great variations arise from changes of water level. Often the method and number of direct measurements must be matters of judgment and practicability as determined by local circumstances and the needs of the work. For limnological purposes it is not sufficient merely to make measurements at convenient and easily workable places. Some system of measurement must be employed which will include all parts of the stream from the narrowest to the widest and which will show the essential variation if such measurements are to have any dependable significance. For results of a statistical analysis of stream-width measurements consult Embody, Goodrum and Edmond (1938), see p. 98. In field work, stream-width measurements may be based upon (a) a selected water level, (b) general channel width, or (c) high-water level. In limnological work, the one first mentioned is usually preferable. The common need is to determine a *mean width* accurate within certain desirable limits.

Measurement of stream width on ordinary maps of small scale is usually out of the question. Dependable measurements can be made only on such maps as show particular sections of a stream on a very large scale and which have been drawn from careful and detailed field surveys.

OTHER MORPHOMETRIC CONSIDERATIONS

Means of making other morphological measurements of streams are easily derivable from methods previously described for work on lakes. To have any value at all, areas, shore lines, volumes, longitudinal and transverse profiles, and similar features must be based upon very careful, detailed surveys and the resulting hydrographic maps drawn on a very large scale. Shore development is computed on the same basis as described for lakes, (p. 93).

Selected References

(For Chapter 6)

Embody, D. R., C. A. Goodrum, and S. D. Edmond: "A Statistical Analysis of Width Measurements on a New Hampshire Stream," Biol. Surv. Merrimack Watershed, Survey Rept. No. 3, New Hampshire Fish and Game Dept., 1938, pp. 198–200.

Gray, F. J.: "The Polar Planimeter." 57 pp. London, St. Bride Press, 1909.

Henrici, O.: "Report on Planimeters," Rept. of the Sixty-fourth Meeting, Brit. Assoc. Advancement Sci., 1894, pp. 496–523.

Welch, P. S.: "The Hatchet Planimeter as a Limnological Instrument." 7 pp. Spec. Pub. No. 14, Limn. Soc. Am., 1944.

Wheatley, J. Y.: "The Polar Planimeter and Its Use in Engineering Calculations." 126 pp. New York, Keuffel & Esser Co., 1903.

Part II

Physical Methods

TEMPERATURE

Since the direct and indirect relations of water temperature to limnological phenomena are both numerous and fundamental, methods of measuring heat content at different depth levels are of prime importance. From the large number of devices now available, those described here are selected because of their general usefulness and dependability.

Surface Temperatures

Measurement of surface-water temperatures is usually a simple matter. Any good grade, simple type of thermometer properly used will suffice. A common chemical thermometer graduated to 0.2 degree C. is a convenient instrument for many kinds of field work. Such an instrument should be provided with a metal case equipped with a clip, chain, or other means of attachment to clothing of the operator to avoid loss in the field.

A good grade maximum-minimum thermometer (p. 118) is not only useful for occasional recording of surface-water temperatures but also is particularly adapted to the measurement of daily maximum and minimum temperatures over a period of time. For routine daily maximum-minimum temperatures, such an instrument should be mounted in a vertical position on a dock post, retaining wall, or other stationary object situated in the desired location. It must also be fixed far enough below the surface of the water so that it will always be submerged irrespective of irregularities of wave action; so disposed that it can be read conveniently; and so protected that it will not suffer damage from storms, floating debris, boats, and accidents in general. However, it often happens that daily maximum-minimum temperatures of the open surface water remote from shore are required and in such an instance it will usually be necessary to install the thermometer on a buoy anchored at the selected station. A useful buoy is illustrated in Fig. 23. It consists of a cylindrical piece of log about 15–20 cm. in diameter, about 1 m. long, and tapered at the lower end. To one side of this block is spiked a wooden staff at the free end of which a red, or a red-and-white, flag about 30 cm. square is attached. On the opposite side of the log and projecting downward is spiked a strong staff, about 1–1.7 m. long, at the free end of which is firmly attached a weight of suitable size and form. This weight serves as a counterpoise and must be adjusted to hold each individual buoy in the proper position and floating at the desired level. Large blocks of cork may be nailed to the sides of the log to prevent waterlogging. Several coats of waterproof paint applied to the log will accomplish the same purpose. Through a large smooth hole in the lower end of the log is tied one end of a 2-cm. (¾-in.) manila hemp anchor rope. The length of this anchor rope should be at least 1.5 times the total depth of the water in which the buoy is to be installed. The anchor may be a piece of scrap iron, concrete block, stone harnessed with strong wire, or any other similar object whose weight will insure against dragging and whose form is suitable for convenient handling. On the side of the log there is stapled a piece of large-size, stiff, electrician's copper wire bent, as shown in Fig. 23, into an inverted U, at the free end of which a maximum-minimum thermometer is securely attached in vertical position. The thermometer is thus held free from contact with any part of the buoy. The life of an anchor rope will be prolonged if it is soaked in hot paraffin or otherwise waterproofed. A thin coat of petroleum jelly will protect the metal parts of the thermometer if they are not rustproof. Tie-on surfaces for both ends of the anchor rope must be broad and smooth, otherwise the almost incessant motion of the buoy will cause wear and, ultimately, its casting loose. A buoy of this sort is much more likely to remain in place during violent storms than those having larger surfaces on which water movement may act. A thermometer so installed may be read whenever occasion demands, but for general purposes once every twelve hours, as for example, 7:00 A.M. and 7:00 P.M., will suffice. The index in each arm of the thermometer must be reset with a magnet after each reading. If it is necessary to lift the instrument out of water for purposes of reading, return it to water quickly to avoid the possible effects of atmospheric temperatures. *Caution:* In selecting a maximum-minimum thermometer for installation on a buoy, care must be taken to determine in advance whether the instrument so chosen is subject to error caused by the motion of the buoy shifting the position of the indices. Such liability of indices in maximum-

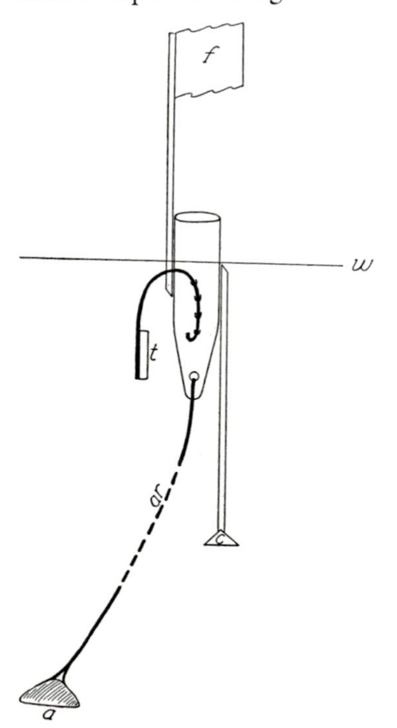

FIG. 23. Structural features of a buoy designed for the support of a recording thermometer. (*a*) Anchor. (*ar*) Anchor rope. (*c*) Weight. (*f*) Flag. (*t*) Thermometer. (*w*) Surface of water. Other features described in text.

minimum thermometers in general to shift position in response to vertical motion is a possible source of error which must be guarded against. However, it is usually possible to select an instrument which can be shown to resist this shifting of the indices. Such an instrument fastened to a buoy can be checked for index shifting by comparing its readings with those of a similar thermometer mounted on an adjacent nonmovable object in the same depth of water.

In ordinary measurements of surface temperature in quiet water, it is only necessary to submerge the thermometer in an inclined position and make the reading, after an adjustment interval, without removing the instrument from the water. If the water is rough, the temperature can be determined by dipping up a container full of surface water and immersing the thermometer in it.

Thermometers used exclusively in surface water need to be corrected for temperature only since for this purpose the hydrostatic pressure is negligible.

Subsurface Temperatures

Measurements of temperature at various depths below the water surface require instruments designed for that purpose. Since temperatures often differ greatly with depth and since most of the positions at which temperature is to be measured are remote from the observer and hence invisible, the construction of these devices must be thoroughly understood and their possible errors known. Such instruments when lowered into water meet rapidly increasing hydrostatic pressure, consequently, they may require correction for pressure as well as for temperature, depending upon the nature of their construction.

REVERSING THERMOMETERS

Of the instruments used for measurement of subsurface temperatures, reversing thermometers are commonly rated as among the best. They are also referred to under other names, such as "deep-sea thermometers," "turnover thermometers," and the names of their makers. Various types are made, all based essentially upon the same principle, namely, the 180° reversal of the thermometer, after the required period of adjustment, in which a column of mercury, whose length is a measure of the temperature at the selected level, is detached and delivered into the opposite end of the capillary tube. No subsequent additions to or subtractions from the detached mercury column can occur until the instrument is brought to the surface and turned back into the original position.

A widely used, high-class reversing thermometer is one manufactured by Negretti and Zambra, London, England, and the following description is based upon that type (Fig. 24). This instrument usually consists of

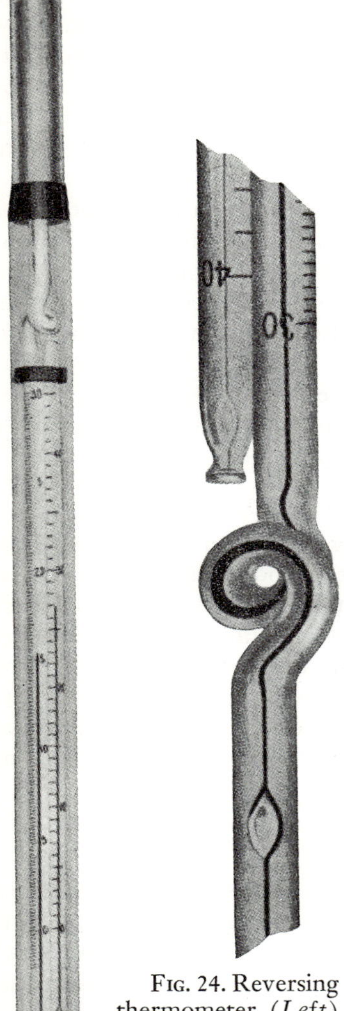

FIG. 24. Reversing thermometer. (*Left*) Complete thermometer in *reversed* position. (*Right*) Portion of main thermometer, in *nonreversed* position, showing details of construction in region of coil. (Courtesy, Negretti and Zambra, London, England.)

two thermometers, a main and an auxiliary thermometer, both in one sealed glass sheath. The main thermometer is the device which makes the temperature measurement at the selected level; the auxiliary thermometer is for purposes of making certain corrections to be described later.

The main thermometer has a long, narrow mercury reservoir at one end which is surrounded on the outside, i.e., the space between the mercury reservoir wall and the surrounding, outer glass sheath, by a layer of mercury, sealed in place, to facilitate the response to temperatures. The stem leading from the main reservoir is coiled. Above this coil the stem is straight and on it are graduations numbered in the reverse direction. A short distance above the mercury reservoir there is a special form of constriction in the bore of the capillary tube. Then follows a short, straight section of uniform bore, leading to the coil in which the bore becomes considerably enlarged. After passing through the coil, the bore becomes uniform, extending to the opposite end of the thermometer where it expands into a small fusiform reservoir. When the thermometer is in a vertical position with the large reservoir down, the mercury column is continuous from the reservoir to its termination in the capillary tube above. As it descends into the water in this position it acts as an ordinary thermometer, the upper end of the mercury column shifting in position in response to the temperature of the various strata through which it passes. When the selected depth is reached, the instrument is allowed to remain at that position for an adjustment period of at least 1 or 2 min., or longer in case of

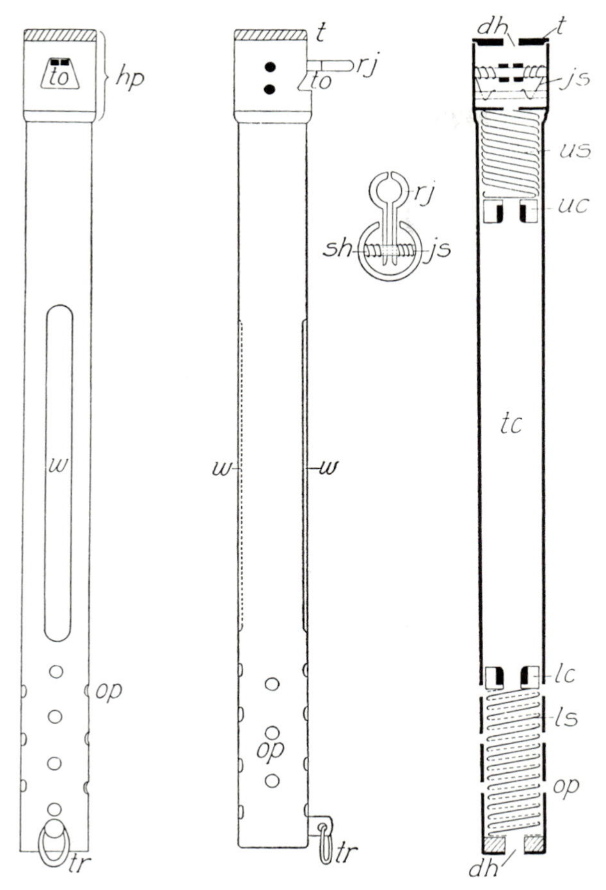

Fig. 25. Metal case for reversing thermometer. Figure to left and figure in middle represent external views of whole case. Figure to right is a median sectional view showing details of internal construction. Small figure shows essential details of releasing mechanism. (*dh*) Drain hole. (*hp*) Removable headpiece. (*js*) Spring operating release jaws. (*lc*) Sliding, circular, rubber-lined collar at lower end of thermometer chamber for support of main-reservoir end of thermometer. (*ls*) Lower coiled spring on which thermometer is supported. (*op*) Lateral openings providing free circulation of water through interior of that part of case which contains principal mercury reservoir. (*rj*) Release jaws. (*sh*) Rotating shaft on which release jaws operate. (*t*) Removable top piece. (*tc*) Thermometer chamber. (*to*) Triangular opening in wall of case through which release jaws operate; when at top position of opening, jaws are closed; when at bottom of triangular opening, jaws are spread apart. (*tr*) Tie-on ring. (*uc*) Sliding, circular, rubber-lined collar at upper end of thermometer chamber which supports upper end of thermometer. (*us*) Upper coiled spring on which upper end of thermometer is supported. (*w*) Elongated window through which thermometer is read.

doubt; then a metal messenger, sent down the rope, releases the tripping device and the thermometer turns over. The effect of the sudden reversal is to separate the mercury column at the constriction point so that it runs into the opposite end of the instrument, filling the fusiform reservoir and a certain portion of the capillary tube. The length of this detached mercury column is a measure of the temperature at the selected level. The instrument is now brought to the surface and read in the reversed position.

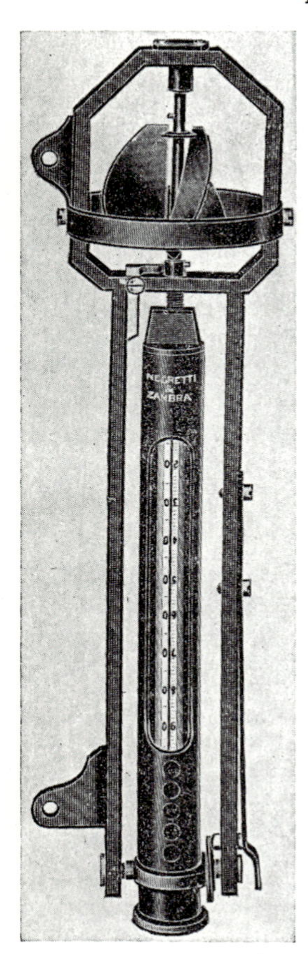

Fig. 26. A frame for reversing thermometers in which release mechanism is operated by a propeller. (Courtesy, Negretti and Zambra.)

In order to function properly, a reversing thermometer must be housed in a special metal case the mechanical construction of which is such that it can be reversed at the selected depth level (Fig. 25) either by a messenger sent down on the line or by a propeller which rotates when it is pulled a short distance through the water (Fig. 26.)

Corrections. Reversing thermometers are subject to two classes of errors, structural and functional.

By structural errors are meant those which arise from failure of the instrument to be true to standard at any of the temperatures and pressures involved in ordinary operations. Such corrections should be secured by submitting the instrument to some well-known standardizing agency, such as the U.S. Bureau of Standards. It is usually possible to secure, at a modest increase in cost, instruments which have certificates of correction already prepared. Usually it is not necessary to consider a correction which is less than one-tenth of the smallest division shown on the thermometer graduations, which on Negretti and Zambra thermometers is one-fifth of a degree C. Instruments made by the company just mentioned are usually certified as satisfactory under pressures up to 3 tons per sq. in.

By functional errors are meant those which occur in the changes in volume of the detached mercury column when it passes through waters of different temperatures on its return journey to the surface, changes which are often so slight as to be negligible for some limnological purposes. In fact certain kinds of reversing thermometers are made without any provision for corrections of this sort. However, it is sometimes desirable to have subsurface temperatures determined as accurately as possible, and to meet this need the auxiliary thermometer, previously mentioned, is included within the instrument. It is a smaller straight thermometer fastened against, and parallel to, the main thermometer, with its graduations in readable position when the main thermometer is reversed. This auxiliary thermometer shows the temperature of the detached mercury at the time of the reading. The correction is calculated from the formula:

$$c = tva$$

in which

c is the correction sought,

t is the *difference* in temperature between the reading on the main thermometer and that of the auxiliary thermometer,

a is the coefficient of apparent cubical expansion of mercury in the glass of which the thermometer is constructed (usually about $1/6000$ per degree Centigrade; value sometimes marked on main thermometer), and

v is the reading of the main thermometer *plus* the volume which would have been detached if the reversal had been made at 0° C. (this "volume at 0°" value is engraved on the main thermometer).

For example: if

main thermometer reads 18°,

auxiliary thermometer reads 24°,

volume at 0° is 79, and

coefficient of expansion of mercury in the glass is $1/6000$, then

$$(24 - 18) \times (79 + 18) \times \frac{1}{6000} = 6 \times 97 \times \frac{1}{6000}$$
$$= 0.09°$$

Corrected temperature reading is $18 - 0.09 = 17.91°$.

The correction, when determined, must be either subtracted or added to the reading made on the main thermometer at the surface. Under ordinary circumstances as they prevail in lakes during the summer, the water becomes cooler with increasing depth; therefore, when the thermometer is reversed at some distance under water, the main and the auxiliary thermometers show, at the reversal level, the same reading if the proper adjustment interval has been provided, but in the passage to the surface through increasingly warmer water the detached mercury column expands; therefore, the correction value must be *subtracted*. In midwinter under the ice cover, the water temperature often becomes warmer with increasing depth (to 4° C.) in the deeper lakes; therefore, the correction value must be *added*. These temperature relations with increasing depth are known to the observer since ordinarily vertical temperature series are made progressively from the surface downward. Special circumstances in which even in summer colder water may exist *above* warmer water because of differences in density should be watched for if only deep water temperatures are taken; in a vertical series such relations are obvious to the observer when the series progresses from the surface downward.

Defects and Adjustments. A defect which seems to be common in all types of reversing thermometers is their liability to develop errors of performance suddenly when a particle of residual gas, displaced from the small reservoir at the top of the thermometer, gets into the mercury column, causing breakage of the latter at the wrong place when reversed. As explained by the makers, great precaution is taken in the manufacture of these thermometers to exclude all gas from the tube. However, a minute quantity of gas is likely to be left in the system in the sealing-off operation or such gas becomes apparent after the instrument has aged. This gas particle is exceedingly minute and not visible in the mercury. If this gas particle remains in the reservoir at the top of the thermometer it not only gives no difficulty but actually is something of an advantage since it tends to prevent the mercury from "locking" in the top of the instrument. If, however, the gas particle gets into the mercury it will cause breakage of the mercury column at that point instead of at the normal breaking point and "locking" of the mercury in the top reservoir after reversal becomes a more frequent occurrence. When the error in breakage is large it can usually be detected at once but if it involves only a fraction of a degree it may be overlooked. For this reason some workers insist that reversing thermometers should be used in pairs mounted together in the same reversing mechanism (Fig. 27). This defect seems to arise usually (a) as a result of improper storage of the instrument when not in use, (b) in transportation, and (c) in instruments which are several years old. To clear this gas particle from the mercury column, some makers advise cooling the thermometer in ice, then heating it up, cooling again, repeating this process several times. This heating-up process must be done with great caution.

When the mercury column "locks" in the upper reservoir it can sometimes be dislodged by swinging the thermometer centrifugally. If this treatment fails to dislodge the column then the makers (Negretti and Zambra) state that it is best to heat the upper reservoir in warm water until the thread of mercury in the capillary tube connecting with the upper reservoir is of sufficient length to assist the down pull when centrifugal force is applied by swinging. After the column is restored, it is very important that the thermometer be cooled in ice and heated several times since the fact that the mercury locked suggests that a particle of gas has been displaced from the upper reservoir and is somewhere in the capillary tube.

Reversing Cases. Mechanisms for the protection and for the reversing of such thermometers are made in many different forms. Two types are common, one of which is operated by a messenger (Fig. 25), the other by a propeller which spins when the whole mechanism is drawn vertically for a short distance (Fig. 26) after the adjustment period. In either type the thermometer is allowed to fall over and come to rest after a change of position of 180°.

Reversing Frames. Detection of abnormal performance in a reversing thermometer can be accomplished in several ways, a quick and convenient one being the simultaneous reversal of two or more thermometers at the same depth level and in the same reversing frame. The frame described here was designed by the writer. Since the operation of reversing thermometers, when used singly, requires that each be housed in a special metal reversing case (Fig. 26), this frame was designed to receive such thermometers without removing them from their cases. A central bar flanked by two rods constitutes the supporting element. At each end is a plate (Fig. 27B, D). Each plate has two cups, the size and form of each adapted to receive the ends of the thermometer cases. On one side, the cups are deeper in order to provide for possible differences in the length of the thermometer cases. Other details are shown in Fig. 27. The length of this frame is such as to hold two thermometer cases snugly and safely in place.

To make ready for use, unscrew wing nuts *wn;* pull out locking staple *ls;* untie line *to;* and remove lower plate. Insert upper ends of two thermometer cases containing two reversing thermometers into cups c_3 and c_4; replace lower plate, making cups c_1 and c_2 engage tie-on ends (main mercury reservoir ends) of thermometer cases; if necessary put soft packing material in bottoms of cups to immobilize thermometer cases completely; screw down wing nuts *wn* and insert lock staple *ls;* test thermometer cases again to make certain that they are safely and firmly installed in frame; then set up frame in position shown in Fig. 27A. The instrument is now ready to be lowered into the water. When it reaches the desired depth level and the proper adjustment interval is concluded, a metal messenger, descending on the line, strikes the top of the trip mechanism, releases the ring *rr*. The frame falls over, reversing the thermometers. When brought to the surface the two thermometers should register exactly the same temperature; if they do not, at least one of the two thermometers is behaving abnormally. Remove one thermometer from the frame; introduce a third one in its place and repeat the procedure. By this means the faulty instrument may be detected. If a third reversing thermometer is not available, the two used in the reversing frame

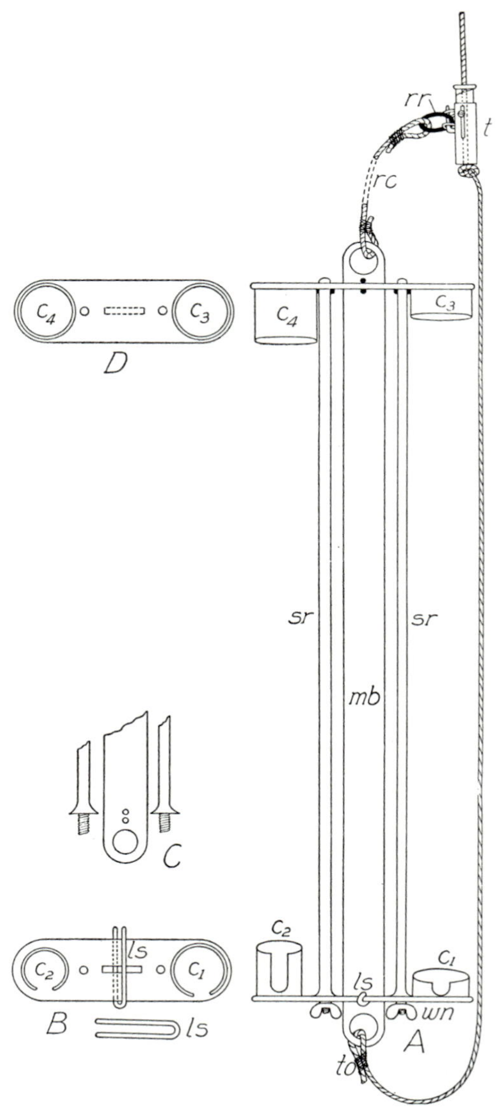

FIG. 27. Frame for reversing two thermometers simultaneously at same depth level. (A) Front view of complete frame. (B) Lower, removable plate showing locking staple in position. (C) Detail of lower end of frame with lower plate removed. (D) Upper, nonremovable plate. (c_1 and c_2) Cups for holding lower ends of reversing thermometer cases, side slots for reception of tie-on posts and rings on thermometer frames. (c_3 and c_4) Cups for holding upper ends of reversing thermometer cases. (*ls*) Locking staple. (*mb*) Middle bar. (*rc*) Release cord. (*rr*) Release ring. (*sr*) Side rod. (*t*) Trip mechanism. (*to*) Tie-on at lower end of frame. (*wn*) Wing nut.

may be tested in the laboratory against another kind of temperature-recording instrument.

Since the operation of two thermometers in this frame is as simple as that of a single one, the regular practice of using a pair simultaneously should be considered seriously by careful workers. Very old thermometers, those which have been long in storage, and those which have been shipped by public transportation must be tested for normal performance before making serious use of them. New instruments must not be accepted blindly on faith. Not all kinds of reversing thermometers are equally good and dependable and those made by manufacturers newly in that kind of production should be tested very carefully.

By means of a few simple structural changes this reversing frame can be enlarged to contain three or four reversing thermometers.

General Considerations

a. When not in use, reversing thermometers should always be stored in a vertical position with the large mercury reservoir down. This procedure minimizes the danger (1) of dislocating a gas bubble from the upper reservoir with resulting break of the mercury column at the wrong place and (2) of the possible locking of the mercury column in the upper reservoir.

b. When not in actual use, these instruments should be kept in a long field case with a strap handle at the upper end. Such a case affords general protection and also a convenient means of hanging the instrument in the correct position.

c. Reversing thermometers may be broken if reversed against rocks on the bottom especially in shallow waters. If the nature of the bottom is not known, a sounding test should be made before a thermometer is reversed at the lowermost position.

d. Every new reversing thermometer should be tested before being put into active use. This test can best be done on a lake where various pressures and temperatures are available in a vertical series. The test can be made by mounting the new instrument in a twin reversing frame (Fig. 27) with another reversing thermometer of known dependability and reversing both simultaneously at various depth levels; or a similar test can be made with some other type of temperature-measuring device, such as a thermophone or an electric-resistance thermometer, by comparing simultaneous readings at the same depth levels.

e. Since in the shallower depths messengers strike tripping devices with considerable force, the battering effect of these impacts is minimized by covering the prongs of the trip on the instrument case with pieces of ordinary rubber tubing of the proper size. It is also desirable that a knot be tied in the graduated rope about midway between the two ends of the thermometer case in order to prevent messengers from striking the tie-on link.

f. A carefully tested reversing thermometer of the highest class construction which has been completely certified by an official standardizing bureau serves as an excellent standard against which to compare other instruments for measuring temperatures at various depths.

g. Reversing thermometers, more or less similar in design to those of Negretti and Zambra but lacking the outer, heavy, protecting glass sheath, are obtainable. Such instruments cannot be used at great depths without risk of crushing and tests for the effects of hydrostatic pressure are imperative.

THERMOGRAPHS

Many of the various forms of thermograph are adaptable to limnological work. Types originally designed for continuous records of temperatures in air and soil are easily turned to the needs for similar records in water. Thermographs have been designed or adapted for the taking of such records in water with installations on hulls of boats or at stations located near or on shore. Recently an instrument known as a bathythermograph (Spilhaus, 1937; 1940) has been devised which appears to have promise of important value in both oceanographic and limnological work. The essential feature of the instrument is the provision whereby a very detailed record of the vertical distribution of temperatures in water is secured by pen tracing against a smoked-glass as the instrument is lowered and raised in the water.

THERMOPHONE

One of the most useful instruments for the measurement of subsurface temperatures is the thermophone designed originally by Warren and Whipple (1895). However, in spite of its obvious merits it is not as extensively used in limnological work as it deserves, possibly in part because of the initial cost.

The thermophone is essentially an electric thermometer which acts upon the principle that the resistance of a conductor to the passage of an electric current changes with the temperature, also that the rate of change in resistance due to temperature differs in different metals. Structurally, the instrument consists of three main portions, (a) an instrument box containing batteries, Wheatstone bridge, galvanometer, and accessories; (b) a 3-wire cable of a length demanded by maximum depths of the waters to be investigated; and (c) the helix-shaped terminal at the free end of the cable which contains the sensitive coils.

The terminal (Fig. 28) is a slender brass tube about 5 ft. long which is wound into a coil about 6.5 cm. in diameter. This tube contains two resistance coils of two different metals which differ from each other markedly in the rate of electric conduction, as for example, copper and German silver. These contrasting coils are united at one end and their other ends are each connected to a wire extending into the cable. A third wire is attached to the junction of the two coils and it also extends into the cable. The brass tube containing the coils is filled with oil and hermetically sealed. The enlargement of the brass housing tube contains the connections with the cable wires.

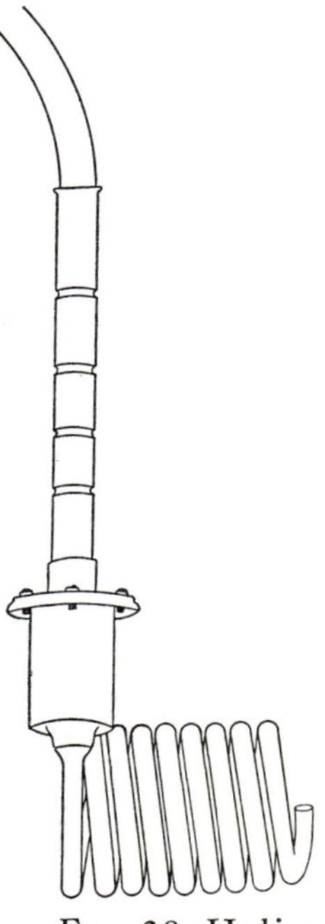

FIG. 28. Helix-shaped terminal of thermophone.

The 3-wire cable is flexible, graduated according to some convenient scale, and usually carried on a revolving reel or drum. At the upper end, the 3 cable wires extend independently from the end of the cable and each terminates in a brass contact number to indicate which binding post on the instrument box is to receive it. The temperatures of the wires in the cable do not affect the temperature reading made on the instrument since two of them are of low resistance and are on opposite sides of the Wheatstone bridge, thus neutralizing each other. The third cable wire connects the junction of the two coils in the terminal with the galvanometer and is not involved in the equation.

The instrument box contains 3 dry-cell batteries in the battery chamber (Fig. 29, *bc*), the galvanometer *g;* the combined Wheatstone bridge and switch *s-w* with dial and movable hand; and 3 binding posts for the attachment of the cable. Fig. 30 is a wiring diagram for the type described here. The coils c_1 and c_2 in the terminal are connected through the cable by individual wires to the two ends of the circular slide wire *xz* (Wheatstone bridge). The two ends of this slide wire are connected in circuit at binding posts *1* and *2* with a battery *b*. The galvanometer *g* is inserted between the upper end of the third cable wire and the movable contact *y*. The galvanometer will indicate no current passing through the No. 3 wire when $\frac{c_1}{c_2} = \frac{xy}{yz}$. Since the resistances of c_1 and c_2 change unequally with different temperatures, there will be a different value for $\frac{c_1}{c_2}$ for each new temperature. This value, measured by $\frac{xy}{yz}$ and expressed in units of temperature, is determined by the position of the movable contact *y* on the slide wire *xz*. In the type being described, the slide wire is

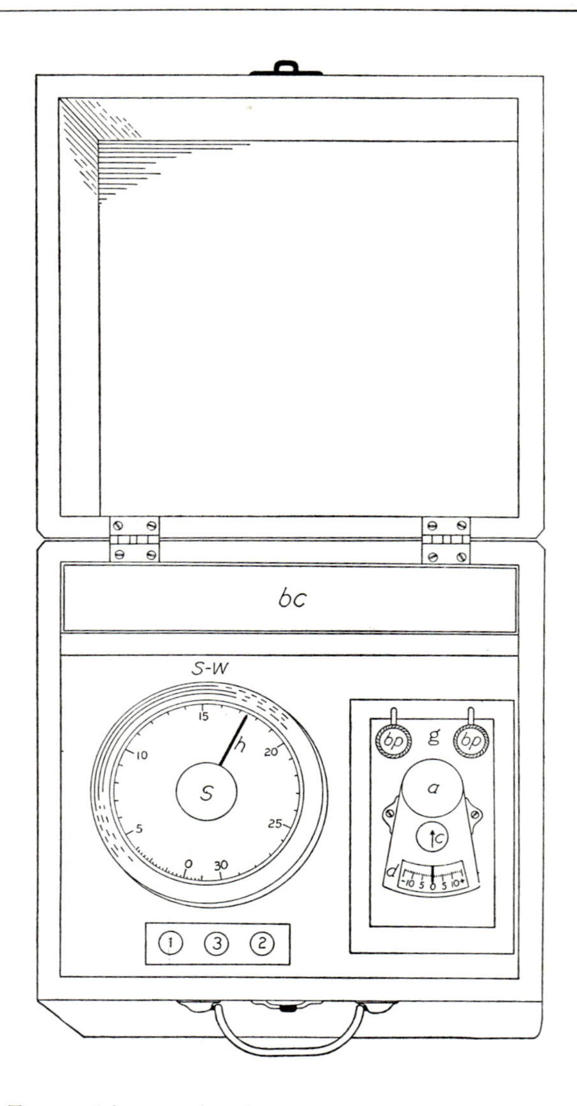

FIG. 29. Diagram showing essentials of instrument box of thermophone. (*a*) Adjusting knob for galvanometer needle. (*bc*) Battery box. (*bp*) Binding posts. (*c*) Sliding clamp for immobilizing or freeing galvanometer needle. (*d*) Scale. (*g*) Galvanometer. (*h*) Movable hand. (*s*) Switch. (*s-w*) Combination switch-Wheatstone bridge dial. (*1, 2, 3*) Binding posts for lead wires in cable.

wound around the edge of a circular disk above which is located a dial graduated in units of temperature. The hand *h* is superimposed directly above the sliding contact, is free to move over the dial, and both hand and sliding contact are moved by turning the knob *s* in the center of the dial.

To operate a thermophone of this type, proceed as follows: Place instrument case on level, firm support in position convenient for work; if in boat, select position where motion of boat will be minimized; connect upper contacts of cable with proper binding posts, *1*, *2*, and *3* (Fig. 29); tie some upper portion of cable securely to boat as measure of safety; release galvanometer needle by sliding clamp *c* into free position and carefully set needle on zero point of underlying scale *d* by means of adjusting knob *a;* lower cable terminal into water to desired depth and allow to stand for adjustment interval of 1.5 minutes; push down knob *s* in center of dial and observe performance of galvanometer needle; in its pushed-down position, turn knob in that direction (clockwise or counterclockwise) which brings galvanometer needle back to zero position; when zero is reached, dial hand *h* indicates temperature of submerged terminal.

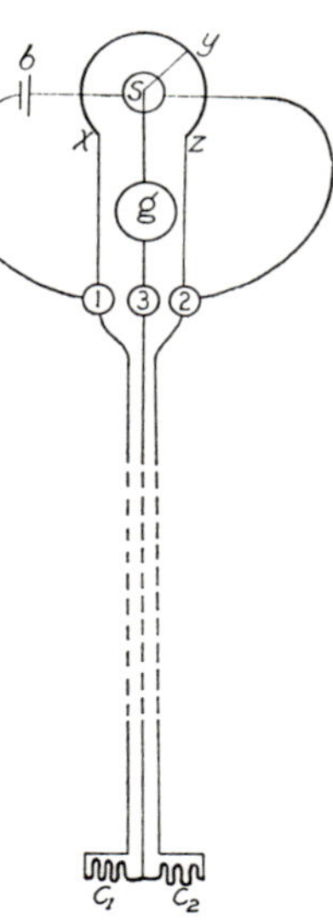

FIG. 30. Wiring diagram of thermophone circuit. (*1*, *2*, *3*) Lead wires. (*b*) Battery. (c_1 and c_2) Coils in terminal. (*g*) Galvanometer. (*s*) Switch. (*x* and *z*) Ends of lead wires attached to circular slide Wheatstone bridge. (*y*) Movable contact.

ADVANTAGES OF INSTRUMENT. The thermophone possesses advantages over most other kinds of subsurface temperature-measuring devices, as follows: (a) It is very sensitive and accurate; (b) its readings are not subject to the effects of water pressures; (c) its adjustment period is small; (d) temperatures are easily and quickly read; (e) it is not necessary to bring the terminal to the surface in order to make a record, thus saving much time; (f) at each selected position the temperature record is made within a water thickness of only about 2.5 in.; (g) it may be used in any of the partly suspended bottom deposits of lakes into which the terminal may be sunk to any depth safe for the instrument; (h) it is particularly suitable for determining the successive fluctuations of water temperature at any one level; (i) with proper care it has a very long life.

The only disadvantages of a thermophone, compared with a reversing thermometer are (a) weight and bulk, especially if a long cable is required, and (b) initial cost.

TESTING AND CORRECTION. Since the thermophone is not affected by pressures, it can be tested and corrected in the laboratory where water of various temperatures can be secured and against any already carefully corrected thermometer.

Modification of Thermophone. The cost of a thermophone may be reduced by substituting for the somewhat expensive galvanometer a current interrupter or buzzer attached to a telephone receiver. In operation, the current is turned on at a switch, the telephone is held to the ear, and the dial hand turned back and forth over the graduated dial. The buzzing sound increases or decreases as the hand approaches or departs from a certain position. The silence point indicates on the dial the temperature of the terminal at the lower end of the cable. This modification has also a certain advantage over the galvanometer in that there is no swinging needle to contend with under conditions of rough water.

Precision. Potentially, the thermophone is a very sensitive instrument. Limitations of precision depend upon the size of the dial and certain other features. A type built to take a dial 10 cm. in diameter is usually graduated to 0.2° C.; a reading of 0.1° by estimation is easy, and one to 0.05° by estimation is possible. Instruments made to use larger dials possess still greater precision.

General Considerations

a. A thermophone should always be tested for good condition in the laboratory before it is taken into the field.

b. The 3 small dry-cell batteries should be replaced from time to time, the interval of replacement depending upon extent of use of the instrument. Mark date of replacement on batteries when installed. In case of doubt, replace the batteries, and if remote from source of supplies, carry an extra set.

c. When not in use wrap both sets of cable terminals in soft, bulky cloth to protect them against damage in transportation. Provide a revolving drum enclosed in a box for the cable (Fig. 31).

d. Establish graduations directly on the cable, either with paint, colored threads, or other permanent means.

e. Ordinarily there is no occasion for disassembling parts within the instrument box. If trouble is serious enough to require disassembling, it will probably be preferable to send the whole thermophone to the maker or to some factory known to be competent and familiar with such instruments. Avoid any changes which alter the position of the dial. If, for any reason, the knob *s* in the center of the dial is removed, be sure that, on reassembling, the vertical rod connecting the knob with the switch below is restored to its proper position; if it is installed wrong end up a short will occur and the batteries will run down.

f. When ready for use in a boat or elsewhere, make certain that the

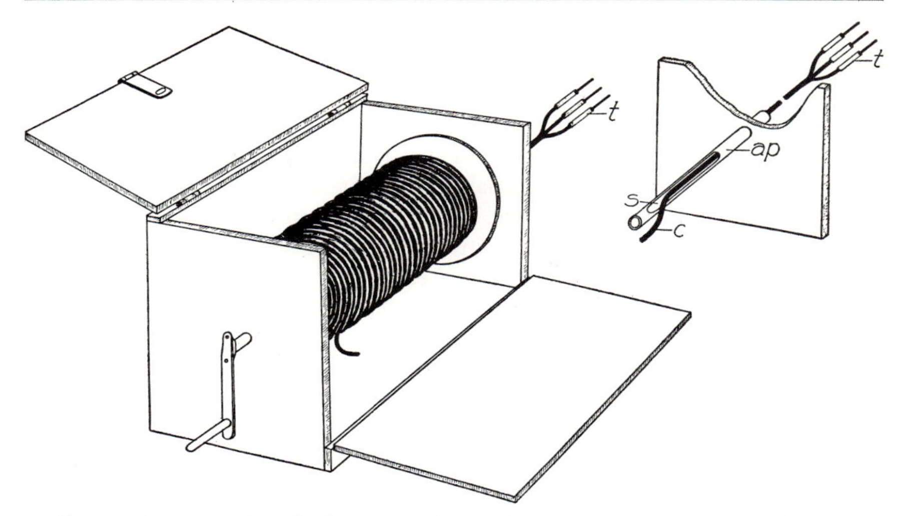

FIG. 31. Drum and enclosing box suitable for storage and field use of thermophone cable. (*Left*) Assembled outfit ready for use. (*Right*) Portion of right end of drum box showing method of carrying terminal end (upper end) of cable through interior of axial pipe to outside of box. (*ap*) Axial pipe which supports drum. (*c*) Cable. (*s*) Slit in pipe for introduction of cable. (*t*) Terminals for attachment to galvanometer.
The spool on which the cable is wound is composed of slats separated by narrow spaces.

cable is firmly tied near its upper end to some fixed object. If this precaution is neglected, there is danger of losing the cable overboard.

g. Every precaution must be taken to exclude water from the interior of the terminal or the cable.

ELECTRIC-RESISTANCE THERMOMETERS

Of recent years other portable types of *electric-resistance thermometers* have been developed which are adapted for the measurement of subsurface temperatures in lakes and streams. These instruments resemble in many respects, both fundamental and superficial, the thermophone just described. In fact, the similarities are so numerous that it is not necessary to describe the instrument in detail here. The following features may be mentioned: (a) The terminal (bulb) which is lowered into the water may be of various forms, a commonly used one being a straight tube about 8 in. long, the lower 3-in. portion of which is the temperature-sensitive region. (b) The potentiometer is so built that the temperature scale may be as much as 17 in. long, thus making it possible to enlarge the temperature units on the scale and provide for much greater precision in subdividing the units (degrees). (c) The adjustment interval is very short (about 30 sec.). (d) Double scales may be provided which make it possible to read results in two scale ranges (e.g., Centigrade and Fahrenheit) simultaneously. (e) Field operation is very similar to that of the thermophone. (f) The advantages and disadvantages of the instrument are essentially those listed on p. 115 for the thermophone. (g) Testing and correcting, see p. 115. These instruments are very accurate and possess high precision.

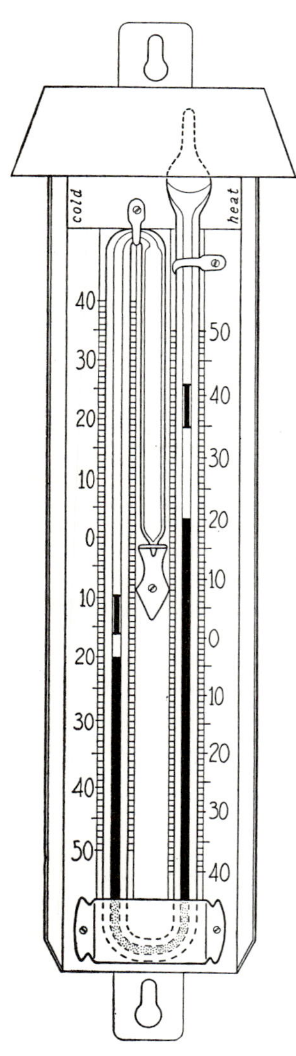

Fig. 32. Maximum-minimum thermometer (Sixes type).

Practically all of the items relating to the thermophone under General Considerations, pp. 116–117, apply to other electric-resistance thermometer outfits.

MAXIMUM-MINIMUM THERMOMETER

If reversing thermometers, a thermophone, or other specially designed instruments cannot be provided, a maximum-minimum thermometer, "Sixes type," can sometimes be used to secure records of subsurface temperatures in inland waters if the limitations of the instrument are understood and certain precautions are observed.

This instrument is a self-registering type of thermometer, made by various companies and in various grades. The instrument consists essentially of a U-shaped capillary tube (Fig. 32) with a slender, elongated reservoir at the upper end of the left arm and a pear-shaped reservoir at the upper end of the right arm. The slender reservoir on the left and a part of the adjacent capillary tube are filled completely with a liquid (creosote). This reservoir acts as the "bulb" of the thermometer since it contracts and expands in response to decrease or increase of temperature. Approximately the lower half of the U-tube is occupied by a mercury column so related to the creosote column on the left that any contraction or expansion of the latter causes a rise of one arm of the mercury column and a corresponding fall of the other. The space in the capillary tube above the right arm of the mercury column is also filled with creosote as is likewise a portion of the pear-shaped reservoir. This partly filled reservoir takes care of the increase and decrease of the liquid volume due to expansion and contraction of the creosote in the other reservoir and the U-tube. In each limb of the tube there is an index above the end of the mercury column. This index is a fine piece of iron wire enclosed in a glass capsule to prevent corrosion. It is equipped with a device (in some types, a minute spring formed of hair or a glass filament) which introduces friction against the sides of the capillary tube and prevents the index from changing position because of its own weight. The U-tube is fastened to a metal frame in such a way that the left limb is the *cold* side and the right limb the *warm* side. Graduations occur on the frame just back of the capillary tube.

To operate the thermometer, proceed as follows: set the two ends of a horseshoe magnet on the capillary tube and over the index slowly stroke downward until the index rests on the end of the mercury column; repeat this procedure on the other index; lower thermometer to the desired depth level and allow a five-minute interval for adjustment; bring instrument to the surface and read the lower end of the index on cold side (see p. 121 for exception); reset with the magnet and proceed to the next selected depth.

The creosote in the left-hand bulb contracts as the result of going into colder, deeper water and the mercury rises in the left hand limb pushing the index *up* the tube to the *minimum reading;* likewise, when the creosote expands as the thermometer comes back into the upper, warmer water, the index remains at the minimum position even though the mercury has retreated to some other position.

CORRECTION. When used for measuring subsurface temperatures, these maximum-minimum thermometers must be corrected for both temperature and pressure. Fortunately, it is possible to correct simultaneously for temperature and pressure in the following way: Choose a lake in midsummer whose depth is such as to provide (a) difference of about 20° C. between surface and bottom waters and (b) considerable water pressure gradient from surface to bottom. Lower simultaneously to same selected depth (near surface) maximum-minimum thermometer to be tested; also already corrected instrument which serves as standard. With both standard and thermometer at same depth, make careful temperature records; lower both to another depth and repeat; continue such records at chosen intervals from surface to bottom. Increasing depth not only provides increasing pressures but also provides colder and colder water thus spreading test over most of operating range of thermometer. From such records a correction table can be made.

Corrections for temperature only, if needed, may be made in the laboratory in the usual way against a standard using adjusted water temperatures. If, for any reason, it is desirable to know the influence of hydrostatic pressure alone, this information can be secured by making records at various depths in a lake at the time of the spring or fall overturns when the water is homothermous. Such a series must be exactly paralleled by one taken with the standard instrument against which the corrections are made.

PRECISION. These instruments are usually graduated in 1° intervals. With experience, readings can be estimated to 0.5°; closer estimates are uncertain.

ADVANTAGES AND DISADVANTAGES. Among the advantages of this instrument may be listed modest cost, small size, light weight, and general adaptability. In subsurface work, several instruments may be attached, one at each of several selected intervals on the graduated rope, thus making records simultaneously at as many different depths as there are instruments and with a considerable saving of time. The following disadvantages should be mentioned: (a) It is fragile; the glass portions are unprotected; (b) it cannot be used in great depths without danger of crushing; (c) it must be brought to the surface for each record; (d) its precision is limited; and (e) it may have a defect which arises from the occasional unreliability of the indices. If the device for holding the index in the tube is too weak the index will tend to slip down of its own weight in the process of handling; if it is too strong, the mercury pushes past the index thus dividing the mercury and spirit columns. Every instrument must be inspected for this kind of trouble. Instruments of best construction and adjustment will give satisfactory and dependable service in ordinary operations, but even they must not be subjected to rough handling or to severe vibration since indices will slip under these conditions, causing erroneous readings.

General Considerations

a. These instruments have one limitation which may or may not be serious depending upon circumstances. If the temperature of the water becomes progressively cooler (or progressively warmer, as just under ice cover) as the depth increases, correct readings will result if the proper side of the thermometer is read, but if, as occasionally happens, an uneven layering of the water occurs in which the temperature change is reversed locally the reading for the still lower position may be a false one. This limitation, which cannot be remedied, has led to the rejection of the instrument by some workers. Certainly, it cannot be used with confidence except on those waters whose temperature behavior is known to be progressive in the vertical direction. It should not be used in any kind of reconnaissance work and it should be avoided in waters manifesting density layering or irregular mixing.

b. Subsurface readings made during the spring, summer and autumn will be taken from the left (cold) side of the thermometer since such temperatures will all be above 4° C. and will ordinarily be progressively colder with increasing depth. However, an exception should be noted if the thermometer is to be read and set in air. On all occasions when the air temperature is *colder* than the surface water, it will be necessary to set and read the right side (warm) of the instrument until water temperatures are reached which are as cold or colder than the air; then read on the cold side (left). Or if preferable, read and set the thermometer just under the surface of the water, i.e., do not bring it out into the air until water is reached which is as cold as the air; then all readings may be made on cold side.

c. To measure subsurface temperatures below an ice cover proceed as follows: (1) if air temperature at time of reading is below zero and if the instrument is to be read in air, read all subsurface temperatures on the right hand side (warm); (2) if air temperature is higher than the temperature of the water just under the ice, submerge thermometer and set indices under water, then read temperatures for successive depths on right (warm) side and do not remove instrument from water but set indices by submerging magnet.

d. Store instruments of this kind in an upright position.

e. If the mercury column becomes separated, grasp the instrument firmly by the upper end and swing vigorously and sharply downward in the arc of a circle until mercury unites.

f. If an index gets into the mercury, follow the directions just given in paragraph (e) above but in addition use the magnet and draw the index up as far as possible after first few swings until it is above mercury column.

g. If the index has been forced into the right reservoir, hold frame upright and manipulate index by tapping the frame gently and by using the magnet until it is drawn into the capillary tube.

Heat Budget of Lakes

Three different things are included under the term *heat budget* of a lake: (a) the amount of heat required to raise the water of a lake from 0° C. to the maximum summer temperature (*gross* or *crude heat budget*); (b) the amount of heat required to raise its water from the minimum winter temperature to the maximum summer temperature (*annual heat budget*); and (c) the amount of heat required to raise its water from 4° C. to the maximum summer temperature (*wind-distributed heat*, or *summer heat income*).

Since lakes do not have bottom temperatures as low as 0° C., the first statement is of least importance. The second statement, representing as it does the two limits of heat content for the lake under consideration, is fundamentally the most valuable of the three approaches. However, it involves the necessity of having available the necessary winter temperature data. The third approach is useful only for *temperate* lakes whose winter temperatures are unknown.

Heat budgets may be computed as follows: (a) the number of calories required to warm a water column of unit base in the deepest part of a lake from a selected minimum (0° C., 4°, winter minimum) to the maximum summer temperature; (b) the number of calories required to warm the whole water mass of a lake from the selected minimum to the maximum summer temperature; and (c) the number of calories required to warm a water column of unit base and a height equal to the *mean depth* of a lake from the selected minimum to the maximum summer temperature.

Heat budgets may be computed, when the lake is considered as a unit, by the following formulas:

$$\textit{Annual heat budget} = Dm\ (Tm^{s} - Tm^{w})$$
$$\textit{Wind distributed heat} = Dm\ (Tm^{s} - 4)$$

in which Dm is the mean depth of the lake *in centimeters;* Tm^{s} is the mean summer temperature; Tm^{w} is the mean winter temperature; and 4 represents 4° C. The results are expressed in *gram calories.*

Mean Temperature of Lakes

A satisfactory method (Birge, 1915) of computing *mean temperature* of a lake is as follows: Make a vertical series of temperature records from surface to bottom at the deepest part of the lake; compute the *mean temperature of each stratum* bounded by each pair of adjacent records by the use of the formula $\frac{T_1 + T_2}{2}$ in which T_1 is the temperature of the upper limit of the stratum, T_2 is temperature of the lower limit of the same stratum. Multiply the mean temperature of each stratum of the lake by the per cent of the volume of the whole lake which that stratum composes. The sum of these products represents the *mean temperature.* The whole computation is represented in the formula

$$T_m = \frac{T_1 + T_2}{2}(P_1) + \frac{T_2 + T_3}{2}(P_2) + \text{-----} + \frac{T_{n-1} + T_n}{2}(P_n)$$

in which

T_m is mean temperature

T_1, T_2, T_3 is temperature at respective levels bounding the various strata of water

T_n is temperature at lowermost level of last stratum

P_1 is that percentage of the entire lake volume which the first stratum constitutes

P_2 is that percentage of the entire lake volume which the second stratum constitutes

P_n is that percentage of the entire lake volume which the lowermost stratum constitutes

According to Scott (1916) the method described above gives satisfactory results for lakes with steep slopes of basin but for lakes with *gently sloping sides* the mean computed below is the true one. Therefore, he proposed the following formula to yield more accurate results:

$$T_m = \frac{(3A + 2\sqrt{AB} + B)\,(T_1 - T_2)}{(A + \sqrt{AB} + B)\,4} + T_2$$

in which this computation is repeated for each stratum of the lake and the sum taken to represent mean temperature, and in which

T_m is mean temperature
T_1 is temperature at the upper level of the first stratum
T_2 is temperature at the lower level of the first stratum
A is area of upper level of first stratum
B is area of lower level of first stratum

Thermal Resistance

Birge (1916) supplied the following formula for the computation of the work done in mixing a column of water of unit base and height and a uniform temperature gradient:

$$W = \frac{AC^2}{12}\,(D_2 - D_1)$$

in which

W is work in ergs
A is area of cross section of water column
C is height of water column
D_1 is density of water at upper surface of water column
D_2 is density of water at lower surface of water column

If A and C are assumed to be constants, as for example, 1 sq. cm. and 100 cm., respectively, the thermal resistance will depend upon the value of $D_2 - D_1$, or, to express it another way, $(1 - D_1) - (1 - D_2)$. In limnological work it is convenient to assume A as 1 sq. cm.; C as 100 cm., 1° C. as the standard temperature gradient; and as the standard unit, the amount of work necessary to mix a $1 \times 1 \times 100$-cm. column of water whose temperature is 5° at the top and 4° at the bottom. Under these conditions W (work) $= 0.0067$ ergs, a value which can be accepted as the unit of work done in mixing (see Appendix, Table 7).

The unit of thermal resistance may also be taken as $D_4 - D_5$ which equals 0.000008. Then in any column of water $1 \times 1 \times 100$ cm. with a uniform temperature gradient the number of units of thermal resistance is equal to $\frac{D_m - D_n}{8}$ in which m and n are the temperatures of the water at the two ends of the column (see Appendix, Table 7, Column IV).

Work of Wind in Warming a Lake

According to Birge (1916) the work of the wind in warming a *stratum* of a lake, by mixing, from 4° C. to any selected temperature may be computed by means of the formula

$$W = RT \times Z \times (1 - D_n)$$

in which

W = work expressed in gram-centimeters per square centimeter of surface of lake

RT = reduced thickness = thickness of stratum if its area is made equal to that of lake and its sides vertical; computed from formula $\frac{V_{n-m}}{A_o}$ in which A_o = area of lake at surface, V_{n-m} = volume of stratum between two selected levels; reduced thickness expressed in *centimeters*

D = density of water; D_n = density of water at any selected temperature; at 4° C., D equals 1, but less than 1 at all other temperatures

Z = distance from surface of lake to middle of stratum expressed in centimeters

Computation of work for an entire lake can be made by adding the results from all of the various strata used from surface to bottom of the lake. If accurate results are needed, the computation should be based upon successive strata, each 1 m. thick, for the entire depth of the lake.

Selected References

(For Chapter 7)

Birge, E. A.: The heat budgets of American and European lakes, *Trans. Wisconsin Acad. Sci.*, **18:** 1–47, 1915.

———: The work of the wind in warming a lake, *Trans. Wisconsin Acad. Sci.*, **18:** 341–391, 1916.

———: A second report on limnological apparatus. *Trans. Wisconsin Acad. Sci.*, **20:** 533–552, 1922.

Olson, R. A.: "A Rapid Response Thermocouple of High Sensitivity for the Determination of Temperature Stratification in Natural Waters." 16 pp. Chesapeake Biol. Lab., Pub. No. 45, Dept. Research and Ed., State of Maryland, 1941.

Saunders, T. J., and P. Ullyott: Thermo-electric apparatus for limnological research, *Intern. Rev. ges. Hydrobiol. Hydrogr.*, **34:** 562–577, 1937.

Scott, Will: Report on the lakes of the Tippecanoe basin (Indiana), *Indiana University Studies*, **3:** 1–39, 1916.

Spilhaus, A. F.: A bathythermograph, *J. Marine Research*, **1**: 95–100, 1937.

———: A detailed study of the surface layers of the ocean in the neighborhood of the Gulf Stream with the aid of rapid measuring hydrographic instruments, *J. Marine Research*, **3**: 51–75, 1940.

Wagler, E.: Die chemische und physikische untersuchungen der Gewässer für biologische Zwecke: I, Abderhalden's Handbuch d. biol. Arbeitsmeth., Part IX, No. 1, 72 pp., 1923.

Warren, Henry E., and G. C. Whipple: The thermophone, a new instrument for determining temperature, *Tech. Quarterly*, **8**: 125–152, 1895.

Welch, P. S.: "Frame for Operating Two Reversing Thermometers Simultaneously." 4 pp. Spec. Pub. No. 13, Limn. Soc. Am., 1943.

Wood, W. P. and J. M. Cork: "Pyrometry." 2d ed., 263 pp. New York, McGraw-Hill Book Co., 1941.

TURBIDITY

Unit of Measure. The standard unit for measuring turbidity is that condition produced by *one part per million* of silica (fuller's earth) in distilled water.

Formerly, the U.S. Geological Survey defined the standard of turbidity as that water containing 100 parts per million of silica the state of fineness of which is such that a bright platinum wire 1 mm. in diameter can just be seen when its center is 100 mm. below the water surface and the observer's eye is 1.2 m. above the wire, the observation being made at midday, in open air but not in direct sunlight, and in a vessel of such size that its sides do not shut out the light so as to influence the results. The turbidity of such water is given the value of 100. This standard is merely 100 times greater than the unit defined in the preceding paragraph and is based upon the same system of measurement (parts per million).

Methods for the measurement of turbidity in natural waters usually fall into three classes: (a) comparison with silica standards; (b) platinum wire method; and (c) turbidimetric methods. The first two classes represent somewhat older practices, have certain limitations, and for many kinds of work are now often replaced by turbidimetric methods.

Comparison with Silica Standards

It is possible to secure from commercial scientific companies permanent turbidity standards already prepared and ready for use. They are available for various graded values and unless a laboratory is supplied with facilities for making standards it is preferable to secure them in this way. However, if standards are to be prepared, proceed as follows: To 1 l. of distilled water add 5 g. of standard fuller's earth; mix thoroughly; repeat this mixing periodically for about one hour; then let stand undisturbed for 24 hrs.; carefully draw off the supernatant fluid, being cautious that settlings at the bottom of the bottle are not brought back into suspension; test this supernatant fluid by the platinum wire method or by a standard turbidimeter (see p. 132) to determine its turbidity value. Take various portions of this liquid and by additions of distilled water adjust each portion to some desired turbidity value in the range of 5–100 p. p. m., testing the accuracy of each dilution by means of the platinum-wire method or by a turbidimeter. For general purposes a series having the following values will be convenient: 5, 10, 15, 20, 25, 30, 40, 50, 60, 70, 80, 90, 100. Nessler tubes of uniform size and glass, having a capacity of 100 cc. and a diameter of 20 mm., are suitable containers for these standards.

In operation, put the sample of unknown value in a glass tube like those containing the standards; shake thoroughly both sample and standard; hold sample and standard side-by-side and observe both sidewise toward the light, looking at some convenient object and noting the relative distinctness of the outlines of the object; match various standards with the sample until close correspondence is obtained. When the two tubes agree, the turbidity value of the unknown is that of the standard which matches it. Experience seems to show that there is some advantage if the object observed is a series of black lines ruled on white paper, if the light used is electric, and if the light illuminates both sample and standard from above, no direct rays reaching the eye of the observer.

GENERAL CONSIDERATIONS

a. It is recommended that silica standards so prepared be made up fresh at least every month.

b. Standards should be put into clean glass containers and kept well stoppered. Leave enough unoccupied space in the top of the container to provide for adequate shaking and mixing.

c. A small quantity of mercuric chloride added to the standards will serve to prevent growths of bacteria and algae.

Platinum-wire Method

The platinum-wire method is operated through the use of an instrument constructed essentially as follows: A piece of straight, bright platinum wire, 1 mm. in diameter and about 25 mm. long, mounted in a detachable screw eye, is set into the end of an aluminum scale, 21 cm. in length, in such a way that it projects at a right angle. At a distance of 1.2 m. above the platinum wire a mark is placed on the tape to indicate the position of the observer's eye when the instrument is being used. At exactly 100 mm. above the center of the wire the graduation mark of 100 is placed on the aluminum scale and other graduations are established above and below the 100 mark in accordance with standard data on the vanishing depths of the wire and their turbidity values. Probably the best known form of platinum-wire turbidity apparatus is that known as the U.S. Geological Survey turbidity rod (Fig. 33) in which the lowermost portion is a flattened bar, 6×17 mm. in transverse section, 21 cm. long, and composed of nonrust metal. On one face of this bar occurs that portion of the graduated scale extending from 3000 to 50. Into the end opposite the platinum wire there is securely fastened a piece of light-colored waterproofed, strong, inelastic cloth tape on which graduation marks 47 to 7 and the eye-position mark occur.

In operation, the end of a slender stick or rod is inserted into the hole of the screw eye. Then holding the tape taut, the scale is lowered vertically into the water to be tested as far as the platinum wire can be seen when the observer's eye is kept at the proper position. When the wire fades from view the surface level of the water on the rod is read on the graduated scale and the value thus obtained is a measure of the turbidity in terms of parts per million. In order to secure as accurate results as possible the

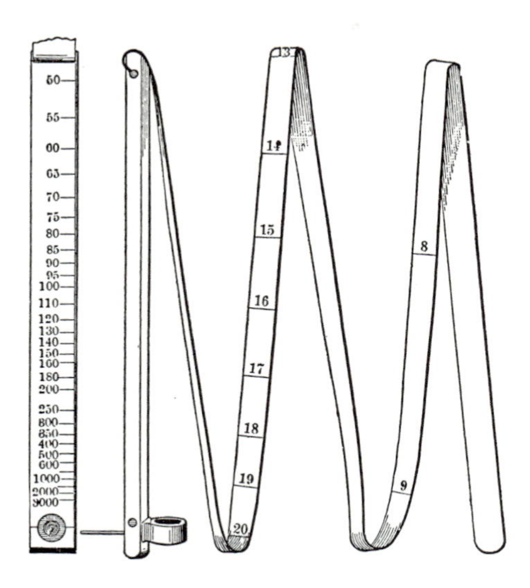

Fig. 33. U.S. Geological Survey turbidity rod. (Reprinted by permission from "The Microscopy of Drinking Water," by Whipple, Fair, and Whipple, published by John Wiley and Sons, Inc.)

following practices should be followed: (1) Make all observations in the open air, preferably near midday, but not in direct sunlight; (2) keep the platinum wire clean and bright; (3) if measurements must be made remote from the natural conditions under which the water normally exists, fill a tank of proper dimensions with the water to be tested and make measurement; stir the water thoroughly before the test is made (use a tank having horizontal dimensions at least twice as great as the depth to which the platinum wire is to be submerged); make measurements under conditions described above; and (4) dilute waters having a turbidity exceeding 500 p. p. m. with clear water; make reading on diluted sample; then compute the turbidity of the original sample from the dilution data.

GENERAL CONSIDERATIONS

a. This instrument has the virtues of low cost, light weight, and very small bulk.

b. This instrument has the following limitations: (1) The working range is approximately 7 to 500 p. p. m.; (2) it is essentially a field instrument; if used in the laboratory very large samples are required, especially in the clearer waters; (3) since the use of the instrument depends upon limit of visibility, different observers may secure somewhat different values for the same sample depending upon differences in eyesight.

c. In field use, waves, ripples, and other forms of surface disturbance interfere seriously with the operation of the instrument and affect the dependability of results. If the rod must be used under these conditions much of the difficulty may be removed by viewing the descending rod through a water telescope (pp. 363–364) so held that surface disturbance is eliminated, or a tank may be used as described previously (p. 129).

d. In its conventional form, the U.S. Geological Survey turbidity rod lacks provision for a safety tie-on. If the rod is to be used in the field over deep water it is imperative that some safety device be added to insure against loss.

e. This instrument may serve a useful function in providing a means of checking turbidity values of silica standards, both in the making up of new silica standards and in checking for changes in such standards with aging.

f. Avoid making measurements under a roof since values so secured are too high, even with very good light.

Jackson Turbidimeter

The Jackson turbidimeter may be used to measure turbidities having a value greater than 25 p. p. m. It operates on the principle that the depth of a column of any turbid liquid necessary to make the image of a flame of a standard candle disappear when viewed vertically through the column is a measure of its turbidity. The instrument (Fig. 34) consists of a substantial base from the center of which a brass candle holder extends vertically. The base also supports in tripod fashion and at the proper height a hollow metal tube holder in which a graduated glass tube is carried. Tube holder and candle holder are both so aligned that the middle long axis of the candle when in place coincides exactly with the middle vertical axis of the graduated tube. Installed in the candle support is a spring device which keeps the top of the candle at the same required level at all times, namely, 7.6 cm. below the bottom of the superimposed glass tube. The glass tubes have flat, polished bottoms, and are graduated to read directly in turbidity units of parts per million. The standard candle is made of beeswax and spermaceti and constructed to burn at the rate of 114–126 grains per hour.

To operate, place a standard candle in the holder in proper position; put a graduated glass tube of length suitable for the particular sample to be measured in the holder above the candle; darken the room; light the candle and pour sample into the graduated tube meanwhile viewing the candle flame from above and through the column or sample; continue pouring until image of candle flame barely disappears. Pour very slowly as end point is approached. If proper care has been exercised in determining disappearing point, the removal of one per cent of the sample should cause the candle image to reappear. The turbidity value in p. p. m. will be read directly on the graduated scale, determined by the top level of the sample in the tube.

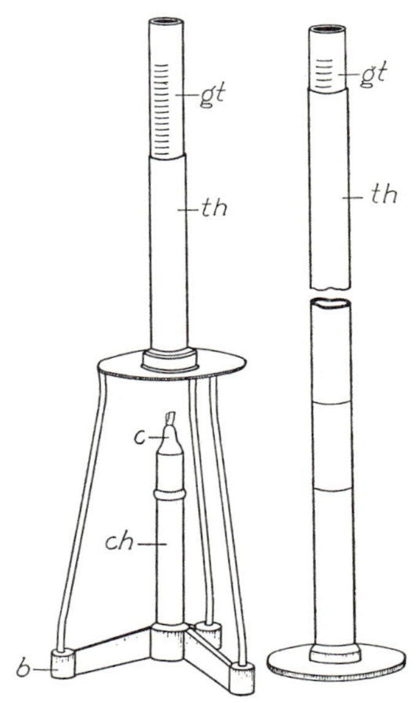

FIG. 34. Jackson turbidimeter. (*Left*) Assembled instrument. (*Right*) Extension tube for the use of a 75-cm. graduated glass tube. (*b*) Tripod base. (*c*) Uppermost tip of candle. (*ch*) Hollow, metal tube holder for candle. (*gt*) Graduated glass tube. (*th*) Hollow metal tube holder for graduated glass tube.

PRECAUTIONS. Keep candle lighted for only a few minutes at a time; the flame tends to increase in size as it continues to burn. Be sure that the spring mechanism is functioning continuously and that the tip of the candle is kept at the proper distance from the bottom of the glass tube. Keep candle well trimmed and all charred portions of the thread removed. Be careful that bottom of glass tube does not become hot before sample is poured into it. Do not permit the water to acquire heat while running test. Keep glass tubes clean both inside and outside. Make certain that soot and moisture are cleaned off the lower surface of the glass tube since any accumulations there detract from the accuracy of the readings. See that air currents in the room are reduced to a minimum.

GENERAL CONSIDERATIONS

a. If a darkroom is not available, a black cloth over the head of the observer will make readings possible.

b. The metal tubes should be black on the inside to prevent light reflection.

c. The standard 25-cm. glass tube, graduated in millimeters on one scale and p. p. m. on another, is used for samples in which the turbidity is greater than 100. For turbidities less than 100, an extra long (75-cm.) graduated glass tube in a metal extension is used.

d. The advantages of the Jackson turbidimeter lie in modest cost, ease of operation, and absence of any requirement for electric current or silica standards.

Hellige Turbidimeter

The Hellige turbidimeter is used to measure directly turbidities of 0–150. Higher turbidities may be measured by diluting the sample with

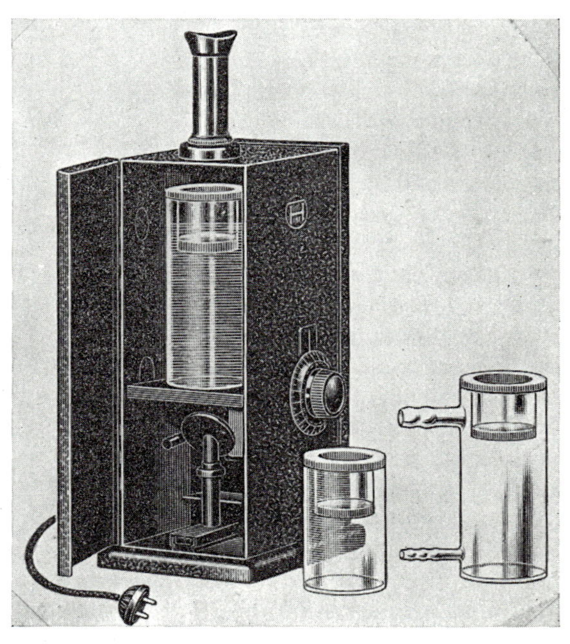

Fig. 35. Hellige turbidimeter. (*Left*) Shown with door of turbidimeter open; tube and plunger in place as in actual test. (*Right*) Tube with side arms for connecting to water line in order to provide continuous turbidity readings (plunger in place). (*Center*) Short tube, with plunger in position, for use with more turbid liquids. (Courtesy, Hellige Inc., Long Island City, N. Y.)

zero-turbid water and calculating the original turbidity from that of the diluted sample. This turbidimeter is particularly useful because of its capacity for measuring very low turbidities with high precision. The principle on which it operates consists of the comparison of a beam of light with the Tyndall effect produced from the lateral illumination of the sample by the same source of light. Thus no standard suspensions are necessary. Figs. 35 and 36 show the essential features of construction of the instrument. Light rays from the opal-glass bulb *b* are reflected from the reflector *r* into the sample in the glass tube *t*. This lateral illumination of the suspended particles in the sample produces the Tyndall effect. Light rays are also transmitted from the lamp *b* through a precision slit *s* operated by a graduated drum knob on the outside of the housing of the whole apparatus. Light rays passing through this precision slit illuminate the milk-glass reflector; thence they are reflected vertically through a filter plate *fp*, then through a circular opening in the horizontal silver mirror *m;* thence through the sample in the glass tube *t*, on through the glass plunger *p* and then into the ocular *o* to the eye of the operator. These light rays appear to the observer as a circular spot in the center of the Tyndall effect of the illuminated sample. This circular spot will appear

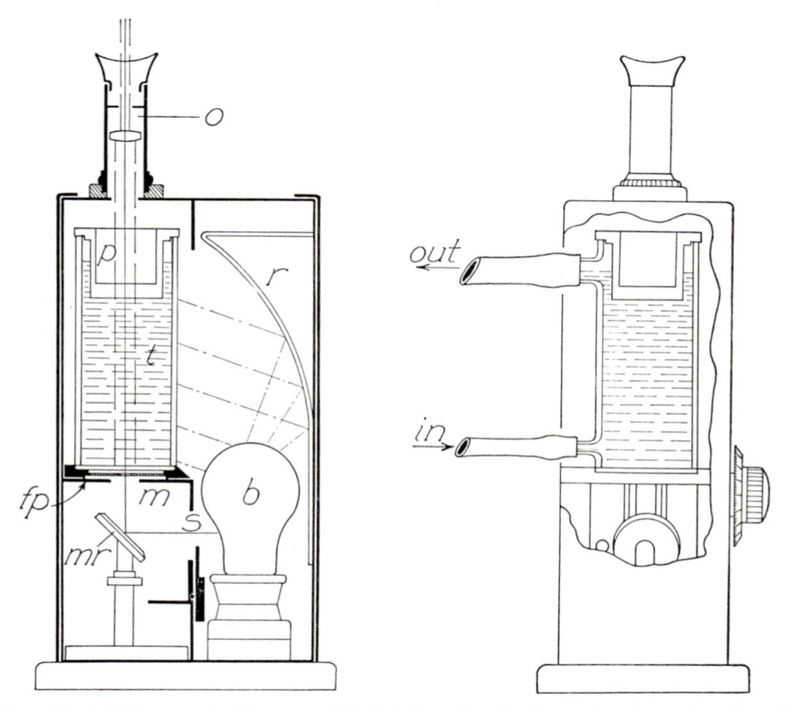

FIG. 36. Hellige turbidimeter. (*Left*) Sectional view showing details of internal construction. (*Right*) Diagram showing use of special tube with side arms for connection with water line so that continuous turbidity readings can be made without necessity of filling tube for each sample. (*b*) Opal-glass bulb. (*fp*) Filter plate. (*o*) Ocular. (*p*) Glass plunger. (*m*) Mirror. (*mr*) Milk-glass reflector. (*r*) Large reflector. (*s*) Adjustable precision slit. (*t*) Specimen tube. (Courtesy, Hellige Inc., Long Island City, N. Y.) (Redrawn.)

lighter or darker than the Tyndall effect depending upon the size of the precision slit. With the operator's eye at the ocular, the graduated dial may be so manipulated (adjusting the opening of the slit) that the circular spot and the surrounding illumination come to have the same degree of brightness, thus obliterating the circular spot. When this adjustment point is reached, its position on the drum scale is read. This value is then applied to a table or curve provided with the instrument and the turbidity is read, without calculation, directly in parts per million.

The glass tubes for holding the samples are made for sample depths of 20, 50, and 100 mm. These tubes have sealed-on plano-parallel bottom plates of optical glass. The upper end of the tube, when filled with a sample, is capped with a short glass "plunger" which automatically adjusts the proper depth in the sample and is usable in all sizes of tubes.

Guidance in the selection of the proper sample tube, turbidity range, and for use or nonuse of filter plates is supplied in the following tabulation.

Tube No.	*Sample Depth of Tube*	*Approximate Range in p.p.m. SiO_2*	*Filter Plate Used*
8010-A or B	100 mm.	0–4	gray glass
8010-A or B	100 mm.	0–15	milk glass
8010-C	50 mm.	0–50	none
8010-D	20 mm.	0–150	none

To operate, select a *thoroughly clean* (outside and inside), glass specimen tube whose length is suitable for the turbidity of the sample; fill tube with sample to level slightly higher than liquid level marked on the tube; close upper end by insertion of *clean* glass plunger; eliminate all gas bubbles in sample or on glass. Place glass tube containing sample in the instrument housing, on the mirror frame and in the circular depression; set filter plate in correct position, close door of housing and switch on light. Set graduated dial at zero. With eye at ocular, revolve dial in direction of increasing values until spot in center matches surrounding area (spot disappears). Read dial and determine turbidity from value thus obtained.

Adjustment of the light fields will probably not be an end point, but rather an *end interval* over which the light fields appear to be balanced. This interval will be larger at higher scale readings. To reduce some of this uncertainty, all adjustments should be made by always turning the dial toward the increasing values until the central spot just disappears, i.e., the reading should be taken when the end interval is first entered.

GENERAL CONSIDERATIONS

a. The whole operation of filling the glass sample tube, the insertion of the sample tube in the instrument, and the adjustment of the dial should be done as quickly as possible to avoid settling of suspended particles to the bottom and convection currents due to heating by the lamp.

b. All parts of the instrument, particularly the glass specimen tube and the plunger, must be scrupulously clean inside and outside. The glass tube and plunger may be cleaned with soap solution but *never* should strongly acid or alkaline cleaning mixtures be used. As the last step in the cleaning procedure, rinse with clean water and a portion of the sample to be tested. Avoid use of cloth; it is too likely to leave lint on the surfaces. One of the best means of removing water from the outside of the sample tube after filling is a cellulose sponge of the type used in cleaning photographic films. Such a sponge should be kept clean and submerged in distilled water in a closed jar during active use; at end of work it should be dried thoroughly in open air. Clean the silver mirror and filter glasses with soft cloth. Clean milk-glass reflector under the mirror with soap solution.

c. In order to avoid fingerprints on the outside of the sample tube at a place where they would interfere, it should be handled by grasping it at the top above the level of the bottom of the plunger.

d. Each opal-glass lamp bulb must be individually gaged and provided with its own set of graphs. This fact must be kept in mind when new bulbs are ordered.

e. Gas bubbles on the lower surface of the plunger may be avoided by putting it into the glass sample tube so tilted that one edge of the plunger makes the first contact with the upper surface of the sample; then as it is introduced further into the water the whole plunger bottom should be settled into final position with a side motion.

f. Be sure that the filter plate is in the proper position when a sample is being tested; otherwise, serious error will occur.

g. Glass sample tubes and the plunger must be handled with great care. Always put them down on some soft, clean material (a rubber pad is recommended) to avoid breakage or scratching. Do not submit the cemented joints to undue mechanical strain; such strains may cause the bottoms to become unsealed.

h. In case it is desired to measure turbidity continuously in flowing water, a sample tube of 100-mm. liquid depth (Fig. 36) can be secured which has two side arms that may be connected by rubber tubing to the water supply in such a way that the water changes continuously in the sample tube. This arrangement avoids the necessity of filling and emptying the tube for each sample.

i. The advantages of this instrument are (1) avoidance of the use of standard turbidity suspensions or standard candles; (2) high accuracy in the low end of the turbidity scale; (3) simplicity and speed of operation; (4) avoidance of need of a darkroom; and (5) compactness and portability.

Other Methods

The Baylis turbidimeter (Baylis, 1926) is now used to considerable extent, especially in sanitary work. A transparency meter for measuring transparency and scattering properties of water *in situ* at various levels in a lake has been developed by Whitney (1938).

Selected References

(For Chapter 8)

Baylis, J. R.: Turbidimeter for accurate measurement of low turbidities, *Ind. Eng. Chem.*, **18:** 311–312, 1926.

"Standard Methods for the Examination of Water and Sewage," 9th ed., 286 pp. New York, Amer. Pub. Health Assoc., 1936.

Whitney, L. V.: Microstratification of inland lakes, *Trans. Wisconsin Acad. Sci.*, **31:** 155–173, 1938.

COLOR

By *color* in water is meant those hues inherent within the water itself which result from colloidal substances and materials in solution. The term *true color* is sometimes used to designate color due to these causes.

Platinum-cobalt Method

Unit of Measure. That color produced by 1 mg. of platinum in 1 l. of distilled water.

Standards. Dissolve 1.245 g. of potassium chloroplatinate (K_2PtCl_6) and 1 g. of crystallized cobaltous chloride ($CoCl_2.6H_2O$) in small quantity of distilled water; add 100 cc. concentrated hydrochloric acid and dilute to 1 l. with distilled water. A solution so prepared has a color value of 500.

Prepare standards having color values of 5, 10, 15, 20, 25, 30, 35, 40, 45, 50, 55, 60, 65, and 70 by diluting 0.5, 1.0, 1.5, 2.0, 2.5, 3.0, 3.5, 4.0, 4.5, 5.0, 5.5, 6.0, 6.5, and 7.0 cc. of solution described above with distilled water to 50 cc. in standard Nessler tubes. Or, if desired, use 100 cc. standard Nessler tubes; double all values 0.5, 1.0, 1.5, etc.; and dilute each to 100 cc. with distilled water and same series of color values mentioned above will result. Label each tube, indicating color value and date of preparation.

PROCEDURE

Put sample through high-speed centrifuge to remove suspended matters. Fill standard Nessler tube with sample to be measured and to same height as that of standards. Compare it with various standard tubes until one is found which has same hue as unknown. Make this comparison by looking down through tubes held vertically and placed side by side above a white or mirrored surface so oriented that light is reflected upward through tubes. Results are read directly in parts per million.

GENERAL CONSIDERATIONS

a. If the sample shows a color greater than 70 it should be diluted with distilled water so that the comparison of hues may be facilitated. The original color is then computed on the basis of the amount of dilution.

b. Do not use filters for the removal of suspended matters since filtering tends to produce a decolorizing effect.

c. Color of an uncentrifuged sample (sometimes referred to as *ap-*

parent color) may be determined if desired but the limitation of such a measurement should be understood.

d. When not in use, standard tubes must be protected from dust and evaporation.

Glass-disks Method

A convenient method for measuring color of water is that in which colored, nonfading glass disks serve as standards. These disks are individually calibrated to correspond to the platinum-cobalt scale. Such a set

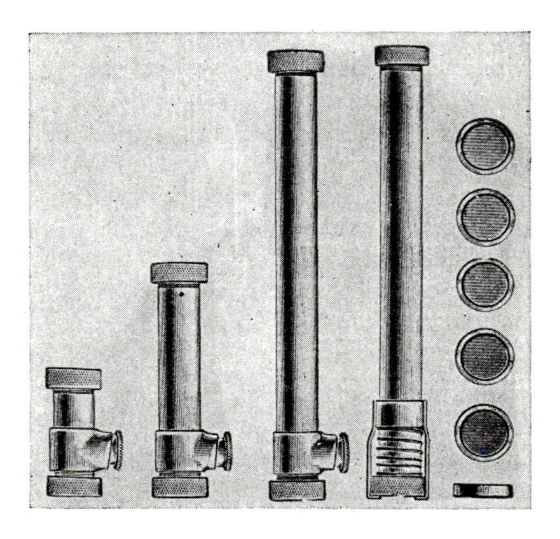

Fig. 37. U.S. Geological Survey outfit for measuring color of water. (Courtesy, Precision Scientific Co., Chicago, Ill.)

devised for the U.S. Geological Survey (Fig. 37) is now recognized as standard equipment for this purpose. Each glass disk is mounted in an aluminum ring. On each ring is stamped the color value of the enclosed glass disk; a serial number is also stamped on this rim. A standard set usually consists of 5 or 6 disks. In addition to the series of glass disks, such an outfit includes three aluminum tubes (sample tubes) having lengths of about 5, 10, and 20 cm., and capacities of about 10, 20, and 40 cc., respectively, and each tube is equipped with removable glass ends and with side filler openings. A fourth aluminum tube (standard tube), in addition to removable glass ends, also has a spring holder at one end for the temporary attachment of the glass disks.

PROCEDURE

1. Either fill *standard* tube with distilled water or, if more convenient, carefully clean inside of tube, removing all water to prevent fogging of glass ends, and use tube empty.

2. Fill one *sample* tube with water whose color is to be measured, selecting a tube of suitable length depending upon color character of sample.

3. Hold sample tube and standard tube parallel and in contact and at such a distance from observer's eye (about 8–10 in.) that inner surfaces of tubes are practically invisible. View tubes through eye-piece ends (glass disks of smaller diameter). Be sure to hold both tubes at such position and at such angles that both can be *looked through simultaneously with one eye.*

4. Interchange tubes in position and determine if light on right and left is approximately equal; equality of illumination is required. On a cloudy day, look through pair of tubes at sky near horizon and away from position of sun; on clear day, look at piece of white paper, white tile, or other similar white surface on which strong light is shining.

5. Slip colored, standard glass disks, singly or in combination, in spring holder on standard tube until by repeated trials, holding tubes as described above, color in both tubes matches. If this result is attained by one disk, color value can be determined at once from color number stamped on metal rim. If disks in combination are necessary, color value is determined either from color ratio of disks used, or from use of serial numbers on rims in transformation table. Some sets of standards are made in such a way that each disk has definite ratio to next one in series so that disks may be used in combination and combination value read directly, as for example, number 10 and number 15 glass used in combination give color value of 25.

The directions given above hold only for waters having a color value less than about 80. Waters having a color value in excess of 80 should not be matched in the 20 cm. tube but instead should be tested in one of the following ways:

a. Use one of shorter sample tubes (10 or 5 cm.) depending upon color depth of water to be tested. Match shorter tube against same standard tube mentioned above. However, do not fail to observe necessary computation step in arriving at true color value, namely, result obtained with the 10 cm. tube must be multiplied by 2; that secured with 5 cm. tube must be multiplied by 4.

b. Dilute sample with number of known volumes of distilled water; match with standard disks as described above; multiply result by number of volumes of distilled water used as dilutant. In preparing dilution of original sample, use tube itself for measuring sample and distilled water, mixing both in any convenient, clean container.

GENERAL CONSIDERATIONS

a. Do not make color measurements by this method under artificial light since dependable results cannot be obtained.

b. When first purchased, standard glass disks should be checked against platinum-cobalt standards.

c. Low turbidity does not interfere with this kind of color measurement. Highly turbid waters require filtration through thick filter paper before color tests are made. However, it is sometimes of considerable value to have a turbidity measurement of the unfiltered sample before it is put into the color sample tube.

d. Because of their size and shape, the color disks must be guarded carefully against loss in the field.

e. Tubes and disks must be kept scrupulously clean; also dry when not in use. Lens paper or a soft cloth free from oil may be used for cleaning glass disks.

f. The glass ends on the tubes must be handled carefully; avoid forcing them on too tightly.

g. In field work this method is very convenient but usable only when color measurements of unfiltered samples are acceptable.

h. If tubes have not been used for some weeks or months the packing materials in all movable joints should be soaked in water for several hours prior to use in order to prevent leakage.

WATER MOVEMENT

Current Velocity

Of the various devices for measuring velocity of flowing liquids, certain ones may be adapted to the needs of limnology. All have limitations which must be fully understood.

PITOT TUBE

In its simplest form a Pitot tube is an L-shaped glass tube (Fig. 38) with both ends open. When it is suspended vertically with the shorter end submerged in flowing water and pointed upstream, water enters the open end at e and rises a distance h above water level outside. If the water flow is steady, height h of the water column within the tube remains practically constant; irregularities of flow cause fluctuations in h. The height h is a measure of velocity of the water at or very near e and assuming stream-line motion of the water outside and ignoring friction, this velocity is computed as follows:

$$v = \sqrt{2gh}$$

in which v = velocity; h = height of column in tube above general water level outside; and g = acceleration of gravity.

Under certain circumstances it is possible to adapt such a simple Pitot tube to limnological needs. It may be made by bending ordinary laboratory glass tubing in a flame and marking a suitable, easily readable set of graduations on the vertical arm. When so constructed the following formula provides an average constant (0.977) which represents more precise results.

$$v = 0.977\sqrt{2gh}$$

If the velocity is desired in terms of feet per second, then v = velocity in feet per second; g = 32.16 ft. per sec.; and h = height of water column, in fractions of a foot, in tube above general water level. If it is desirable to take capillarity into account, the upper end of the tube may be dipped into water, the rise of water due to capillarity measured, and this value subtracted from h.

In its simplest form, as described above, the Pitot tube is sometimes difficult to use. In quietly flowing water with not too great velocity, h may be measured with some ease and accuracy. However, in waters of

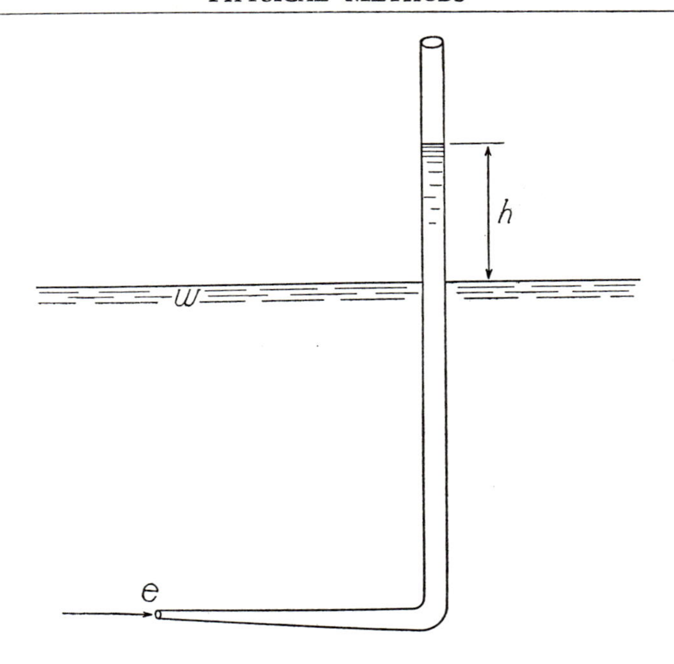

Fig. 38. Pitot tube in simplest form. (*e*) Point of entrance of flowing water. (*h*) Distance between water level inside of tube and outside water level. (*w*) General water level. Arrow at *e* represents direction of water flow.

high velocity or currents with much surface disturbance, the measurement of *h* is difficult, uncertain, and sometimes virtually impossible.

IMPROVED PITOT TUBES

Many modifications in and additions to the simple Pitot tube have been made in order to overcome difficulties and extend its use to flowing waters in general. An early improvement appears in Darcy's instrument which consisted of two duplicate L-shaped tubes built together as shown in Fig. 39. One tube is pointed upstream; the other downstream. Both are connected at their upper ends to a single tube which is provided with a closing valve *s* and which terminates at the top in a small plunger pump. When *s* is opened and the lower portions of the instrument are submerged in water to the desired depth the air in the upper portion of the tubes is partly pumped out, allowing the water to rise above the outside water level. The water will rise higher in the upstream tube than in the downstream one and the difference in level *h* has a constant relation to the actual velocity of the water at the point tested. This device eliminates the effects of surface disturbance by putting both levels to be read above the water line. However, it should be noted that the downstream arm does not provide a

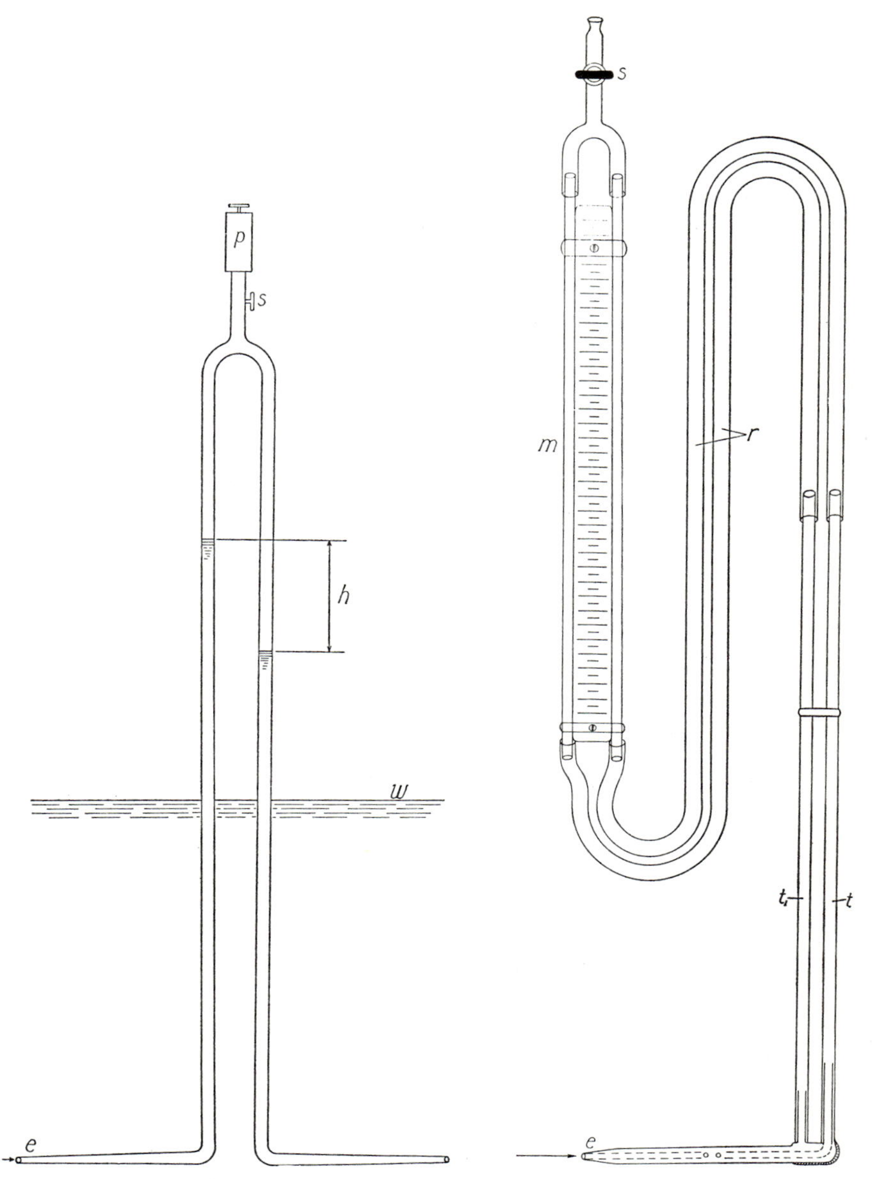

FIG. 39. (*Left*) Darcy's modification of the Pitot tube. (*e*) Point of entrance of water. (*h*) Difference in height of water levels in two arms of instrument. (*p*) Pump. (*s*) Stopcock. (*w*) General water level.

FIG. 40. (*Right*) Diagram of the essential features of a modern, simple form of Pitot tube. (*e*) Point of entrance of flowing water. (*m*) Manometer. (*r*) Rubber tubing connecting two arms of Pitot tube with manometer. (*s*) Stopcock. (*t*) Tube registering rate of flow of water (opening at *e*). (t_1) Tube registering hydrostatic pressure, opening to outside through two small lateral pores. Arrow indicates direction of water flow.

true measure of the general static pressure in the water and hence h cannot be used directly in the formula mentioned above. In order to secure values for determining current velocity, each instrument must be tested in a rating canal or by some other experimental means.

An improved but relatively simple Pitot tube is illustrated in Fig. 40. It consists of two tubes so arranged that one, t_1, indicates the static pressure (normal water level) and the other, t, the sum of the static pressure and the velocity pressure, the difference between the two being the height h in the formula stated above. On each side of the lower horizontal arm two small openings lead into a space continuous with that in t_1. The opening at e leads directly into a closed tube which in turn leads into t. When in operation the two tubes are connected to a manometer by means of two rubber tubes of suitable length. The manometer consists of two glass tubes of equal diameter and length mounted on the two sides of a graduated scale. These two tubes unite at the top into a single tube which terminates in a stopcock. When ready for use, the lower, horizontal arm is lowered into the water at the selected location and depth, and fixed in position by some stable form of support which will not interfere with the action of the instrument; the manometer is fixed above water to some convenient support; the stopcock at the top is opened and by means of suction applied at the upper end water is drawn into the manometer tubes until both are partly filled. It may be necessary to determine by various trials the most favorable height of the manometer above the water surface and the most favorable height of the water levels in the tubes. The stopcock is then closed and if the instrument is free from leaks, a reading of difference in height h of the water columns in the manometer can be made at once. When so measured, the height h may be substituted directly in the formula $v = \sqrt{2gh}$. However, manufacturers usually supply a set of curves for these instruments by means of which the velocity may be read directly as soon as the manometer reading is known.

When properly used, this instrument gives reliable results. It has the disadvantage of being somewhat cumbersome for one person to use unless stable installation is provided by means of which both tube and manometer are immobilized. Consequently it is slower than certain other methods. It has the advantage of simplicity in construction and can be built in any laboratory from simple materials. Obviously, such an instrument is usable only in shallow water. It also shares with other forms of Pitot tube the limitation of being less effective in very low velocities. It measures velocity at any instant, although the average velocity over a period of time may be secured by making a series of manometer readings at selected intervals during the total time period. It also has the virtue of measuring velocity at a small area, i.e., the size of the tube at e, and this aperture may be directed into various situations within a stream which are inaccessible to some other kinds of current-measuring devices.

BENTZEL VELOCITY TUBE

The Bentzel velocity tube overcomes some of the disadvantages inherent in other current-measuring devices. It consists essentially of (a) a long metal tube in the shape of the letter U, with both ends bent through 90° and directed opposite each other; (b) a tapered glass or plastic tube with accompanying graduated scale in the upper part of one arm; (c) a priming device (usually a small air pump); and (d) an index (either a float or a sinker, depending upon the form of the instrument) in the tapered glass tube. These features may be assembled in various ways.

The principle on which this instrument acts is as follows: When the instrument is held vertically, with the two oppositely directed orifices of the long U-shaped tube submerged in running water, the whole tube filled with water, and one aperture directed upstream and the other downstream, a flow of water through the instrument is established, entering one aperture and emerging from the other, due to the difference of pressure at the orifices. Near the upper end of one arm is a tapering glass tube, a *dilator*, whose diameter increases in the direction of flow. Within the dilator is a floating index; outside it is a half-cylinder rotating sleeve which bears on its concave surface a graduated scale. In one type of instrument, the dilator is in the downstream arm, its expansion is downward, and the index is lighter than water; in the other type, the dilator is in the upstream arm, its expansion is upward and the index must be heavier than water. Fig. 41 represents an instrument of the latter type. The description which follows relates to this type.

When in correct operating position, the water flowing through the dilator exerts a force on the index lifting it vertically. The index comes to rest at some level, determined by the velocity of water flow, where the tractive force on the index just balances the weight of the index. This level, read on the adjacent scale, is a measure of the water velocity. If the scale is an arbitrary one, actual velocity is determined by referring the scale reading to a rating curve supplied with the instrument; if the scale is graduated in terms of velocity units, the reading is direct.

The velocity range of types now commercially available is 0.3–4.0 ft. per sec. (0.09–1.21 m.). However, with special indices, adjustments, and precautions, velocities as low as 0.15 and as high as 7.0 ft. per sec. (0.04–2.13 m.) are measurable. While depths to which velocity measurements may be made are usually limited to about 1 m., it is possible to build such an instrument so that greater depths may be studied.

The advantages of the Bentzel velocity tube are as follows: (a) general convenience; (b) ease of operation by one person; (c) mechanical ac-

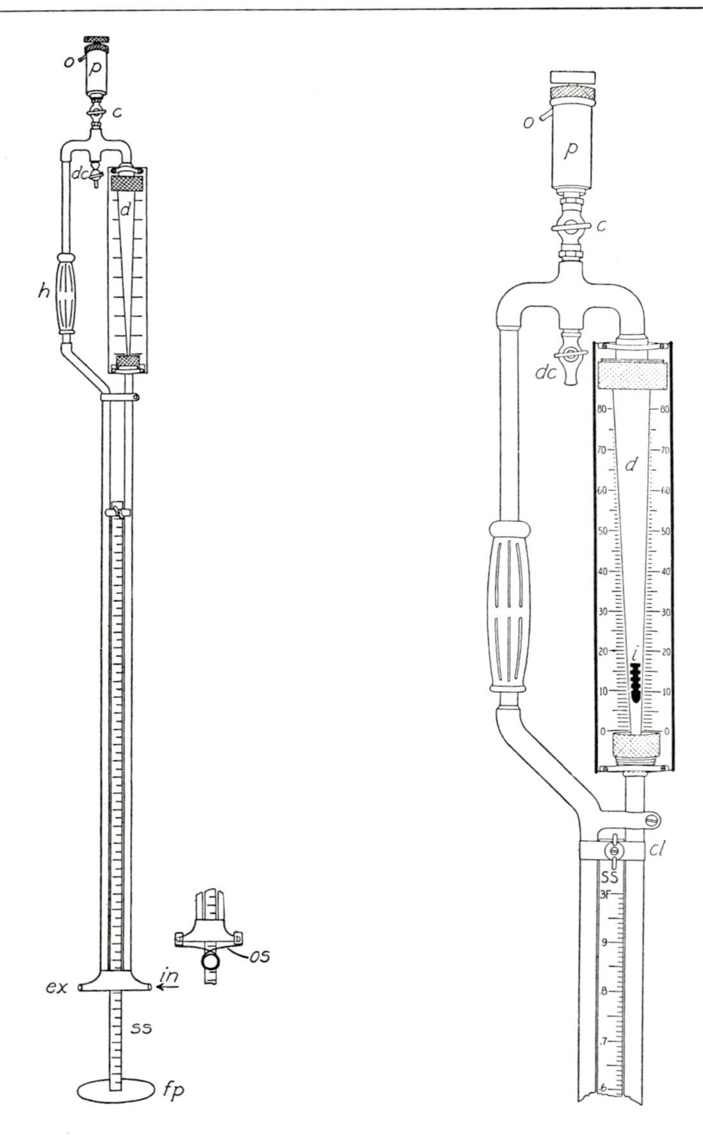

FIG. 41. (*Left*) Bentzel velocity tube; type with priming pump and sliding scale with foot plate; detail to right represents orifice stop in position, closing incurrent and excurrent openings. (*Right*) Enlarged view of upper portion of velocity tube showing structure of dilator and associated scale. (*c*) Stopcock. (*cl*) Clamp for immobilizing sliding scale. (*d*) Dilator. (*dc*) Drain cock. (*ex*) Excurrent opening. (*fp*) Foot plate. (*h*) Handle. (*i*) Index. (*in*) Incurrent opening. (*o*) Pump overflow. (*os*) Orifice stop. (*p*) Priming pump. (*ss*) Sliding scale.

curacy to about 2 per cent; (d) sensitivity to low velocities; (e) measurement of both velocity at any instant and mean velocities over a period of time; (f) rapidity of measurements; (g) mechanical supports not required; (h) measurement of velocity at very small spot and at situations very near the bottom, and (i) usability in vegetation beds, regions of coarse debris, or in streams full of rocks and obstructions.

General Considerations

a. Since it is not possible at present to manufacture dilators which are exact duplicates, every dilator must be calibrated independently together with the index to be used in it.

b. Temperature of the water produces an effect on the rating but the change is so slight that it can usually be disregarded within the velocity range mentioned above.

c. All air bubbles must be removed from the index before records are made since they affect the rating.

d. The index will give correct readings only if the proper end is up. If the index becomes accidentally reversed in the dilator it may be restored to its correct position by pumping the dilator partly full of water and then turning the instrument into a horizontal position and rocking it in an up-and-down fashion.

e. Each instrument is provided with an orifice stop which can be inserted under water thus making it possible to move from position to position without repriming.

f. A sliding scale and foot plate, provided in certain types of instrument, make it possible to clamp the recording orifices at any height above the level of the foot plate. This scale and foot plate can be removed in case they interfere with some special use of the instrument.

g. For work in the field, a substantial case for this instrument is necessary to protect it against accidents in transportation.

MECHANICAL CURRENT METERS

Various kinds of mechanical current meters have been devised. One of the best known and most dependable is the Price pattern meter the principal feature of which is a bucket wheel rotating on a vertical axis, the rotations of which are transmitted to a recording device. Fig. 42 represents one of the superior types. In this instrument the carefully balanced bucket wheel is mounted on a pivot providing minimal friction. The vertical wheel shaft extends into the superimposed enclosed contact box where the rotations of the shaft are transmitted through a set of simple gears to a make-and-break mechanism connected to two exposed binding posts. The meter yoke is attached to a guide vane which, when the instrument is in use, keeps the bucket wheel headed into the current. The bucket wheel assembly and its attached guide vane are carried either on a cable suspension or on a wading rod suspension. The instrument may be fastened to the lower end of the wading rod (see Fig. 42), or it may be attached to a rod adapter which can be adjusted to any height along the rod. In the latter instance, the wading rod has a foot plate at its lower end. These suspensions are interchangeable. When cable suspension is employed a 15-lb. streamline weight attached immediately below the meter gives added stability of position. A set of flexible, insulated wires, a dry-cell battery, and an earphone with flexible headband complete the outfit.

When the instrument is completely assembled and in position in a stream, the flowing water causes the bucket wheel to rotate. If the upper binding post of the contact box is connected in the electrical circuit, each rotation of the bucket wheel will be transmitted to the earphone as one click; if attached to the lower binding post a click occurs in the earphone for every five revolutions. The latter arrangement is used for the higher velocities. The speed of revolution of the bucket wheel is a measure of water velocity. Manufacturers provide a rating table which converts the revolutions per unit of time into velocity of flow. Such instruments have a velocity range of 0.1–6.5 ft. per sec. They measure velocity over a period of time, not at a particular instant. The operator must have a timepiece, preferably a stop watch, as a part of the equipment since the time interval

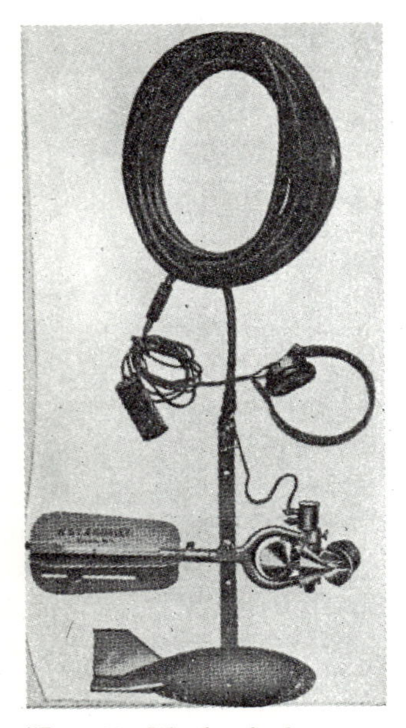

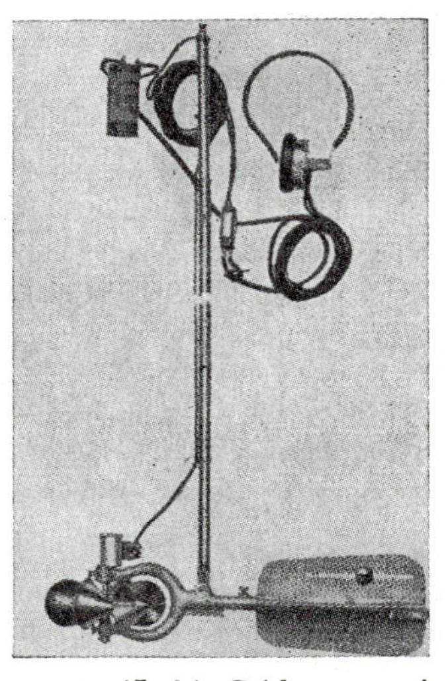

Fig. 42. Mechanical current meter, Price pattern. (*Left*) Cable-suspension outfit. (*Right*) Rod-suspension outfit. (Courtesy, W. and L. E. Gurley, Troy, N. Y.)

of each observation must be known. The operator may count clicks and watch the time interval simultaneously, or if his funds permit, he may use an electric counter which registers the number of clicks.

This instrument has the following advantages: (a) Cable suspension permits measurements in deep water; wading-rod suspension facilitates measurements in shallow waters; (b) worn or damaged parts are easily replaceable by the operator; (c) rating of the instrument is not altered by disassembling it.

Certain disadvantages are inherent in the instrument: (a) Initial cost is considerable; (b) preparation for operation is slow; (c) wires and headphone offer some inconvenience; (d) it cannot be used very close to the bottom or in very small space; (e) it does not measure velocity at any one instant; and (f) it cannot be used among floating vegetation or debris.

This instrument, though sturdy in most respects, requires careful treatment in the field, in transportation, and in storage.

EKMAN CURRENT METER

The Ekman current meter is commonly used in oceanographic work but limnologists may also have need for such a device in the very large inland waters such as the Great Lakes. This instrument records simultaneously both current rate and direction of current. The reader is referred to other works for a full description of this instrument; but the essential features are as follows: (a) a mounting in ball bearings on a vertical axis; (b) a carefully balanced propeller the revolutions of which are recorded on a set of dials and are a measure of current rate; (c) a compass box and an associated device for recording direction of current; (d) a vane which faces the propeller into the current; and (e) various accessories for starting and stopping the recording of the instrument, for maintaining its vertical balance in the water, and for suspending the meter at the proper level.

THRUPP METHOD

The Thrupp method makes use of the fact that if the surface of a smooth-flowing stream is cut by a small vertical object a definite set of ripples is formed provided the stream velocity exceeds a certain threshold. The angle of divergence of the ripples formed is related to the velocity of the water. Thrupp utilized this relation by devising an arrangement which employs two duplicate obstructions so located that they produce intersecting ripples. A Thrupp apparatus can be made as follows: Select a straight, flat bar (Fig. 43, *ss′*) about 1 m. long and graduate it in millimeters; at its zero end attach firmly at right angles another flat bar *bb′* of similar cross section and about 25 cm. long, the top surface of both bars being on same level. Attach two pegs (nails with 3 mm. diameter are suit-

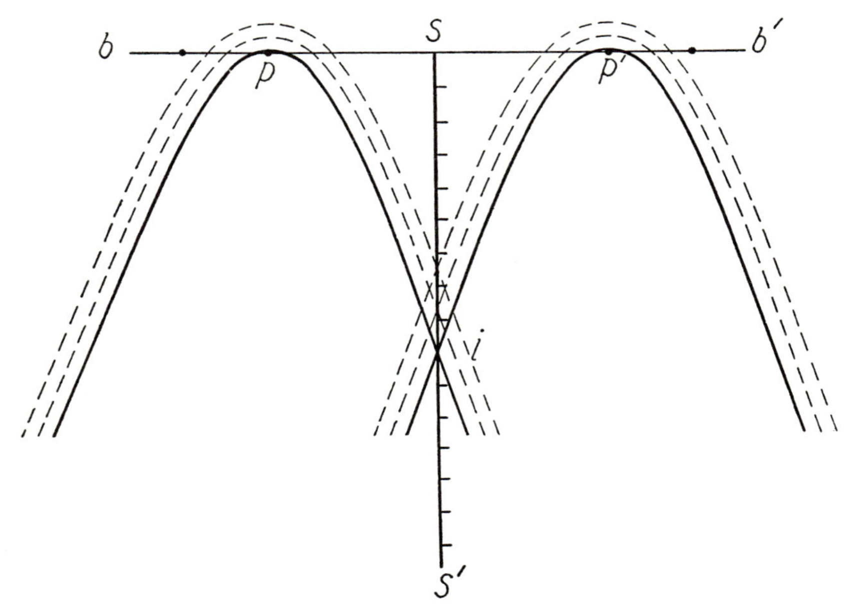

Fig. 43. Diagram showing essentials of Thrupp method. (bb') Transverse bar. (p, p') Two vertical pegs which cause surface ripples in water. (pp') Distance between pegs. (ss') Scale. (s) Mid-point between b and b'. (si) Distance from s to intersection of inner ripple of one set with that of the other set.

able) to the crossbar pp' on exactly the same transverse level of, and equidistant from, the zero end s of the longer bar. Make the pegs adjustable to different distances apart, convenient distances being 10 and 15 cm. respectively. This apparatus may be made wholly of metal, preferably brass, but if metal supplies are lacking, a crossbar properly fastened to the zero end of an ordinary meter stick serves well, particularly if the whole device can be so loaded that the flotation tendency of wood in water is overcome.

In order to operate, hold apparatus in flowing water, peg end upstream, the supporting bars parallel to the water surface and barely but completely submerged. Hold instrument as steady as possible and note point on submerged graduated bar at which innermost ripple from one peg intersects the corresponding one from the other peg and measure distance si. This datum substituted in the following formula will give surface velocity in *meters per second.*

When distance pp' is 10 cm., $v = 0.1555l$
When distance pp' is 15 cm., $v = 0.1466l$
in which v = velocity in meters per second
l = distance si in centimeters

General Considerations

a. Obviously this method is limited to measurements of surface velocity.

b. Accuracy of results depends much upon care in using the instrument. Even when all personal error is reduced to a minimum, this method is not as satisfactory as those described previously.

c. Turbulent surface flowage renders this method difficult or impossible to use.

d. This method is not adapted to the measurement of very low velocity.

FLOAT METHODS

In the absence of more satisfactory methods, velocities of the uppermost stratum of water may be measured roughly by means of floats set adrift in the current and the velocity of flow determined by the time required to drift them a known distance. Such floats may be of various materials, forms, and sizes. It is important that they be loaded so that they are almost entirely submerged thus reducing the effect of wind action. Even so simple a thing as a flat cork with a stone of appropriate size and weight tied to its lower surface will serve in an emergency. Various disadvantages are inherent in this method most of which are too obvious to justify mention. Wind, surface disturbances, and eddies are always possible sources of error.

Surface Floats. These floats indicate roughly the surface speed of the stream at the place of measurement only. If such a float passes along the region of highest velocity of water flow, the *mean velocity* of the whole stream at that section may be calculated as 0.8 of the float value. It must be understood, however, that such a calculation is only a very rough approximation, since mean velocity is affected by various factors such as nature of bed, size of channel, and amount of discharge per unit of time. The coefficient for transforming observed surface velocity into mean velocity will be highest numerically for large deep streams with smooth, uniform channels and smallest numerically for small streams with irregular, rough beds.

Tube or Rod Floats. A float which gives a different type of information from that of the surface float may be made by selecting a wooden or metal tube of uniform diameter and so weighting it at its lower end that it will float in a vertical position with only a very small portion of its length above the surface of the water. Tubes may be loaded properly by pouring lead shot or sand into the lower ends and closing the top. A small flag installed in the upper end aids in making the float more visible when used in the water. Wooden rods may be used instead of tubes provided that they are properly loaded at their lower ends. Such tubes or rods may be made with adjustable lengths if desired. The length of such floats must be determined by the user in adapting them to his needs. If it is the purpose of the operator to secure a direct measurement of the mean velocity of the water in that particular vertical plane in which the rod moves, then the rod should be made as long as free flotation will permit, that is, the clearance between the lower end of the float and the objects on the bottom should be as small as practicable. However, any dragging of the lower end on the bottom will invalidate the results. Data resulting from rod- or tube-float measurements should always be accompanied by information as to the length of the float, the mean depth of the stream, and the position of the plane of flotation. Obviously rod or tube floats are not suitable for streams in which submerged vegetation or other kinds of obstructions are prevalent.

When a rod float is released in a stream, the velocity of its movement will be approximately that of the vertical thin stratum or "filament" of water in which it drifts. Hence, the velocity of filaments in different positions in the cross section of the stream can be measured. If the immersed length of the rod is about 0.9 of the total depth of the stream, then, according to some authorities, the mean velocity of the filament of water in which it drifts may be calculated by means of the following formula:

$$V_m = V_r [1 - 0.116 (\sqrt{D} - 0.1)]$$

in which

V_m = mean velocity of water filament
V_r = observed velocity of rod float
D = ratio of depth of water below bottom of rod to total depth of water

The mean velocity at any cross section of the stream may be determined by introducing the float several times at different positions and observing the velocities.

In the operation of rod floats definite arrangements must be set up for determining the *time* and *course* of each float. Where applied to small streams, a wire or rope graduated with tags may be stretched across the stream at each end of the section of stream to be measured, the rod float released above the upstream line, the time required for the float to pass from one line to the other noted, and the length of the course of the float determined. In a large stream, some other arrangement will be required, as for example, the following one: Establish two ranges, one at each end of the section of stream to be measured, and at right angles to it; determine distance between ranges; set up a transit on each range; set transits to read zero on each other; start rod float just above upstream range; transitman on upstream range signals transitman on downstream range (a) when float is *about* to cross range and (b) when it *does* cross range; downstream transitman reads an angle on float just as it crosses upstream range; repeat this procedure when float crosses downstream range, the upstream transitman reading angle on float; each transitman records time of crossing range at which he is located. This procedure makes it possible to determine (a) *course* of the float, and (b) *time* required for float to drift from one range to the other. These operations may be repeated either at the same position along the range or at different positions and the results recorded. In the latter instance, results are secured which make possible the construction of a cross-section diagram of the stream at the position of the downstream range with velocities of the various filaments of water indicated.

Volume of Flow of Streams

In nature, surface streams running in nonartificial, open channels ordinarily present so many departures from a uniform condition that the formulas of the hydraulic engineer for open-channel computations cannot be applied with any dependable accuracy. If the character of the data desired and the features of the stream are such that the installation of weirs is justified, then the volume of flow can be measured with considerable accuracy and the mathematical formulas appropriate for weir computations may be found in almost any treatise on hydraulics. Commonly, however, the limnologist is forced to be content with less accurate results secured by means of simpler approaches. Embody's formula (1927) is expressed as follows:

$$r = \frac{wdal}{t}$$

in which

r = rate of flow in cubic feet per second
w = average width of channel section tested
d = average depth in feet
a = constant
l = length in feet of channel section tested
t = average time (three tests) in seconds required for float to traverse channel section

In this formula it is assumed that rate of flow is measured by timing the passage of a float over the selected channel section. If a current-meter reading in feet per second is available and the average velocity over the selected channel section is determined, then the formula becomes

$$r = wdav$$

in which v = average velocity of water over section tested. The value of a depends upon the nature of the bottom. If the bottom is rough and composed of loose rocks and coarse gravel, *a* has a value of 0.8; if smooth and composed of such materials as mud, sand, hardpan, or bedrock, *a* is given a value of 0.9.

Current velocity measured by timing a float in mid-current is upper-stratum velocity only. In the absence of current meters, a somewhat more accurate value for current velocity can be obtained as follows: if the channel is not more than 2 ft. deep, multiply the surface current rate by 1.33; if 10 ft. or more in depth, multiply the surface current rate by 1.05; for intermediate depths, interpolate between these two values.

Average depth is computed by measuring the depth of the water at uniform horizontal intervals across the stream and dividing the sum of such depths by the number of intervals plus 1.

GENERAL CONSIDERATIONS

a. Streams having a volume flow of less than 1 cu. ft. per sec. are sometimes described in terms of gal. per min.; 1 cu. ft. per sec. is approximately equivalent to 450 gal. per min.

b. Records of volume of flowage should be accompanied by information concerning the water level, since high or low water conditions alter the flowage markedly.

c. It must be remembered that any values secured by means of Embody's formula are approximate only.

STREAM DISCHARGE FROM ROD-FLOAT DATA

From results secured by the rod-float method described on pp. 151–153, it is possible to calculate approximately the stream *discharge* by multiplying the mean velocity of each water filament by its cross-sectional area and adding the products so secured. Improved results are obtained if two or more sections within the entire stretch between ranges are measured in order to compute a mean section.

Direction of Water Currents

SCOTT'S VANE METHOD

Direction of water currents, particularly in shallower water of the littoral region, may be determined by the use of an instrument modified from Scott's (1916) vane method. This modification consists of a galvanized iron vane 75 cm. long and 25 cm. wide mounted vertically in the end of a graduated steel rod 2 m. long and counterbalanced in such a way that it hangs vertically when suspended in water. The upper end of the rod is attached to the axle of a front-wheel bicycle hub which serves as a handle and allows the suspended vane to swing freely on the ball-bearing support. On the upper end of the axle a small pointer is fastened thus providing means of determining current direction when the vane itself is not clearly visible.

Ordinarily, this instrument can operate only in very shallow water. By the use of sectional extension rods, Scott found it possible to use his instrument to depths of 30 ft. or more with satisfactory results.

STREAMER METHOD

A simple method of studying current direction in shallow water may be devised as follows: A rod of convenient dimensions is installed in a vertical position at the desired location by thrusting it into the bottom. At regular intervals along the submerged portion of the rod, pieces of red silk ribbon or silk thread are tied at one end, leaving the other end free. Currents may be detected and their direction determined by watching the behavior of the ribbons at the various levels above the bottom. Obviously, this method is not usable in very turbid water. Its use in clear but rough water may be facilitated by the aid of a water telescope (pp. 363–364).

Detection and Measurement of Seiches

Methods for the measurement of ordinary changes of water level have been described on pp. 61–64. Such methods may be adapted also to the detection and measurement of seiches if general information only is desired. For a highly precise and detailed study of seiches, various complicated and costly types of recording equipment have been devised, a discussion of which falls outside the province of this book. For information concerning these refined instruments the reader must consult some of the more extensive research papers dealing with investigation of seiches. However, certain simpler outfits yield general information and, while their limitations must be fully understood, results so secured may be useful for some purposes.

INDEX LIMNOGRAPH

An example of simple equipment is represented in Fig. 44. It is a modification of a device, known as an *index limnograph*, first employed by European workers many years ago. It consists essentially of a very sturdy tripod; an immobilized small table *t* with a vertical back piece *vb* of similar size attached to it at one edge; a float *f* from which a flexible, inelastic cord *c* passes vertically over a pulley installed on the table, and vertically downward to a counterpoise; and an index or pointer, attached to the pulley *p*, which operates over a graduated scale *s* on the vertical back. This outfit is mounted over an improvised still well *w* at the margin of the lake. The graduated scale may have any form and pattern which convenience dictates.

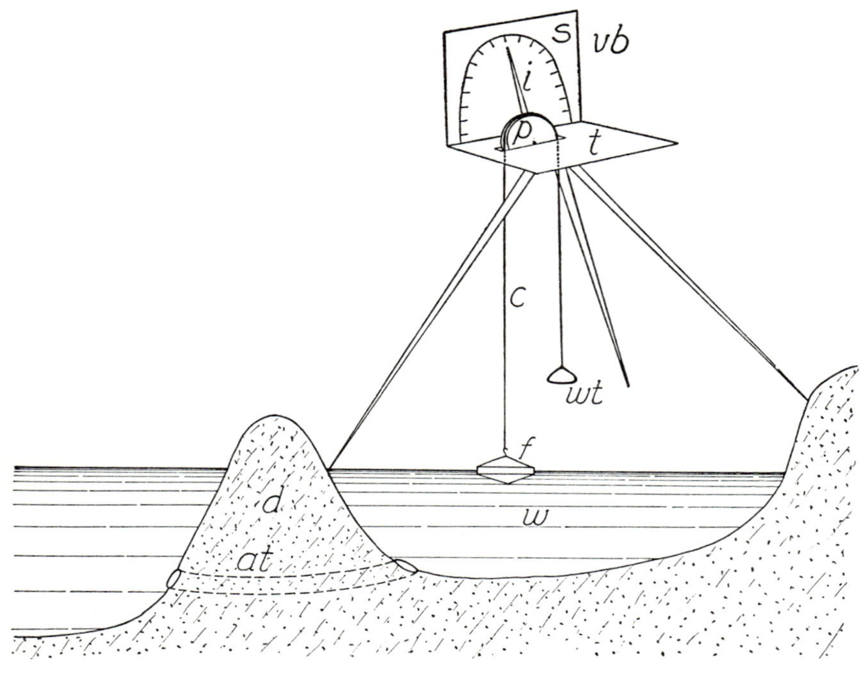

FIG. 44. Diagram of essential features of a simple index limnograph. (*at*) Subterranean access tube connecting lake with improvised still well. (*c*) Cord passing from float, over pulley, to counterpoise. (*d*) Dam separating still well from lake. (*f*) Float. (*i*) Index or pointer fastened to back surface of pulley. (*p*) Grooved pulley mounted in table over which passes cord connecting float and counterpoise. (*s*) Scale mounted on vertical back piece. (*t*) Horizontal table. (*vb*) Vertical back piece. (*w*) Improvised still well. (*wt*) Counterpoise.

Procedure

1. Set up limnograph in laboratory and calibrate graduated scale by raising and lowering float known distances.

2. Adjust weight of counterpoise to provide for proper performance of float.

3. In field, select suitable position at lake margin; construct improvised still well as indicated in Fig. 44, excavating basin to such depth that all possible changes of water level are provided for. Install access tube *at* to connect still-well basin and lake basin.

4. Set limnograph over still well and make as stable as possible. Adjust pulley position so that all changes of position of float will register on vertical scale.

5. Make records of index position at regular intervals of time so chosen that the size of the time interval is appropriate to the demands of the work.

General Considerations

a. This method requires that the limnograph be under constant observation during the period of work since usually the time interval for recording must be short.

b. If desired, records may be plotted as they are taken by the use of cross-section paper mounted on a board of convenient size. By plotting *time* along the abscissa and the index values along the ordinate a curve can be constructed as the work progresses.

c. Adequate protection against rain and windstorms must be provided.

EQUIPMENT FOR CONTINUOUS RECORDS

In the study of seiches it is often desirable to have a continuous record extending over several hours and possibly during times when the instruments cannot be under constant observation. Elaborate and costly instruments, designed especially for this purpose, fulfill such requirements, but since they are not discussed here it should be mentioned that certain simpler devices may yield continuous records satisfactory for some purposes. An example of such a simpler instrument is that sometimes designated as the *wagon recorder*. Changes in water level are transmitted by a float, line and counterpoise, similar to those of the index limnograph, but the line passes over two pulleys between which it drags, back and forth, a little carriage carefully mounted on three wheels that run, one on one side and two on the other, on two parallel tracks. This carriage supports a stylographic pen or pencil, so located and controlled that its movements are recorded on a long strip of paper moved horizontally and perpendicularly to the movement of the pen or pencil on rollers operated by a clockwork mechanism. Thus the rise and fall of the water surface is recorded on the paper, in a continuous record. Such an apparatus is easily constructed at modest cost in any laboratory from materials readily obtainable. Obviously, the best records are provided when it is operated in connection with some form of still well. The apparatus is easily calibrated in the laboratory.

SELECTED REFERENCES

(For Chapter 10)

Chrystal, G.: Seiches and other oscillations of lake-surfaces, observed by the Scottish lake survey, in Murray and Pullar: "Bathymetric Survey of the Scottish Fresh-Water Lochs," Challenger Office, Edinburgh, 1910, Vol. 1, pp. 29–90.

Davis, H. S.: "Instructions for Conducting Stream and Lake Surveys." 55 pp. Fishery Circular No. 26, Bur. Fisheries, U.S. Dept. of Commerce. 1938.

Hellström, B.: Wind effect on lakes and rivers. 191 pp. *Ing. Vetenskaps Akad., Handl.* **158**, 1941.

King, H. W., C. O. Wisler, and J. Woodburn: "Hydraulics," 4th ed., 303 pp. New York, John Wiley & Sons, 1941.

Pierce, C. H.: Methods of stream gaging, in National Research Council: "Physics of the Earth," Vol. 9, "Hydrology," ed. by O. E. Meinzer, New York, McGraw-Hill Book Co., 1942, pp. 486–498.

Scott, W.: Report on the lakes of the Tippecanoe Basin (Indiana). 39 pp. Indiana University Studies, Study No. 31, 1916.

LIGHT PENETRATION

Methods for the measurement of light penetration in water fall into two general groups, namely, limit of visibility tests, and light measurements.

Limit of Visibility Methods

SECCHI DISK

Secchi disk, as now used, is a circular metal plate, 20 cm. in diameter, the upper surface of which is divided into four equal quadrants (Fig. 45) and so painted that two quadrants directly opposite each other are black and the intervening ones white. A staple fixed at the center of the upper surface provides attachment for the graduated rope. Opposite the staple on the lower surface is a weight which facilitates the sinking of the disk in proper position. The lower side of the disk is painted black in order to eliminate reflection of light from that surface. Use of Secchi disk consists in lowering it into the water on a graduated line, noting the depth at which it disappears, then lifting the disk and noting the depth at which it reappears. The average of these two readings is considered to be the *limit of visibility*. For more exact results the disk, as it sinks, should be viewed vertically through a water telescope and under a sunshade.

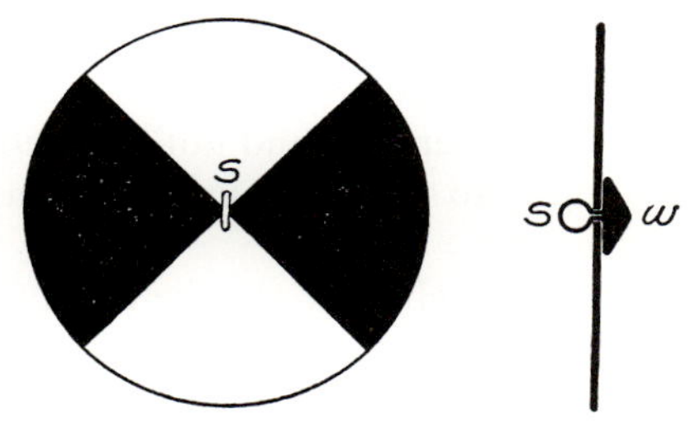

Fig. 45. Diagrams showing construction of Secchi disk. (*Left*) View of upper surface showing alternate black and white quadrants. (*Right*) Transverse section through center. (*s*) Staple at center. (*w*) Weight at center on lower surface.

This method has come into wide use as a routine means of comparing different waters. Obviously it is not an actual measure of light penetration, but merely a useful, rough index of visibility. It is also useful in making comparisons of the same water at different times. Readings made at the same date on the same water will vary with different observers, with the time of day, with the degree of roughness of water, and with the clearness of the atmosphere. Therefore all Secchi-disk records should carry full information on these items.

Standard conditions for the use of the Secchi disk are as follows: clear sky; sun directly overhead; shaded, protected side of boat; under a sunshade; minimal waves or ripples (or use water telescope). Any necessary departure from these conditions should be specifically stated in the records.

General Considerations

a. The upper surface of the disk must not be allowed to become dull or scarred. Frequent painting or enameling is recommended.

b. If the water is rough the protected side of the boat must be used.

c. If a water telescope is not available, the observation should be made on the shaded, protected side of the boat with the observer's eye at some fixed distance above the surface of the water. When small boats are used, a distance of 1 m. is suitable.

d. Observations should be made during the middle of the day; early morning and late afternoon hours must be avoided.

e. The rope must be accurately graduated and the graduations made as fine as practicability will permit. A steel tape is an excellent substitute for a rope but may rust unless it is kept dry when not in use.

f. An incandescent light, substituted for a Secchi disk, is more complicated, cumbersome and not justified in ordinary practice. Such equipment is subject to the same limitations as the Secchi disk and has no essential advantages.

g. A Secchi disk may be made up with a suspension composed of three cords attached to the edge of the metal disk, equidistant from each other, and converging upward to union with the graduated line; also on the opposite surface a similar suspension may be provided for the weight. Such suspensions hold the disk more steadily in a horizontal position when it is sinking, but the uppermost suspension interferes with visibility and the lower one may become entangled with objects on or near the bottom. The disk as described in this section avoids both disadvantages.

Light-measurement Methods

Satisfactory methods for the qualitative and quantitative measurement of light in water are developments of very recent years. Improvement and reduction in cost of the photoelectric cell are the outstanding events in the history of such methods. At the present time, rectifier cells of the selenium type appear to be the most satisfactory means for measuring subsurface daylight. Improvements in photoelectric cells and the accessories necessary for subsurface measurements are constantly appearing and some of the present devices for securing high-precision results are complicated, highly technical, and exacting in time and experience. So rapidly is the picture changing that no attempt will be made here to do more than indicate the general features of an outfit which, at the present time, is considered suitable for routine limnological work.

PHOTRONIC CELL OUTFIT

Assembled outfits for measurement of subsurface illumination are not yet available on the market. Hence each worker must arrange individually for the construction of his equipment. He must understand the essential details of the device and will probably require the assistance of some physicist familiar with photoelectric cells and their uses. This account follows the recommendations of the Subcommittee on Light Measurements of the International Council for the Exploration of the Sea (Atkins, Clarke, Pettersson, Poole, Utterback and Ångstrom, 1938). Only essential features are discussed here. Construction and assembling details are omitted since they may vary widely with different types of materials, and with various uses which the outfit is to serve.

PHOTOMETERS. Photronic cells manufactured by the Weston Electrical Instrument Corporation have been found suitable and reliable when operated under proper conditions. Two such cells, properly mounted, are needed, one to be used as the *deck instrument* and the other as the *subsurface photometer.*

The deck photometer requires no special mounting unless it is expected that the instrument must be exposed at times to rain or spray. Ordinarily it may be mounted in any convenient type of container so constructed that the instrument window is horizontal and that the necessary glass plates and color filters may be properly installed. A gimbaled support is advantageous although not necessary.

The subsurface photometer requires special mounting. It must be installed in a case which is absolutely watertight under all conditions of use. The case must be compact, sturdy, much heavier than an equivalent volume of water, and capable of suspension with the cell window horizontal. It must also be so constructed that the opal-glass disk and color filters, to be mentioned later, can be installed or removed quickly; also so built that the upper rim of the case does not produce any appreciable shading of the low-angle light. Other details are described in the reference already mentioned (Atkins, *et al.*, 1938). Particular attention should be given to the method of union of the cable to the photometer. The case should be provided with a supporting bridle by means of which the photometer in its heavy case can be attached to the galvanized supporting rope. This bridle should be composed of two strands of strong, small-diameter galvanized steel cable, long enough to produce a minimum of shading on the photometer.

MICROAMMETERS. Two suitable, duplicate microammeters are required (see p. 167 for possible use of but one microammeter), one to measure the current from the deck photometer and the other to measure current from the subsurface one. The choice of an instrument for the measurement of these currents involves a number of considerations and the decision must be determined by such things as the nature of the proposed measurements, degree of precision, cost, and adaptability. It is probable that, at this point also, most limnologists should seek the advice of a competent physicist. For routine purposes where high sensitivity is not expected, certain microammeters which serve the purposes are now obtainable at relatively low cost. However, highly sensitive microammeters are required for very precise work.

Diffusing Glass. Since a photronic cell mounted beneath a clear glass window will not give an accurate measure of light falling on a freely exposed horizontal surface because of the fact that the rim of the case interferes with reception of low-angle light, it is necessary to reduce this error by use of a *diffusing glass*. This diffusing device is in the form of a disk of flashed opal glass mounted above the photronic cell and in contact with the water. If this glass meets certain specifications, it combines desirable diffusing action with minimum loss of light. Both surfaces of the glass are polished and the diffusing layer is internal thus avoiding any interference with the diffusing action by submersion in water. When used, this diffusion disk is always the outermost part of the window structure. Both deck and subsurface photometers must be supplied with diffusing disks.

Color Filters. Selenium rectifier cells have a sensitivity over the visible spectrum and beyond it at both ends; have maximum sensitivity at about 5900–6000 Å.; and have rapidly declining sensitivity toward both ends of the spectrum (drops off to 10 per cent of maximum at about 7000 Å. and at about 3400 Å.). For well established reasons which need not be reviewed here, it is usually advisable to limit the range of sensitivity by the use of color filters of which the following are likely to be most useful for routine purposes: (a) If but one set of measurements is to be made, a Schott green filter VG9 (maximum sensitivity near 5400 Å.; range about 4800–6000 Å.) will probably be most suitable; (b) if other readings are to be made, a Schott red filter RG1 (transmits freely above sharp cutoff at 6000 Å.; upper limit determined by falling sensitivity of cell and for daylight is about 7200 Å.), and a Schott blue filter BG12 (maximum sensitivity near 4300 Å.; range about 3600–4800 Å. under diffusing glass and thick glass window) are recommended.

Accessory Equipment. Galvanized strand steel rope of least diameter which will provide necessary tensile strength is recommended. It should be graduated in convenient units, if not used in connection with a recording meter block, and used on some form of winch or hoist. The lifting device should be of such construction and occupy such a position that a minimum of shading is produced.

This cable must have a certain margin of excess length over the maximum depth of the water in which it is to be used. If it is to have a length of more than a few meters it should be used on a drum.

The connections between the deck photometer and its microammeter should be of convenient length, well insulated, equipped with adequate terminals, and have a resistance as nearly as possible that of the connecting cable for the subsurface photometer in order that the two sets of instruments will yield very similar readings when under the same light exposure.

Since the subsurface photometer is housed in a heavy brass case, additional weight may not be necessary to insure prompt sinking and maintenance of vertical position of the supporting steel rope. However, in case of doubt, a sinker weight of elongated streamline shape and equipped with a tail fin should be suspended from the lower part of the photometer case, care being taken not to interfere with the horizontal position of the photometer window. Such a sinker arrangement is also useful in preventing rotation of the photometer and twisting of cables when it is being hoisted on board the boat.

STANDARDIZATION OF PHOTRONIC CELLS. Measurement of subsurface illumination may be expressed in two general ways: (a) percentage of surface illumination; and (b) definite units.

PERCENTAGE OF SURFACE ILLUMINATION. For many kinds of work adequate results are secured if the subsurface illumination is expressed as a percentage of the surface illumination in the same spectral band. From such results it is possible to determine the relative penetration of the light of one kind of color in the water concerned. In this simpler method, it is only necessary to know the relative sensitivity of the deck and subsurface photometers and the curvature corrections of each. Current generated by photronic cells is not exactly proportional to the illumination producing it but departs from a linear relationship. The greater the illumination the greater the curvature from the linear; also the greater the external resistance the greater the curvature relation. Therefore, it is clear that such cells, their cables and microammeters must be standardized against some standard light source and the proper curves drawn from values determined for different illuminations. If a microammeter having a series of different resistances is used, then each resistance scale must be standardized. The different glass color filters must be standardized in the same fashion.

DEFINITE UNITS. For some purposes it may be necessary to express light-penetration results in terms of definite units. At this point it is probable that the limnologist should engage the assistance of a competent physicist. It has been suggested by certain authorities that it may be desirable to express these units in terms of radiant power, such as milliwatts per square centimeter, or gram-calories per minute per square centimeter in the particular spectral bands concerned. For the present at least and until certain problems relating to the standardization of color filters are solved, the Subcommittee on Light Measurements of the International Council for the Exploration of the Sea makes the following recommendations:

1. Employment of some definite although arbitrary unit to measure the illumination as shown by the photometer with and without various color filters.

2. Unit may be called the *lux* when the photometer is used with plain opal diffusing glass but without color filters.

3. Readings of photometer with the opal glass and color filters used simultaneously may be expressed in red units, green units, blue units, etc., respectively.

4. One *red* unit in this system would be defined as representing the red light (6000–7200 Å.) in a total illumination of one lux as supplied by a standard source. Other color units (green unit, blue unit, etc.) are defined in a similar way.

5. In the choice of a standard source the following requirements should be observed: (a) If standardization is expressed in visual units, the spectral composition of the standard source should closely resemble that of average daylight; if expressed in radiant power units the color temperatures must be accurately known; (b) the standard source should provide an intensity of illumination corresponding to that to be measured by the photometer, thereby reducing the curvature correction which must be determined for the whole range of use.

6. The standard filament lamp, the artificial mean noon sunlight secured with such a lamp equipped with the Davis-Gibson double liquid filter, and the carbon arc are suggested as standard sources. The second is regarded as the most satisfactory for standardizing selenium cells in visual units.

PROCEDURES FOR MAKING MEASUREMENTS

1. On open, level, unshaded portion of deck, set up instruments, making proper connections between one microammeter and deck cell and between other microammeter and cable leading to subsurface cell.

2. If "lux" measurements only are to be made (a) install plain opal diffusing glass in place over deck cell; (b) similarly, install opal glass over subsurface cell, *completely filling space between window and opal glass with water.*

3. If measurements are to be made in terms of color (a) install selected color filter over window of deck cell and superimpose opal diffusing glass; (b) similarly, install selected color filter over window of subsurface cell and superimpose opal diffusing glass, *completely filling the spaces between window, color filter, and opal glass with water.*

4. With both cells under same conditions of illumination, make readings on both microammeters and compare results. Interchange microammeters to determine if reading is essentially the same—a condition which should prevail if the two microammeters are duplicates and if the cells are in good condition. A constant difference in reading of the two duplicate microammeters indicates difference in resistance in the two systems and such difference must be taken into account in interpretation of results. If desired, the two systems may be equalized by installation of a suitable shunt.

5. When preliminary arrangements are being made under unobscured sunlight, a reading of diffuse light by screening off direct sun by holding small opaque object between sun and cell at distance of about 2 m. from latter should be made. A convenient index of relative intensity of sunlight is provided by ratio of total to diffuse light. It is usually recommended that this measurement be made at the beginning and the end of each vertical series.

6. When all preliminary preparations and records are completed, lower subsurface cell to first desired depth, observing following precautions: (a) Be sure subsurface cell is free from direct shadows created by boat; (b) make certain that subsurface cell is so suspended that its window is horizontal; and (c) if possible, insure that supporting line is vertical in position; if this is not possible, then note approximately that angle from vertical made by supporting line.

7. Make simultaneous readings of two microammeters and record results. Also record time and other circumstances which bear upon problem of light penetration at that moment.

8. Lower subsurface cell to next selected depth and repeat readings and records. Continue until whole vertical series is completed. Make series as rapidly as possible in order to minimize possible changes in light or in water. If changes are occurring due to shifting light at surface, to effects of water flowage, or to other circumstances, the series may be repeated in reverse order as subsurface cell is brought to top and results of two readings at each level averaged, or results treated in any other desirable way.

9. At conclusion of each series, determine ratio of direct to diffuse light at surface as outlined in paragraph 5.

CALCULATION AND RECORDING OF RESULTS

Results secured in ordinary routine work should be entered on some record form such as the one suggested in Form 6. The calculations and the units used will depend upon the kind of work in progress. For the simpler information concerning light penetration, both with the opal diffusing glass and some selected color filter, calculation of the per cent of the light transmitted to a selected depth in terms of the total amount registered by the deck cell may be sufficient. Results more detailed than the computation of per cent transmitted will depend upon the proper standardizations and calibrations of both the microammeters and of the photronic cells and their filters. It is likely that at this point the aid of a physicist should be secured, and no attempt will be made to introduce the more elaborate methods here.

For the determination of the per cent transmitted, all that is necessary is to translate the two microammeter readings into terms of actual illumination by use of the illumination curves resulting from the standardization of the cells. From these transformed values the calculation of the per cent transmitted at the selected depth is simple.

FORM 6

LIGHT PENETRATION RECORD

State	Wind	Boat
County	Sky	Operator
Water	Air Conditions	Recorder
Station	Air Temperature	Apparatus
Date	Surface Illumination	Opal Glass
	Deck Cell	
Time	Subsurface Cell	Filters
	Exposure	
Water Surface		

No.	*Depth (m.)*	*Temp. (C.)*	*Deck Cell Reading*	*Subsurface Cell Reading*	*Per Cent Transmitted*	*Remarks*

GENERAL CONSIDERATIONS

a. Photronic cells must be protected against high temperatures. When exposed to direct sunlight for any considerable interval, an infrared filter should be used.

b. Both deck and subsurface cells must be kept dry at all times. The subsurface cell must be watched with particular care since it is submerged when in use and under rising water pressures as the depth is increased. Frequent inspection of the photometer case and of its various packings is imperative.

c. Continuous exposure of Weston cells (type 1) to high illumination does not, *per se*, damage the cells.

d. Measurements of light penetration should be made as near midday as possible.

e. Some investigators have found that, for most purposes, one microammeter provided with a double-throw switch can be substituted for the two instruments described above.

Other Methods

An instrument known as a pyrlimnometer was designed by Birge and Juday (1931) and in their hands has yielded much valuable information concerning light penetration in lakes. Thus far its use has been almost exclusively confined to the Wisconsin laboratory.

Selected References

(For Chapter 11)

Atkins, W. R. G., G. L. Clarke, H. Pettersson, H. H. Poole, C. L. Utterback and A. Ångstrom: Measurement of submarine daylight, *J. conseil permanent intern. exploration mer*, **13:** 37–57, 1938.

Birge, E. A.: A second report on limnological apparatus, *Trans. Wisconsin Acad. Sci.*, **20:** 533–552, 1922.

———, and C. Juday: A third report on solar radiation and inland lakes, *Trans. Wisconsin Acad. Sci.*, **26:** 383–425, 1931.

Shelford, V. E.: "Laboratory and Field Ecology." 608 pp. Baltimore, The Williams & Wilkins Co., 1929.

Zinn, D. J., and J. D. Ifft: A new limnophotometer with a special adaptation for the measurement of the penetration of light through ice under natural conditions, *Ecology*, **22:** 209–211, 1941.

WIND VELOCITY

Many limnological phenomena are related, directly or indirectly, to air movements and wind-velocity data often become a necessity. Far too commonly the limnologist must work in regions remote from weather-bureau stations, astronomical observatories, and other institutions making routine wind measurements; hence he is obliged to provide his own means of securing information. In all situations an accurate, local measure of air movement is obtainable only by an instrument operated at the locations concerned.

Anemometers

Anemometers may be classified (Middleton, 1941) into five principal groups on the basis of method of performance, namely, (a) rotation; (b) pressure-plate; (c) bridled; (d) pressure-tube; and (e) cooling. Of these, the rotational, pressure-plate, and pressure-tube types are more likely to be used for limnological purposes, although obviously any standard type may be employed if available and practicable for the work in hand. Rotation anemometers are commonly of two kinds: (a) the *cup* type with vertical axis; and (b) propeller (windmill) type with horizontal axis and with blades either flat or helicoidal.

CUP ANEMOMETER

The well-known cup anemometer is a standard rotational instrument in general meteorological practice and no description is needed here. Certain of the simpler forms may be used on temporary installations and supply data useful for some purposes. The ideal types are those mounted in elevated situations and equipped with continuous recording mechanism, but such outfits are very expensive and require considerable attention. So many types are on the market and so readily available is the detailed information about them that the limnologist will have little difficulty in seeking those best suited to his needs. The cup anemometer gives wind velocity over a selected period of time.

PROPELLER OR WINDMILL TYPE

The well-known Biram anemometer (Fig. 46) is particularly useful for some limnological purposes. It has the advantage of being portable. The blades of the propellerlike wheel are usually flat. Several dials record the number of revolutions of the wheel. Calibrations and corrections are commonly supplied by the makers. Like the cup anemometer, this instrument gives wind velocity over a selected period of time only. A correct record depends upon the operator's facing the instrument into the wind current. However, an installation can be improvised by suspending the instrument on a cord, holding it in vertical position by a counterweight suspended beneath, and by attaching a very lightweight vane to the instrument in such a position that it will not interfere with the air flow. This vane will hold the receiving side of the instrument (side opposite the dials) in the face of the wind.

Fig. 46. Biram anemometer. (Courtesy of Eugene Dietzgen Co., Chicago, Ill.)

PRESSURE-TUBE ANEMOMETER

Pitot Tube. Certain forms of Pitot tube may be adapted to the measurement of air velocities. The type described on pp. 142–145 may be used for that purpose. If it were possible to use air in the manometer when determining wind velocities in the same way as water is used in the manometer when measuring water velocities, the height of the air column would be the value needed in order to compute the air velocity. But since air cannot be so used in the manometer it is necessary to employ water or some other suitable liquid. Some makers supply calibrated Pitot tubes and a set of curves from which the velocity of the air can be ascertained directly when the manometer is read. Such curves are made for the particular instrument used, stated conditions of temperature, barometric pressure, humidity, and the kind of liquid used in the manometer. Fig. 47 is a diagram illustrating the principal features of a Pitot tube used for measurement of air velocities.

When the air velocity is being measured, particularly the lower velocities, difficulty may be experienced in reading the small level differences in the manometer tubes when the manometer is in a vertical position. This difficulty may be avoided by inclining the manometer to that angle which increases the reading over the normal one by some definite amount. For example, if the manometer is inclined 30° from the horizontal the readings are twice the normal value; if inclined at an angle of 5.75° from the horizontal the readings are 10 times the normal value.

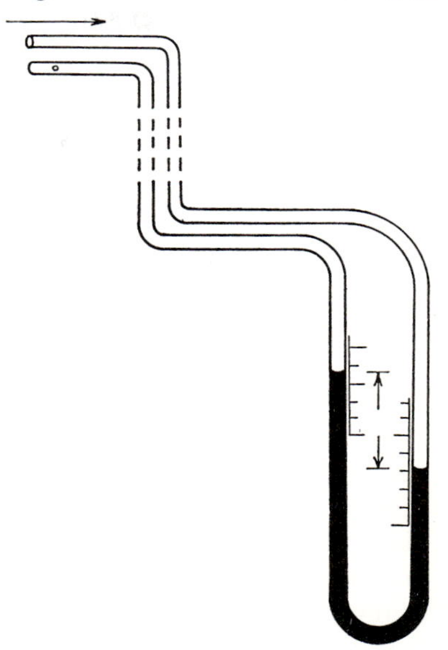

FIG. 47. Diagram showing general features of a Pitot tube as used for measuring wind velocity. Horizontal arrow indicates direction of air flow. Upper, *pressure* tube is open ended; lower, *static* tube is closed at the end but has holes in the side. Vertical arrows indicate difference of level in the two columns which is a measure of wind velocity. (Modified from Negretti and Zambra, London, England.)

The pressure difference in a Pitot tube is proportional to the square of the velocity and the following formula indicates the relations:

$$P = \frac{9992.24\, a\, C}{gd} V^2$$

in which

P = pressure difference in millimeters of water
a = density of air in grams per cc.
d = density of water in grams per cc.
g = acceleration in centimeters per sec.^{2} due to gravity
V = air velocity in miles per hour
C = constant

Under ordinary conditions of temperature and pressure the formula is

$$P = 0.012504\, V^2$$

A very useful and convenient accessory to the Pitot tube for air-velocity measurements is the differential inclined-tube liquid gage which enables the operator to read velocities directly in conventional units.

PRESSURE-PLATE ANEMOMETERS

WINDIKATOR. Various other means for measuring wind velocity have been devised. A simple, inexpensive instrument now on the market and known as a "Windikator" serves many routine limnological purposes reasonably well. The mechanism consists essentially of a pressure vane delicately geared to a pointer rotating over a graduated dial, the units of which express wind velocity in miles per hour. To operate, the base of the instrument is held at chin level in either hand with the wind velocity dial toward the face of the operator; then the operator turns his body, swinging the windikator in the arc of a circle, until that position is reached in which the wind stream meets the pressure vane to the fullest extent, thus securing the maximum reading on the wind velocity dial. This apparatus is also equipped with a compass and reference lines by means of which wind direction is also determined. It is a serviceable instrument for securing routine information in local situations.

WILD'S PLATE

A simple inexpensive means of measuring wind velocity which is known under the name of Wild's plate is a modification of the ancient Hooke's pendulum anemometer. It consists of a sheet-iron plate, 30 × 15 cm. in dimensions, weight 200 g., loosely hinged at one end to a suitable support. When set up for use, the support faces into the wind and the plate hangs from one edge in a vertical position. Air movement exerts pressure against the surface of the plate and lifts it, the amount of lifting being a measure of wind velocity. Some convenient means of measuring the angle between the plate and the perpendicular when influenced by the wind is provided. This device should be mounted on a wind vane in order that it may automatically adjust to the varying direction of the wind and insure that the plate faces the direction of maximum air-mass movement. It must be understood that Wild's plate, under the best of conditions of mounting and operation, will yield results of only approximate accuracy. Obviously such a device must be calibrated against some standard in order to make its results usable. According to Landsberg (1941), Wild's plate with dimensions and weight mentioned above will have wind velocity relations as indicated in Table 13, Appendix.

MAXIM PRESSURE-PLATE ANEMOMETER

Another pressure-plate anemometer of simple construction is illustrated in Fig. 48. The pressure plate against which the wind blows is 13 in. in diameter and wind pressure is transmitted directly to an index operating over a graduated scale. Wind velocity is read directly in miles per hour.

WIND VELOCITY ESTIMATIONS

Long ago, the need for some means of estimating and describing various wind velocities gave rise to numerous proposals known as *wind scales*, prominent among which was the well-known Beaufort twelve-point scale.

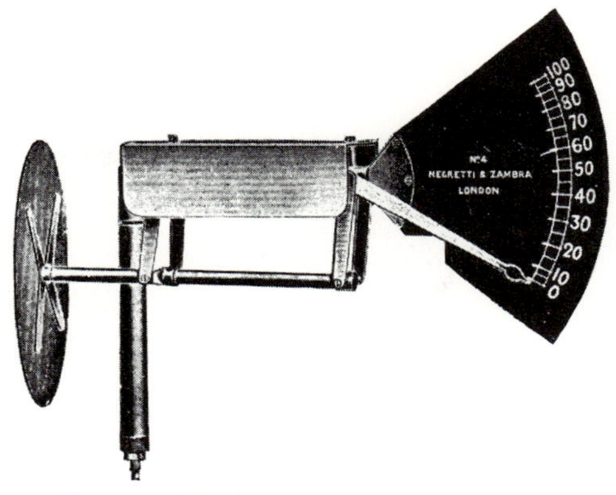

FIG. 48. Maxim pressure-plate anemometer. (Courtesy, Negretti and Zambra, London, England.)

The later development of instruments for measuring air movements caused these wind scales to be largely superseded. However, for certain descriptive, routine, and substitutional purposes a scale, such as the Beaufort scale, may be of some value to the limnologist (see p. 355). In almost any standard treatise on meteorology will be found a discussion of the Beaufort and other scales.

SELECTED REFERENCES

(For Chapter 12)

Hewson, E. W., and R. W. Longley: "Meteorology, Theoretical and Applied." 468 pp. New York, John Wiley & Sons, 1944.

Landsberg, H.: "Physical Climatology." 283 pp. Gray Printing Co., Du Bois, Pa., 1941.

Middleton, W. E. K.: "Meteorological Instruments." 213 pp. Toronto, Can., Univ. of Toronto Press, 1941.

Thornthwaite, C. W., and Benjamin Holzman: "Measurement of Evaporation from Land and Water Surfaces." 143 pp. Tech. Bull. No. 817, U.S. Dept. Agr., 1942.

BOTTOM MATERIALS

Bottom Samplers

The requirements of bottom sampling are so diverse that no one sampler has been devised which will serve all purposes. For preliminary or reconnaissance work certain simple forms of sampler may be used in most situations, but for more precise work they must be chosen in accordance with the particular needs of the program and the structure of the bottom materials.

EXPLORATORY SAMPLERS

A combination sounding weight and bottom sampler, illustrated in Fig. 8 is a useful exploratory instrument. One serviceable type weighs about 5 lb. The conical cup at the lower end acts as a bottom sampler. As it descends to the bottom the water pressure holds the cover *v*, composed of two layers of sole leather, above the edge of the cup. On reaching the bottom the cup sinks into the deposit and fills; then when drawn upward the water pressure acts in the other direction pressing the cover down upon the top of the cup and keeping it closed until arrival at the surface. If the sampler is in good condition a sample so taken is delivered at the surface with no significant loss or mixing. In soft bottoms such a sampler sinks for a certain distance into the deposit, usually securing a sample below the uppermost surface of the mud.

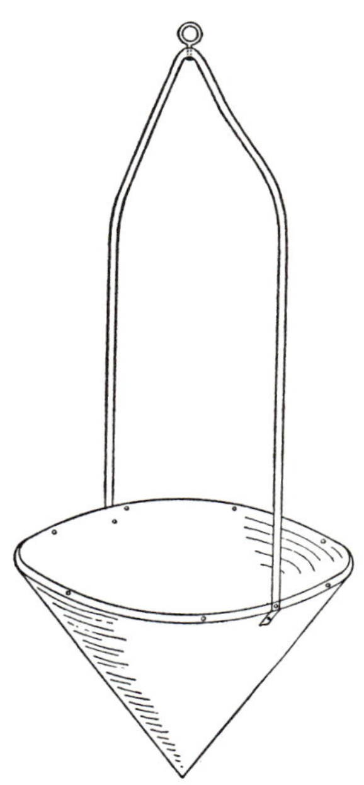

Fig. 49. Scoop sampler designed for U.S. Public Health Service.

A simple scoop sampler, designed for the U.S. Public Health Service, is illustrated in Fig. 49. It consists of a cone-shaped cup fastened to a strap-iron handle. The rim of the cup bears a cutting edge. In operation, a rope is attached to the free end of the handle and the sampler thrown into the water. It is dragged along the bottom for a selected distance; then brought to the surface. The cup may be of any convenient size. Obviously, such a sampler must be used in shallow water, and it secures a sample suitable for general purposes only.

EKMAN DREDGE

Of all the various bottom samplers, the Ekman dredge is so preëminently successful and so widely used for soft bottoms that it has become the standard instrument. In its present form (Fig. 50) it is modified from the original Ekman design by being adapted to the use of a messenger for closing. It is commonly built in two sizes, one having a cross section of 6 × 6 in. (15.2 × 15.2 cm.) and the other 9 × 9 in. (22.8 × 22.8 cm.). The body of the dredge consists of a square or rectangular box of sheet brass. The lower opening of this box is closed by a pair of strong brass jaws so made and installed that they oppose each other and, when shut, close tightly; when fully pulled apart, they leave the whole bottom of the box open. Two strong external springs, when released by the messenger,

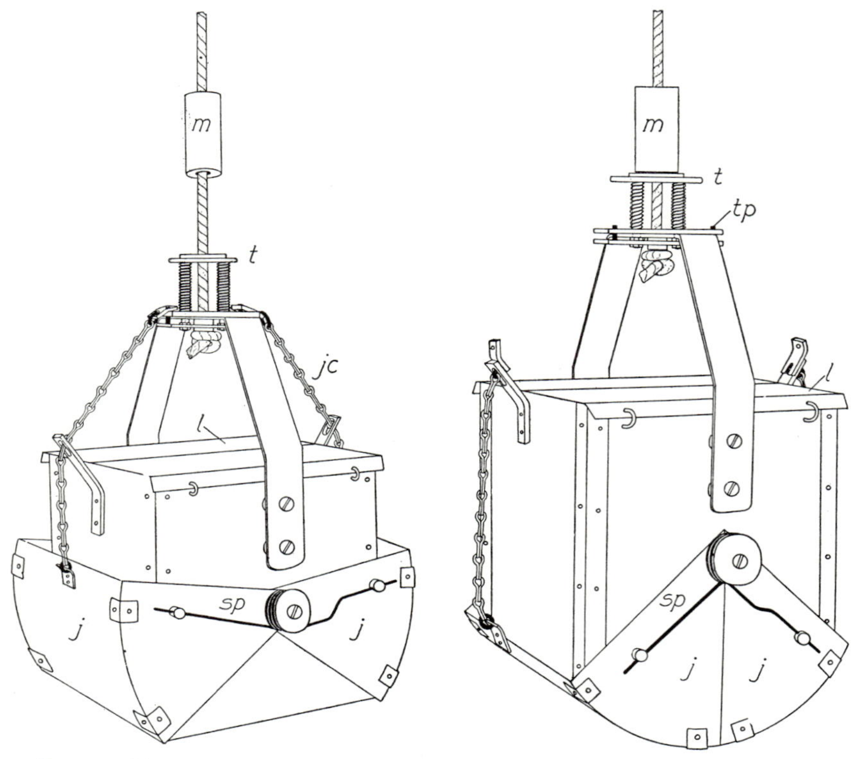

Fig. 50. Ekman dredge of usual type. (*Left*) In open form and ready to be lowered into water. (*Right*) In closed form after messenger has released trip mechanism and jaws have closed. (*j*) Jaw. (*jc*) Jaw chain. (*l*) Top lid. (*m*) Messenger. (*sp*) Spring which operates jaws. (*t*) Trip mechanism. (*tp*) Trip pin.

snap the jaws shut. The top of the box is closed by two thin, hinged, overlapping lids which open easily when the dredge is descending through the water but which close and are held tightly shut by the water pressure while the dredge is lifted to the surface. A spring mechanism at the top of the sampler provides a means of releasing the jaws by a messenger when the sampler is at the proper place. Two short chains extend from the upper edges of the jaws to two pins in the spring mechanism when the jaws are set at the open position. Other details of construction appear in Fig. 50.

In operation, the sampler is attached to a strong rope by passing the latter through the trip mechanism and knotting it securely below the underlying plate. The sampler is then lowered into the water until it rests on the bottom. Its own weight is usually sufficient to sink it in the mud for much or all of its height. After a short time to allow for settling, the messenger is sent down on the rope causing the jaws to close and bite out a sample. If the sampler is in good condition this sample can be drawn to the surface without loss and without any contact with overlying water. On arrival at the top the sample is delivered into screens, containers or elsewhere by merely pulling up the jaw chains.

This sampler is especially adapted for use in soft, finely divided mud, muck, ooze, submerged marl and fine peaty materials. It will not function on sand bottoms since the springs are not strong enough to force the jaws closed; also fine grains of sand get between the sides of the box and the closely opposed sides of the jaws, preventing the latter from closing. This difficulty may appear even in muds having some intermixture of fine sand. The presence of hard objects (sticks, partly decayed leaves, clams, stones) may cause difficulty by getting between the jaws. Ordinarily the instrument is useless on hard bottoms.

This sampler has been modified in various ways to suit special purposes. For example, it may be made in a tall form in order to provide a better chance of securing the uppermost, fine, bottom materials. Such a tall form is sometimes equipped with a series of regularly spaced, horizontal slits through which thin metal sheets resembling the shutters on a camera plate holder can be inserted thus dividing the sample into horizontal strata which can be delivered one at a time, thus making possible a study of stratification of materials and vertical distribution of organisms. Both the standard and the tall forms are sometimes equipped with a brass screen of selected mesh which covers the upper end of the instrument just below the lids. Such a screen is looked upon by some workers as a means of preventing the overspill loss of organisms and coarser materials by the sampler sinking into the mud deeper than its own height. However, when a sampler is so screened, it must be used with care since if lowered too speedily into a bottom the passage of watery mud through the screen may not equal the speed of lowering and then there is a danger of underspill at the bottom opening. The size of mesh of such a screen must be selected with care and preliminary tests of a sampler so equipped should be conducted in the laboratory before it is used in the field.

General Considerations

a. All nuts must be sealed into place with solder to prevent loss in the field.

b. In attaching the rope to the sampler, care must be taken to make a close, secure knot below the supporting plate with the rope end worked into the knot, otherwise the loose end may interfere with the closure of the upper lids.

c. When lowering the sampler into the water, it should be started down in a vertical position and then not dropped too rapidly, otherwise the open sampler will tend to descend in some diagonal direction. When lowered properly, the sampler will meet bottom in the correct position.

d. Ekman samplers of larger sizes (9×9 in. or larger) when loaded are heavy and ordinarily should be used with the aid of a hoist (Fig. 51).

e. While this sampler, as usually built, is sturdy and will function indefinitely, damage in the form of bent or dented sides, jaws, and lids is likely to result in leaks, seriously impairing its usefulness. Hence in use, impacts with the boat and other objects should be avoided.

f. An Ekman dredge may be used at any depths in inland waters.

g. The Ekman dredge is suitable both for qualitative and quantitative work.

PETERSEN DREDGE

The Petersen dredge is now widely used for taking samples on hard bottoms, such as sand, gravel, marl, clay, and similar materials. This sturdy dredge is usually built of iron and so constructed that both by its own weight and by the leverage exerted by its closing mechanism it bites its way into hard bottoms deep enough to secure satisfactory samples. The principal features of construction appear in Fig. 52. It may be made in any desired size and weight. For ordinary uses one which weighs 35 lb. empty and which may be loaded with weights bolted to the outer surface of the jaws to a total weight of 70 lb. is recommended. Since the total weight is markedly increased when the instrument acquires a sample it is usually necessary to operate it with a hoist (Fig. 51). A dredge of the size just mentioned will enclose an area, when open, of about 0.08 sq. m. The weight of its load depends upon how nearly it fills and upon the character of the bottom materials. One of its virtues is simplicity of construction and operation. Barring accidents, such a dredge will last indefinitely. The tripping device consists only of a horizontal locking bar which holds the dredge open until it reaches the bottom and the tension is taken off the cable. Then the upper bar drops at its free end and the locking bar falls out of the notch in the end of the upper bar. Tension on the cable now exerts a closing motion on the jaws by pulling on the ends of the crossed bars, a motion which not only closes the jaws but tends to force them deeper into the bottom.

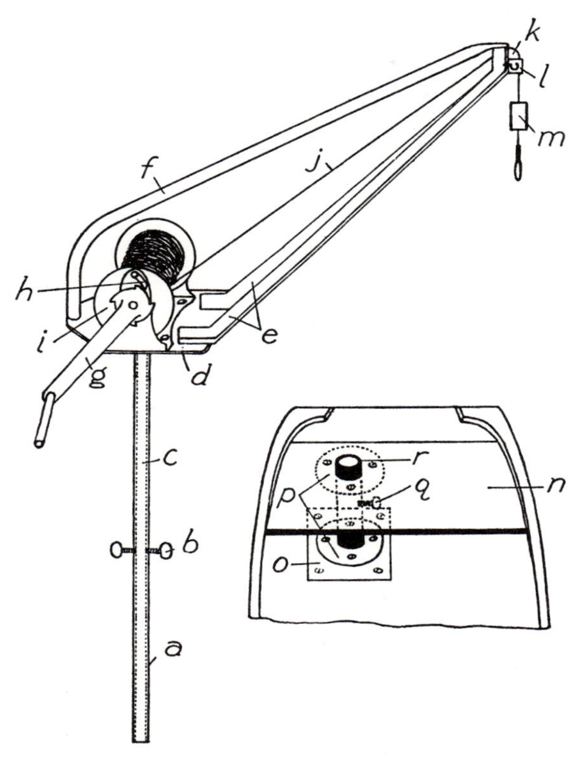

FIG. 51. A crane suitable for lifting limnological sampling apparatus, designed by Brown and Ball, 1940. Originally devised for use on rowboats, it can also be adapted for launches, scows, and other larger craft. (*Left*) Side view of crane. (*Right*) Stern of rowboat showing attachments for crane. (*a*) 1¼-in. pipe, 30 in. long, not fastened at either end. (*b*) Set screw to lock davit to fixed vertical rod. (*c*) Pipe, 30 in. long, attached to plate *d*, is turned to fit inside *a* and acts as bearing for *a*. (*d*) 3⁄16-in. steel plate, 7 in. wide and 9¼ in. long, with two truncate corners. (*e*) ¾ × ⅞-in. angle irons, 27 in. long. (*f*) ½-in. pipe, 37¾ in. long. (*g*) Lever handle, 10¼ in. long. (*h*) Dog. (*i*) Reel, 3 × 6-in. drum, known as "Sasgen small puller, capacity 300–600 lbs." (*j*) ⅛-in. steel cable, with 6 twisted strands and cord center. (*k*) 1¼-in. ball-bearing pulley. (*l*) Guard with hook to suspend messenger when latter is not needed. (*m*) Messenger. (*n*) Rear boat seat. (*o*) ¼-in. plate, 8 in. square. (*p*) Pipe flange to fit 1½-in. pipe *r*. (*q*) Set pin to hold pipe *a* in position. (*r*) 1½-in. pipe which acts as bearing for vertical arm *a*. (Explanation largely quoted from Brown and Ball.)

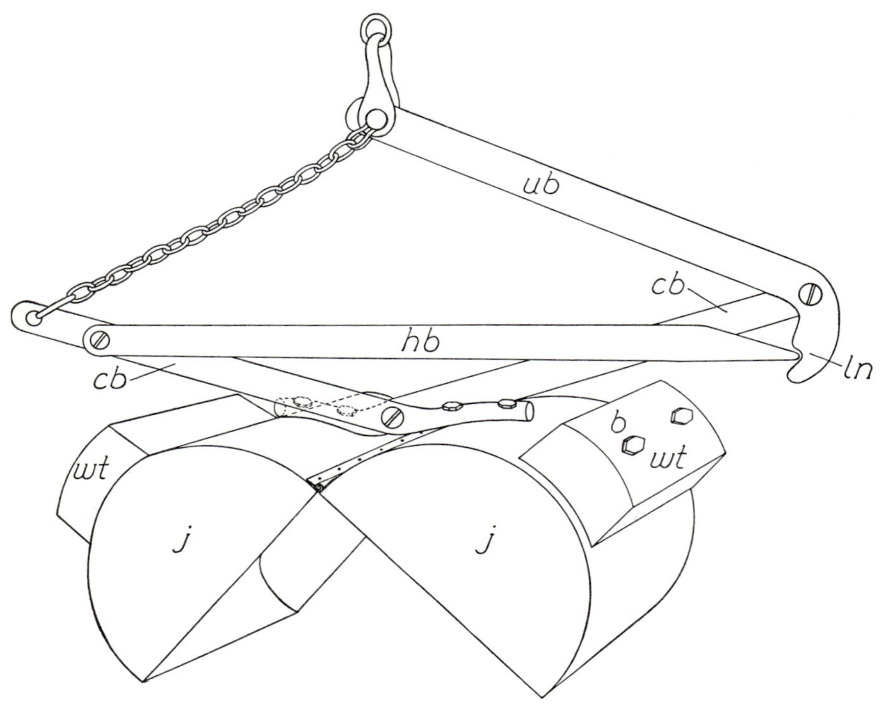

Fig. 52. Figure showing principal features of construction of Petersen dredge. (*b*) Stud bolt for fastening weight to jaw of dredge. (*cb*) Cross bar. (*hb*) Horizontal locking bar. (*j*) Semicylindrical jaw of dredge. (*ln*) Locking notch. (*ub*) Upper bar. (*wt*) Weight bolted to dredge jaw.

General Considerations

a. Most Petersen dredges are of such weight that they should be operated on wire cables of suitable size and tensile strength.

b. In spite of its vigorous bite there are limits to the sampling capacity of any Petersen dredge. It must not be expected to cut through sizable sticks and similar objects or to secure samples from rocky bottoms.

c. Provided sufficient lifting power is available, metal weights of various magnitudes may be added to the dredge.

d. If the dredge is constructed of iron, it should be kept well painted to prevent rusting. When stored for considerable periods of time, a thorough coating of oil or petrolatum is recommended. This treatment should include all surfaces, joints, bolts, and stud-bolt holes if the latter are left open.

e. This dredge is satisfactory for both qualitative and quantitative work.

OOZE SUCKER

A useful instrument for securing samples of that fine ooze layer which usually constitutes the uppermost part of the bottom deposits is an ooze sucker modified by Moore (1939) from an original design by Rawson. The principal structural features of the instrument may be seen in Fig. 53. The brass ring at the base of the instrument is about 15 in. in diameter and has two functions: (a) to insure that when the instrument arrives at the bottom it will stop sinking in the proper position, and (b) it aids in keeping the instrument in the upper part of the ooze layer. A bridge composed of two parallel rods extends from one side of the ring to the other and supports the sampling mechanism, the tripping device, and the attachment ring. The lowermost part of the sampling mechanism is a shallow funnel the broad end of which is directed downward and covered by a brass, 20 meshes-to-the-inch screen the function of which is to exclude very coarse materials that might clog the collecting tube. Six small holes, about 3 mm. in diameter, through the wall of the funnel provide for the passage of water through the funnel while it descends to the bottom. The funnel is attached to and opens into the brass tube which in turn bears at its upper end a large ovoid rubber bulb. The funnel tube and bulb mechanism can be disassembled and reassembled in a short time. Two large flat plates are

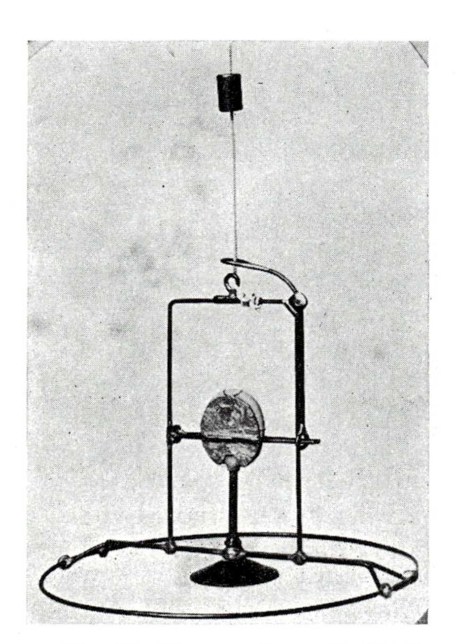

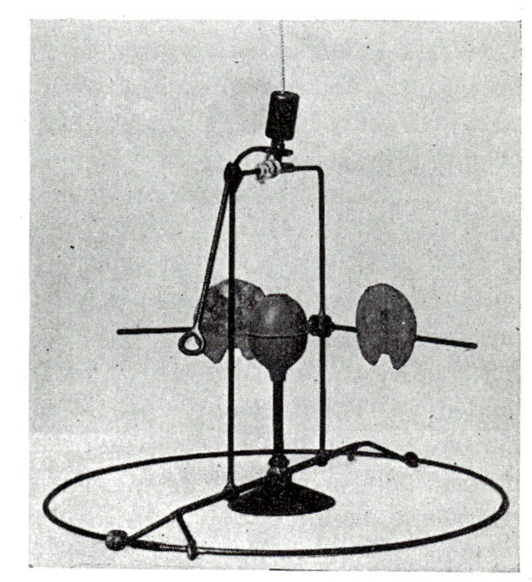

FIG. 53. Ooze sucker. (*Left*) Sampler shown with compressor arms closed. (*Right*) Sampler with compressor arms released by messenger. (From Moore, 1939.)

carried on two swinging horizontal arms. When the rubber bulb is in place these flat plates, brought together, squeeze the bulb and temporarily eliminate its inner space. The ends of the two horizontal swinging arms are then caught within the ring of the tripping device, the rope is passed through the other ring of the trip and tied into the supporting ring, and the sampler is ready to be sent to the bottom.

In operation, the sampler is lowered to the bottom and allowed to rest for a moment; then a messenger, sent down the rope, strikes the upper end of the trip releasing the ends of the horizontal swinging arms and allowing the rubber bulb to expand to its normal shape. The bulb thus sucks up a load of ooze through the funnel and tube. On arrival at the surface, the bulb is gently twisted off the tube, its load of ooze delivered wherever desired by squeezing it out, and all residues are thoroughly washed out by repeated squeezings of the bulb partly submerged in water.

General Considerations

a. This instrument finds its greatest usefulness in qualitative work.

b. The sampling mechanism is so constructed that funnels of different sizes, screens of different mesh, and bulbs of different capacities may be used interchangeably.

c. The general delicacy of the instrument demands that it be carried and stored in a specially made case.

VERTICAL CORE SAMPLER

A simple, effective core sampler of the type illustrated in Fig. 54 is especially suitable for bottom sampling which involves consideration of bottom stratification. It consists essentially of a brass tube with a detachable steel nose on the lower end and a weight and valve on the other. The valve permits the free passage of water through the sampler as it descends but as it is hauled up the water pressure automatically keeps it securely closed. The over-all length of the sampler is 60 cm. When the steel nose at the lower end is removed, a glass tube, 48.4 cm. long, inside diameter of 2.2 cm., and open at both ends, is inserted into the brass tube and the steel nose replaced. In operation, the sampler, attached to a rope, is allowed to fall rapidly through the water and imbed itself in the bottom. It is then hauled to the surface, a cork is placed in the bottom opening of the sampler, and the glass tube with its enclosed and undisturbed core of bottom deposit is removed by unscrewing the steel nose. The area of the transverse section of the sample so taken is 3.8 sq. cm.

The general usefulness of this instrument has been extended by additions and accessories designed by Moore (1939). In soft bottoms the sampler imbeds itself properly and without difficulty, but on harder bottoms additional weights must be provided. Such weights may be con-

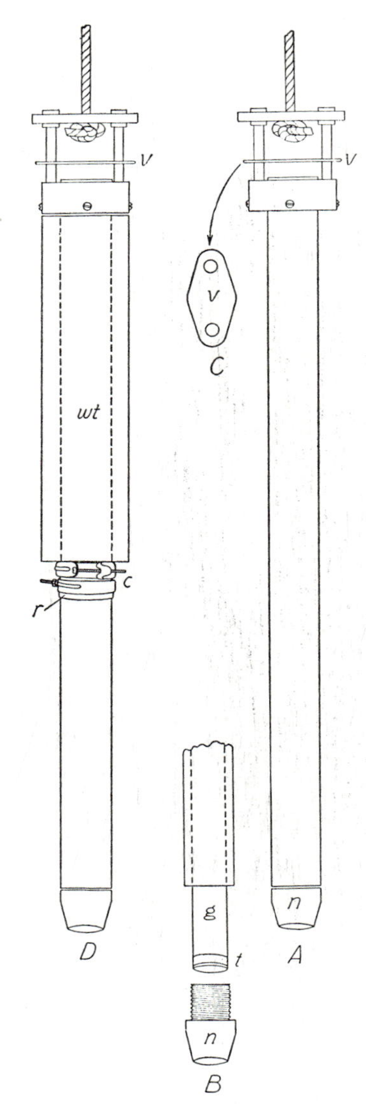

Fig. 54. Vertical core sampler. (A) Side view of sampler as usually used. (B) Lower end of sampler with steel nose unscrewed and dropped down; also showing lower end of glass tube dropped down out of main tube of sampler. (C) Top view of valve. (D) Side view of sampler with weight in place for use in more resistant bottoms. (*c*) Hose clamps holding weight in place. (*g*) Glass tube. (*n*) Steel nose of sampler. (*r*) Sheet of rubber between clamps and glass tube. (*t*) Strip of adhesive tape about end of glass tube. (*v*) Valve. (*wt*) Cylindrical, hollow weight which slips over main tube of sampler and is held in place by hose clamps.

structed as follows: a collar consisting of two pieces of iron pipe, one inside of the other and the space between them filled with lead, is slipped over the brass tube of the sampler (Fig. 54), pushed to the topmost position and held in place by hose clamps. Properly weighted, the sampler will secure samples even from hard sand bottoms. Empty glass tubes and tubes containing cores may be carried by a field case as shown in Fig. 55.

FIG. 55. Field case for extra glass tubes used in vertical core sampler. (From Moore, 1939.)

When brought into the laboratory, the cores may be treated in any desired way. Moore's method for studying the stratification and vertical distribution of materials in the cores is as follows: A close-fitting piston *p* (Fig. 56) is inserted into the lower end of the glass tube containing the core. A rubber cup *c*, 6 cm. in diameter, 3 cm. high, and with a hole 2.2 cm. in diameter in the bottom, is slipped over the upper end of the glass sampling tube until its rim is about 2 cm. below the top of the tube. This provides a watertight trough around the upper end of the tube. Any clear water above the bottom deposits in the glass tube is gently siphoned off.

By means of the piston the core is then pushed up the tube until its upper surface is just at the level of the upper end of the tube. Then the first centimeter of core, or any other desired unit of depth, is pushed up by the piston and spills over the end of the tube into the rubber trough. The top is leveled off by passing a knife or flat blade across the end of the tube. This material is then removed from the cup by means of a small bulb syringe and the cup washed with tap water. The next vertical unit of the core is then collected in the same way, and so on until the sectioning of the core is completed. The base which serves as a support for the glass tube may be made from a plumber's rubber suction cup.

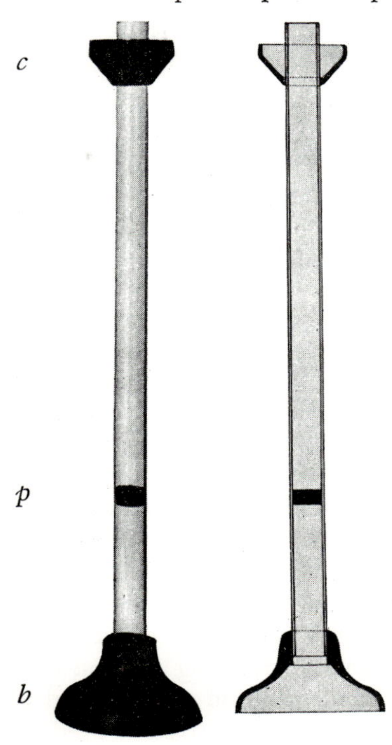

FIG. 56. Apparatus for sectioning cores of bottom deposits. Figure to right shows longitudinal section of glass tube and supports. (*b*) Rubber base for supporting glass tube. (*c*) Rubber cup on top end. (*p*) Rubber piston (stem not shown). (From Moore, 1939.)

General Considerations

a. It has not yet been satisfactorily determined to what extent this kind of a sampler compresses the core linearly, but it is believed that any compacting of cores, if and when it occurs, is not serious for most limnological purposes.

b. Only unusually watery cores fail to remain in the sampler when hauled to the surface.

c. There is no evidence that when the sampler is used properly there is any consequential mixing of materials along the sides of the glass tube when the sampler sinks into the deposit or when the core is displaced in the glass tube by a piston.

d. A strip of adhesive tape about each of the two ends of the glass tubes insures better fitting within the brass tube of the sampler and also offers greater protection against breakage.

OTHER BOTTOM SAMPLERS

PEAT BORERS. Various types of peat borers are available on the market. Certain ones are useful to the limnologist in exploratory work on bog lakes, marl lakes, and the soft bottoms of the littoral and sublittoral regions of lakes in general.

DEEP-CORE SAMPLERS. For a description of equipment and technique for deep-core sampling in lake and stream deposits, the reader is referred to Wilson (1941).

BOTTOM SAMPLING

In the collection of bottom samples for physical analyses, the plan used must take into account the following features: (a) general purpose of whole project; (b) nature of bottom, i.e., kind of materials and degree of uniformity; (c) kinds of analyses to be made; and (d) utilization of results.

It is imperative that every effort is made to insure that samples are as *truly representative* of the whole area as possible. This will depend largely upon the good judgment of the operator since by the very nature of the problems involved no definite rules can be formulated in advance. Size of the individual sample must also be determined by the nature of the plan and the physical character of bottom materials. Unfortunately there seems to be no specific answer at present to the time-honored question as to exactly how many samples must be taken to make the sampling satisfactory. Ordinarily one sample has little or no value and as many as practicable must be provided.

Horizontal sampling over the selected area may be accomplished by the random sampling method executed in a completely random way or by the distribution of individual samples along transects which cut across the area according to some predetermined plan.

Vertical sampling, *seasonal sampling*, and *special sampling* must be planned in accordance with the objects of the work.

Ordinarily, the time interval between sampling and analysis is not as important in mechanical analyses as in certain others although prompt analyses are recommended as the best general procedure since different bottom materials may not behave uniformly in storage.

SCREENS

Suitable screens are standard requirements in many mechanical, chemical, or biological methods for analyzing bottom materials. While screens may be made up in various forms, the following types are suitable for many limnological purposes.

GENERAL UTILITY SCREENS. One of the best screens for general purposes is built in circular form with flaring sides and having the following dimensions: top diameter, 17 in.; bottom diameter, 15 in. The sides are constructed of heavy galvanized sheet iron having a height of 4 in. above the level of the screen and an extent of ¾ in. below the screen level. The top and bottom rims (Fig. 57) are each turned over to surround an iron ring having a diameter of ¼ inch. The topmost, 4-in. portion and the lowermost ¾-in. part are made separately, and when assembled, the circular screen is installed between the two as shown in the sketch. The screen is brass and of whatever mesh is desired. Common and convenient sizes are 10, 20 and 30 meshes to the inch. Across the lower surface of the screen and in contact with it are two ¼-in. brass rods installed at right angles to each other, intersecting at the center of the screen and fastened to the lower rim of the frame. These rods act as supports for the screen when it is loaded. In some kinds of work in which the screen must take unusually heavy loads, these cross rods should be of larger size. A small ring is installed in the upper rim to which a strong cord can be attached for tie-on purposes. This type of construction provides a very sturdy screen, suitable for heavy-duty work, and can be made in any tin shop. All parts are soldered securely into place.

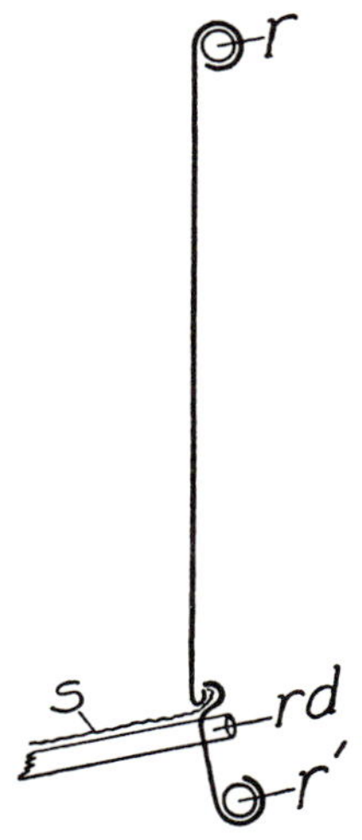

FIG. 57. Cross section of side of circular metal screen, showing construction. (*r*) Upper iron ring. (*r'*) Lower iron ring. (*rd*) Transverse supporting rod. (*s*) Screen.

Another form of general utility screen is made by building a wooden frame in the form of a square or a rectangle with vertical sides 4–6 in. high. Across one end of this open frame the screen of desired mesh is securely fastened. Mechanical support, if needed, may be supplied by installing large-mesh screening immediately below the brass screen. Handles on the outside of each end of the frame facilitate manipulation. Such screens can be constructed in any desired size and made to nest if different meshes are to be used simultaneously.

Screening may be done by means of canvas or muslin bags into which are sewed large windows of strong grit gauze of suitable mesh. A sample of bottom mud is delivered into one of these bags, the bag closed, the bag and contents submerged in water, and the bag so manipulated that a thorough washing out of the finer contents is accomplished.

TESTING SIEVES. When precise results are required, standard testing sieves must be used. Such sieves are now made for high accuracy and exact screen analyses. In the past, several bases have been adopted for the initial point in screen scaling, some makers using 1 in. as the base, and others 1 mm., the scales ranging above and below these units. The U.S. Bureau of Standards standardized the 200-mesh sieve constructed from 0.0021-in. wire and having an opening of 0.0029 in., and some manufacturers now adopt this as base for their screens. Of the various ratios between the different sizes of the screen scale which have been proposed, that of using the square root of 2, or 1.414, has apparently proved to be the most satisfactory. The convenience of this system appears in the fact that in such a series the area or surface of each successive opening in the scale is just double that of the next finer or half that of the next coarser sieve. Thus the widths of the successive openings have a constant ratio of 1.414, while the areas of the successive openings have a constant ratio of 2. Another advantage in this ratio is that by skipping every other screen a ratio of width of 2:1 is secured, by skipping two sizes a ratio of 3:1 is attained, and by skipping three sizes an approximate ratio of 4:1 results.

Table 11, Appendix, presents features of certain well-known systems. In most series a 200-mesh screen represents the finest sieve. When special circumstances demand, a 400-mesh screen may be obtained. However, it is recommended (Twenhofel and Tyler, 1941) that for the usual purposes sieves be used for materials the particles of which have diameters which exceed $1/16$ mm.; for smaller particles either elutriation or subsidation methods should be employed.

Rating of Sieves

Manufacturers of high-grade precision sieves supply the necessary information concerning their products and it may not seem necessary to rate them, but as a safety measure, it should always be done. When it is desirable to confirm the rating of such sieves or to rate less precise ones, proceed as follows:

Screen a sample of sand which has been dried in an oven to constant weight; pass sample through a single sieve, or a nest of sieves if more than one is to be rated; then take each sieve individually and shake slightly, collecting the grains which pass through the mesh. Count a number of these grains and then weigh accurately on an analytical balance those counted and express results in milligrams. Calculate the weight of each grain from this result. Determine the diameter of each sand grain by the formula

$$\text{Diameter} = 0.9\sqrt[3]{w}$$

in which w is weight of one grain in mg. and the resulting diameter is expressed in millimeters.

This formula gives satisfactory results for sands having a specific gravity of 2.65, an average value commonly used. If, however, sands of some other specific gravity (*sp. gr.*) are used the diameter of the grains may be computed from the following formula

$$\text{Diameter} = \sqrt[3]{\frac{1.91w}{sp.\,gr.}}$$

From the procedures outlined above it appears that the diameter of the sand grains represents the diameter of the sieve meshes through which the grains passed.

Care of Screens

1. Clean screens thoroughly as soon as the task is completed. Use a stiff brush to dislodge sand grains. If a screen becomes dry before cleaning, it must be thoroughly soaked in water, then scrubbed and rinsed. Certain bottom materials when dried in the meshes are resistant to cleaning.
2. Avoid passing through screens any collections which may have been treated or preserved in reagents which might cause corrosion.
3. Avoid using the supporting cross rods as handles in transporting or in emptying screens.
4. Screens should never be overloaded.

Mechanical Analyses of Bottom Materials

Assuming that samples have been taken with an adequate sampler and that they are truly representative of the areas under consideration, materials so secured may be treated as described below.

PREPARATION OF FIELD SAMPLES

For gross limnological analyses in which large sieves equipped with the coarser mesh screen are used it is often possible and even desirable to use the entire field sample. Even samples of large volume may be screened in their entirety with advantageous results. In such instances no special preliminary preparation is required.

When graded precision sieves are used, the volume of the field sample determines whether it can be used in its entirety or only in part. If the field sample is small in volume compared to the capacity volume of the sieve the whole sample may be screened; but if the reverse is true, practicality may demand that some kind of subdivision of the sample be made. Likewise, in the instance of several samples taken at same time and place, all collections should be assembled into a composite and then subdivided in the manner indicated below.

Subdivision by Quartering. For this kind of subdivision, proceed as follows:

1. Mix thoroughly in same container all materials representing same field sample.

2. Deposit on flat surface and spread out into circular, flat mass.

3. Divide mass into quarters; reject two opposite quarters.

4. Mix together two quarters retained; if volume of sample so secured is still too great, again spread out into circular flat mass, quarter, reject two opposite quarters, and mix together other quarters, and so continue to subdivide by this same quartering procedure until final sample of proper size is secured. This procedure is suitable for both wet and dry field samples.

Size of Final Sample. The size of the final sample must be governed by several considerations, namely, character of the material composing the sample; size of meshes used in the sieves; capacity of the sieves; time available for the test, and degree of accuracy desired. As a general principle, the use of as large a sample as is consistent with practicalities of the situation is recommended.

Wet Samples. Wet samples usually require no special treatment unless the water content is so high that subdivision by quartering described above cannot be accomplished in which instance it is necessary either to pass the whole field sample through the sieves even at the added cost in time and effort, or to resort to subdivision by some other method, as for example, thorough mixing followed immediately by cup sampling.

Dried Samples. If dry samples are demanded, wet field samples should be placed in an oven with temperature maintained at about 105° C. and dried to a weight which is practically constant before they are put through the sieves. This treatment is satisfactory for bottom materials containing a large component of sand. However, it may happen that some bottom materials, on being dried, tend to change their physical character, perhaps tend to consolidate, or otherwise undergo changes which make any subsequent dry analyses by graded sieves unsatisfactory. In such instances wet screening is recommended.

GROSS ANALYSIS

If only a gross analysis is needed, the collected wet materials may be passed through a vertical nest of three sieves of sizes No. 10, 20, and 30. This passage of materials is facilitated by directing a stream of tap water into the upper sieve (No. 10) in such a way that the finer materials are washed down into the lower sieves. When this washing is completed, the materials in the uppermost two sieves are removed and are ready for whatever form of examination is desired. If the material which accumulates in the No. 30 screen does not appear to be well washed, washing may be concluded by treatment with a jet of tap water. If screening must be done in the field, lower the sieve with its contents until it is about one-half submerged in the lake or stream water; then give the sieve a vigorous rotating motion by grasping the frame with the two hands 180° apart and swinging the sieve first one-half turn in one direction and then a like swing in the other, at the same time imparting to the screen a slight vertical up-and-down motion. By this method all of the material passing through the third screen is lost. Should this material be desired it must be collected by the addition of another screen of still finer mesh. If all solid materials are to be examined, materials escaping through the finest sieve must be collected and passed either through a filter of some suitable type, or through a high-speed centrifuge.

For most limnological purposes, final results may be expressed in terms of percentage of the total sample, either by weight or by volume.

Analyses with Graded Sieves

Graded sieves are commonly made so that the extended bottom rim of each fits snugly into the top of any other, forming a vertical series. Thus sieves may be selected in any desired combination and nested together with the coarsest mesh at the top and the others arranged in a progressively finer series. A whole set so assembled may be operated simultaneously, or the sieves may be used individually and successively.

Different limnological needs demand different procedures in the use of graded sieves and the operator must adapt his practices accordingly. Under all circumstances the underlying purpose is to provide a means of sorting mechanically and differentially the physical components of a sample on the basis of size groups as determined by the sizes of mesh used. The following procedures are presented for their general utility, but it must be understood that still others may be required, depending upon the character of the bottom materials and other conditions.

PROCEDURE NO. 1—WET SIEVING

1. Assemble nest of sieves including desired sizes of mesh graded progressively from coarsest at top to finest at bottom; leave top and bottom open; arrange so that outdrainage from lower screen is properly provided for.

2. Allow excess water, if present, to drain from field sample; weigh final sample and place in top (coarsest) sieve.

3. By means of jet of tap water under low pressure, gently wash sample through underlying screens until material no longer leaves lowermost screen; remove uppermost screen and again pass water through underlying screens, making certain that accumulated material in each sieve has been adequately washed; remove next sieve and proceed as before until all are so treated.

4. Carefully remove all material retained by each sieve; allow any excess water, if present, to drain out; weigh material from each sieve.

5. Calculate per cent by weight of material passing each sieve as follows:

$$p = \frac{s - (w + W)}{s}$$

in which

p = per cent of material passing individual sieve
w = weight of material retained by individual sieve
W = combined weights of all materials retained on all coarser screens above
s = total weight of original sample

Because of difficulty of reducing water content in all screened material to same level, this method does not yield results of high precision. For general purposes the results are useful. Some bottom materials are not readily handled in this manner.

PROCEDURE NO. 2—DRY SIEVING

1. Assemble nest of sieves as in Procedure No. 1; weigh dry sample and place in top sieve.
2. Give whole nest of sieves an initial shaking of about 3–4 min.
3. Remove topmost (coarsest) sieve from nest and shake separately but directly over sieve next below to complete screening; continue shaking until no material passes through screen; proceed in same fashion with other sieves.
4. Weigh materials retained by each sieve and calculate per cent by weight of material *passing through each sieve* by the use of the same formula given in Procedure No. 1.

PROCEDURE NO. 3—WET-AND-DRY METHOD

1. Assemble selected sieves in graded series. Deposit sample in *finest* sieve to be used in analysis; place under small, low-pressure stream of tap water and wash until fine materials no longer pass through sieve; collect all material passing through sieve, dry, and weigh.
2. Dry and weigh material retained in sieve; add this weight to dry weight of that material which passed through sieve.
3. Place dried material (material *retained* by finest sieve) on coarsest sieve; shake until material no longer passes through screen; and collect all materials passing through. Transfer collected material to next finest sieve and proceed as before. Complete sieve analysis by using successively and in proper order all sieves in series, including that one used in preliminary washing process. Weigh separately materials retained by each sieve.
4. Calculate per cent by weight of material passing through each sieve, using same formula as given in connection with Procedure No. 1. (Note

that s in formula is here the sum of (a) the dry weight of the material passing through, and (b) the dry weight of material retained by, the finest sieve used in the preliminary washing process.)

If, in sieving, a significant amount of fine material passes through the finest screen in the last step of the analysis, it may be collected, weighed and properly accounted for in the final computation.

GENERAL CONSIDERATIONS

a. The size of a sample for sieve analysis should not be so large that the sieve cannot perform properly. Interference with free passage of material through the sieve may result from a sample which is too large.

b. Shaking is commonly performed by hand. Special machines for shaking samples have been devised.

c. Manufacturers of graded sieves commonly supply detailed technical information concerning their products.

d. Dried samples of materials which tend to cake or consolidate may require dispersal before sieving.

e. For information concerning highly precise methods for the study of bottom sediments, the reader is referred to Twenhofel and Tyler (1941) and similar treatises.

Ripple-mark Records

Ripple marks on lake and stream bottoms are indices of certain limnological processes. Under some circumstances they may be studied *in situ* in the field; under others, direct study may be difficult or impossible. In either instance, permanent records are desirable and sometimes necessary. The methods described below were originally designed by Kindle (1917).

Shallow-water Method. In water having a depth of 0.3 m. or less, plaster of Paris casts may be made as follows: The bottom of a wooden box, about 46 × 25 × 15 cm., is removed. One edge is beveled all around the periphery. This frame is then pressed firmly, beveled edge down, into the bottom surrounding the ripple-mark area to be recorded. Plaster of Paris is sifted through a screen until an accumulation of 2.4 cm. or more in thickness has been deposited inside the frame. Complete protection against any disturbance for one hour must be provided. Then the cast may be removed.

Deep-water Method. When greater depth of overlying water prohibits the use of the method just described, casts may be made with an apparatus the essential features of which are as follows: a base in which the cast is to be formed is constructed. This base has a diameter of about 40.5 cm., a form like that of an inverted dishpan, and the lowermost edges sharp. Just above the edge of the base, short nails are inserted through holes, most of the length of each nail extending into the interior of the base. On top of this broad circular base is mounted a 2.5 quart cylindrical tin reservoir the top end of which is closed by a cap resembling that of a milk can. In the opposite end of this reservoir is a wide short tube which fits snugly into a hole in the center of the base and contains a valve which opens upward. Through a small hole in the reservoir cap a string passes from the valve at the bottom of the reservoir to the surface where it can be manipulated by the operator. Two or more valves in the upper surface of the base, so constructed that they will automatically be open during the descent and then closed when on the bottom, facilitate the lowering of the apparatus. The reservoir is filled with a mixture of plaster of Paris and water (equal parts of each) and the reservoir cap put into place. Then the apparatus is lowered to the desired place by supporting cords and allowed to settle for a moment. The valve at the lower end of the reservoir is then opened by pulling the controlling string, and the plaster of Paris mixture pours into the base forming a layer over the area of bottom so surrounded. The plaster of Paris mixture must be delivered into place within 5–6 min. after it is first prepared since it begins to show signs of setting after 6 or 7 min. One hour is allowed for setting of the plaster; then the cast may be lifted, the small nails in the base preventing its escape. If the apparatus must be left during the hour required for setting it should be marked with a buoy.

Another device useful in deeper waters gives information on the spacing or amplitude of ripple marks. A thin piece of sheet iron or zinc, about 50 cm. square, is suspended by four cords attached at the four corners. These cords are united into a single cord by means of which the plate is lowered after being suitably weighted. The lower surface of the plate is coated with petroleum jelly. When lowered to the bottom, allowed to rest there momentarily, and then raised to the surface, the plate will show lines of adhering sand representing the crests of the ripple marks.

Firmness of Bottom

No method of measuring firmness of bottom has yet been adopted as a standard procedure. The degree to which bottom materials pack under natural conditions has limnological significance and some method of measure is needed. The method described here makes use of the principle that this firmness (degree of packing) of bottoms in which the predominating materials are sand, silt, organic deposits, gravel, marl, and clay may be expressed by that force which is necessary to cause the penetration to a measurable depth of an object having known weight and cross section in a selected unit of time. Equipment for this purpose may be assembled as follows:

Choose a straight iron rod, 1.2 cm. in diameter, and round off one end to the form of a hemisphere; using rounded extremity as the zero end,

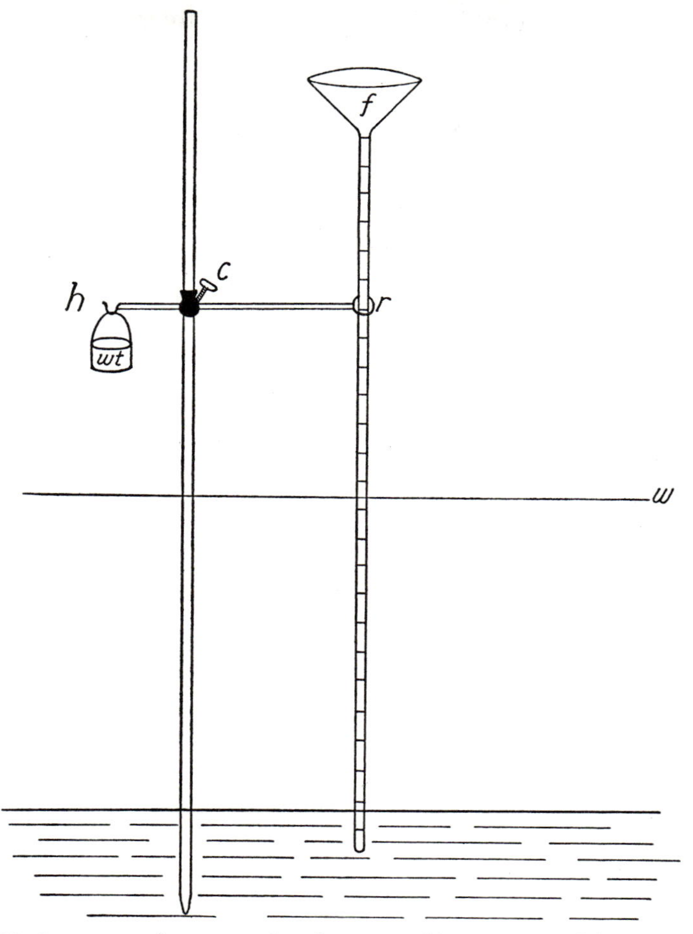

FIG. 58. Apparatus for measuring firmness of bottom. (*c*) Ring-stand clamp. (*f*) Funnel for receiving weights. (*h*) Hook on horizontal bar for holding supply of weights. (*r*) Ring on horizontal bar. (*w*) Water line. (*wt*) Pail holding weights.

graduate the rod in some system of convenient, linear units by cutting shallow rings around the rod with a pipe cutter or some similar instrument. Fasten a metal funnel securely to the top of rod as shown in Fig. 58. Select another piece of rod of similar size and having length greater than the graduated rod; taper one end to a point; by means of an adjustable ring-stand clamp, secure to this rod, and at right angles to it, a section of rod about 1 m. long in the free end of which is a metal ring having a diameter of about 2.4 cm.; insert a hook in opposite end of this transverse rod. In operation, the supporting rod is sunk firmly into the bottom at the point to be tested and in a vertical position; the transverse rod is adjusted to the proper height above the water surface *w* and the pail containing the weights hung on the hook *h;* the graduated rod is thrust vertically through the supporting ring *r* and allowed to come to rest with lower rounded tip in bottom materials. Weights are added to the funnel *f* until the sinking (penetration) is considered to be a sufficient test. For ordinary purposes the results may be expressed in *distance of penetration under the weight used* per unit of time for a rod of stated size. The *weight used* is computed as $wts + (R - h)$ in which wts = the weights added in the funnel; R = weight of the graduated rod; and h = buoyancy of water exerted upon the graduated rod.

GENERAL CONSIDERATIONS

a. If desired, the rods may be made sectional and thus adapted to deeper water. However, the method is obviously limited in general to shallow depths.

b. The supporting rod may be built with a metal tripod foot thus avoiding the necessity of setting the rod into the bottom.

c. A sturdy transit tripod with a large opening through its metal head serves very well as an improvised support for the graduated rod. However, if the buoyant effect of the water on the wood legs of the tripod is too great it will be necessary to load the tripod.

d. Keep rods well coated with oil, if of iron, to prevent rusting. Brass is preferable if it can be provided.

SELECTED REFERENCES

(For Chapter 13)

Brown, C. J. D., and R. C. Ball: "A Rowboat Crane for Hauling Limnological Sampling Apparatus." 3 pp. Spec. Pub., No. 3, Limn. Soc. Am., 1940.

Jenkins, B. M., and C. H. Mortimer: Sampling lake deposits, *Nature,* **142:** 834, 1938.

Kindle, E. M.: "Recent and fossil ripple-mark." 121 pp. Can. Dept. of Mines, Geol. Survey, Mus. Bull. No. 25, Geol. Ser., No. 34, 1917.

Moore, G. M.: A limnological investigation of the microscopic benthic fauna of Douglas Lake, Michigan, *Ecol. Monographs,* **9:** 537–582, 1939.

"The Profitable Use of Testing Sieves," 1940 ed., 51 pp. Cleveland, W. S. Tyler Co.

Twenhofel, W. H., and S. A. Tyler: "Methods of Study of Sediments." 183 pp., 17 figs. New York, McGraw-Hill Book Co., 1941.

Wilson, Ira T.: A new device for sampling lake sediments, *J. Sediment. Petrol.,* **11:** 73–79, 1941.

Part III

CHEMICAL METHODS

In this section, only a limited group of chemical methods is presented. The methods chosen are those needed for an initial entry into limnological work; also those commonly used in routine limnological procedures. Inclusion of other chemical methods is made unnecessary by the existence of an authoritative and readily available work, the well-known "Standard Methods for the Examination of Water and Sewage," 9th ed., 1946, American Public Health Association, which covers the field adequately and extensively. Several other similar works also exist. Hence more extensive consideration of chemical methods here would merely result in unjustified repetition. The less experienced worker in limnology should be warned that, while "Standard Methods" mentioned above is a leader among works of its kind, it is written for the professional chemist with consequent omission of much detailed and elementary material.

SELECTED CHEMICAL METHODS

Samplers

Scrupulous care must be exercised in the choice of means whereby samples of water for chemical analyses are secured since the water (a) must be taken at a positively known depth, and (b) must be brought to the surface in strictly unmodified condition. Any sampler used for this purpose must, if properly operated, guarantee these results, otherwise serious error will result. In the past, many samplers of various designs have been constructed, some of which are still used with satisfactory results.

KEMMERER WATER SAMPLER

The modified Kemmerer water sampler, illustrated in Fig. 59, is now so generally used that it has virtually become a standard instrument. When in good condition and properly used, it meets the two general requirements mentioned above. In addition it has the virtues of simplicity of construction, convenience, small bulk, slight weight, and modest cost. All parts save the rubber valves are made of brass and the number of working parts is very small. It is commonly constructed in three different sizes, namely, 1200-, 2000-, and 3000-cc. capacities, the first being best suited for general limnological purposes. With the exception of differences in the tripping mechanism, improvements made upon the original Kemmerer design have now produced an instrument of one standard type. In operation, the open sampler is lowered on a graduated rope to the desired depth. A messenger is sent down the rope causing the release and closure of both the upper and lower valves. These valves when seated are watertight; furthermore, the instrument is so constructed that when closed the entire weight of sampler and contents is carried upon the lower valve, further insuring its complete closure. In spite of the fact that the instrument does cause a slight disturbance of the water as it sinks, practical experience has shown that such influence is negligible.

General Considerations

a. The chief disadvantage of the Kemmerer sampler is its relatively small capacity thus requiring repeated trips to the selected depth whenever large quantities of water are desired.

b. When in storage the sampler should be hung from a hook in vertical

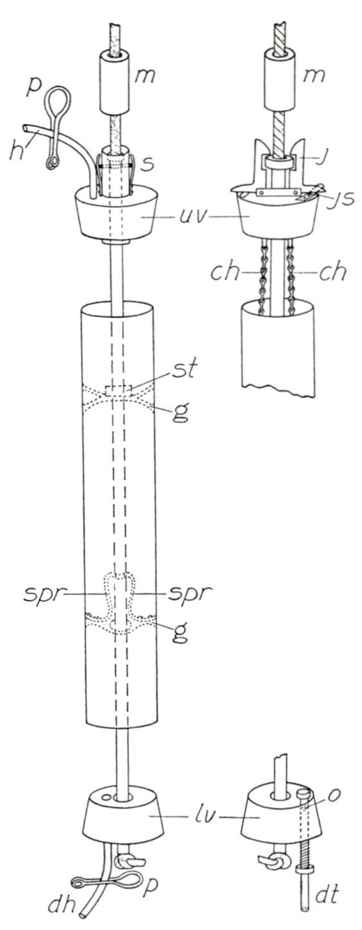

Fig. 59. Diagram showing structural features of modified Kemmerer sampler. (*Left*) View of complete sampler with valves open. (*Top right*) Another type of construction of upper valve and tripping device. (*Bottom right*) Another type of construction of lower valve and drain tube. (*ch*) Chain which anchors upper valve to upper interior guide. (*dh*) Rubber drain tube. (*dt*) Brass drain tube. (*g*) Interior guide fastened to inner surface of body of sampler. (*h*) Rubber tube. (*j*) Jaw of release. (*js*) Jaw spring. (*lv*) Lower valve. (*m*) Messenger. (*o*) Opening into interior of drain tube. (*p*) Pinch cock. (*s*) Upper release spring operating on horizontal pin one end of which fits into groove on central rod. (*spr*) Spring fastened to lower internal guide and operating in groove on central rod to provide lower release. (*st*) Stop on central rod. (*uv*) Upper valve.

position with the valves open. Samplers stored horizontally are more subject to accidents, and storage in the closed form may injure the seating surfaces of the valves.

c. This instrument must be protected against blows on the ends of the brass cylinder which may cause them to be knocked out-of-round. This is a common cause of sampler leakage and leakage leads to serious errors.

d. Leakage may also arise from a bent central rod. When this rod is bent, it is usually best to replace it with a new one.

e. It is necessary to replace the large rubber-stopper valves whenever they become worn or hardened from aging. As these samplers are now made this is not a difficult operation.

f. In the field such samplers should be carried in suitable field kits in order to avoid accidents.

g. For special purposes these samplers can be made in forms either shorter or longer than the conventional ones.

h. One type of tripping device (Fig. 59, *left*) is of such a construction that when all springs are carefully adjusted no messenger is required but the instrument may be closed at the selected depth by giving a sharp, sudden jerk on the rope. Such an arrangement may be convenient particularly in the shallower waters.

PUMP AND HOSE METHOD

Samples of water from various selected depths and satisfactory for chemical analyses may be secured by the pump and hose method provided certain conditions are fulfilled and the limitations are understood. This method requires the use of a suitable pump of the type which, when fully primed, has no leaks and contains no air spaces. Such a pump may be any form consistent with the demands of the particular kind of sampling desired. Suitable pumps vary from the small plunger pumps which are held in the hand, through medium sized "clock" pumps, to the larger power-driven types. The hose must be appropriate to the size of the pump used, must be free from leaks, and should be graduated, if inelastic, to facilitate depth readings. It may be necessary to weight the lowermost end of the hose in order to insure that it descends vertically through the water.

In addition to cumbersomeness, weight, difficulties of handling, and cost, the pump and hose method has one fundamental weakness, namely, it may fail to meet the requirement of sampling water from one narrow, definitely known stratum. When drawn into the open lowermost end of the hose, the water moves into that region from all directions and continues to do so until pumping ceases. There is no way of knowing from what distances, radially from the end of the hose, the water has come. This might not appear serious if only small quantities of water are needed but it must be remembered that before any samples can be taken pumping must proceed until the whole system is filled with the water from the selected level; also that in taking series of samples at successively lower depths the water already filling the system must all be pumped out and the new water take its place whenever the hose is lowered to the next sampling depth. These pumpings necessitate the removal of a considerable quantity of water from each depth. How much water has come from other positions and levels in the sphere of inflowage about the end of the hose is uncertain. If sampling from a wider stratum presents no difficulty the method will be adequate.

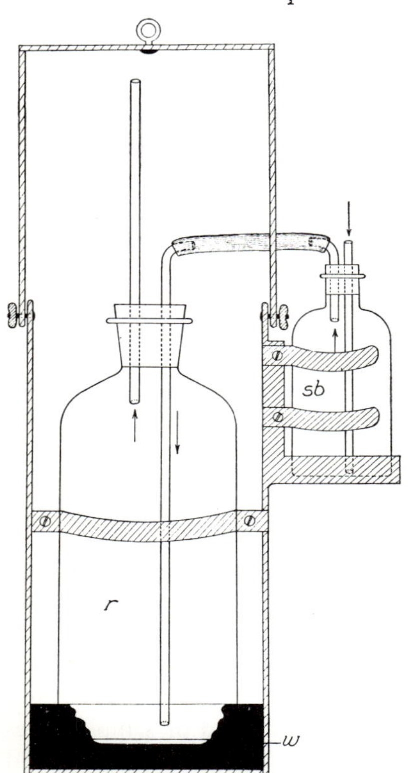

FIG. 60. Side view of Hale's water sampler. (*r*) Reservoir bottle. (*sb*) Sample bottle. (*w*) Weight at bottom, shown partly cut away in front. Arrows indicate direction of flow through sampler when lowered into water.

HALE'S WATER SAMPLER

In case it is necessary to improvise means of securing samples for chemical analyses, Hale's water sampler can easily be assembled from ordinary laboratory materials and, if its limitations are understood and if it is operated properly, samples suitable for ordinary purposes may be secured. Fig. 60 indicates the general features of construction. The reservoir bottle (*r*) must have a capacity several times (preferably six or more) that of the sample bottle (*sb*). The top of the outlet tube in the reservoir bottle must be at a higher level than the top of the inlet tube in the sample bottle —a condition which insures that the water will flow through the system in the proper direction. At the bottom of the frame which carries the reservoir bottle is a base of lead (*w*) or some other suitable heavy substance, shown partly cut away in Fig. 60, which facilitates the rapid sinking of the sampler to any desired depth.

When this sampler is submerged, water enters the intake tube of the sample bottle, filling it from the bottom and displacing the air; then water passes over into the reservoir bottle until it also is completely filled. In this process the sample bottle has been overflowed several times and, at the cessation of flow, contains the last portion of water to enter the apparatus, thus providing a good sample. Care must be taken to insure that the apparatus remains at the sampling position long enough to completely fill the reservoir bottle, otherwise flowage will continue while the outfit is being brought to the surface with resultant entry of water from other levels. The time interval necessary for the filling of any apparatus of this kind should be determined in a tank in the laboratory before it is used in the field. To the interval so determined should be added a certain amount of additional time as a margin of safety.

Because of the construction of the apparatus, water begins to enter the system as soon as it is lowered below the surface. If samples are being taken at considerable depths, it is necessary to lower the outfit as fast as possible so that the amount of water entering at different positions above the desired sampling level will be minimal. However, since the sample bottle is overflowed several times by water entering from the proper sampling level, that which entered on the way down is displaced in the first overflow and in no way affects the final sample. Once the system is completely filled, no water can enter while the apparatus is being brought to the surface, thus no error is introduced on the return trip.

The sample bottle must be one of the regular sample types selected for water analyses so that when detached from the Hale apparatus the analysis can proceed without any transfer of the water to another container. A narrow-neck bottle with about 250-cc. capacity and with ground-glass stopper is a good type since, on detachment from the frame of the apparatus and the necessary withdrawal of the first stopper and its glass tubes, the water level in the bottle is such that the glass stopper can be dropped into place with a small overflow and no inclusion of air bubbles. In rapid transfer of the two stoppers momentary exposure of the small amount of water surface in the neck of the bottle leads to no appreciable error.

As usually set up, the Hale apparatus carries but one sample bottle. However, if it is desirable to secure more than one sample from the same level and at the same time, it is possible to add a limited number of other bottles by providing a much larger reservoir bottle, and by supplying the necessary additional supports. The sample bottles must be coupled *in series*, so that the water flows through all of them before entering the reservoir bottle. Additional time must be allowed for completing the filling at the sampling position. This multiple arrangement must be tried out carefully in the laboratory before taking it into the field. It may be necessary to increase the difference in height level between the intake opening of the first sample bottle and the top of the outlet tube in the reservoir bottle in order to provide enough difference in hydrostatic pressure to insure prompt and certain flowage through the series of bottles.

General Considerations

a. Inside surfaces of bottles, tubes and stoppers must be of such character that there is no opportunity for air bubbles to become trapped.

b. Sample bottles of smaller capacity may be used if the occasion warrants.

c. In order to avoid loss of time due to breakage of glass parts, extra glass tubing, rubber tubing, and a file should be carried in the field.

d. Proper time allowance for filling at the sampling level is of prime importance. Since the process of filling cannot be observed by the operator, except in very shallow water, the only certainty that the sample will be trustworthy is that which comes from leaving the apparatus in the

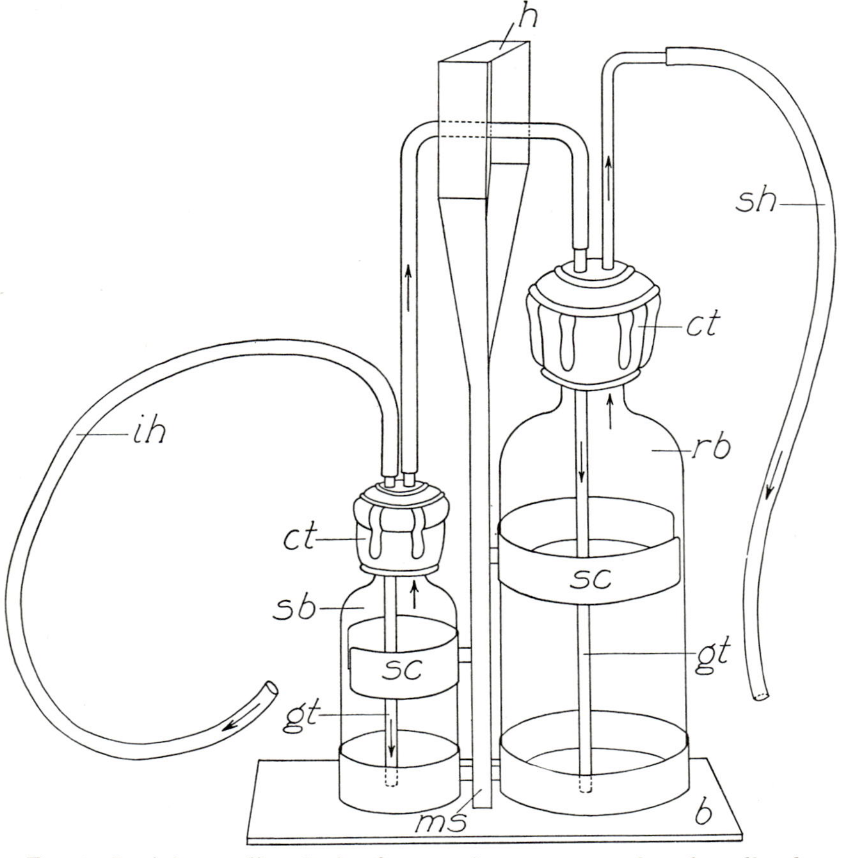

FIG. 61. Irwin's sampling device for securing water samples of small volume. (*b*) Base. (*ct*) Rubber crutch tip. (*gt*) Glass tube. (*h*) Handle. (*ih*) Intake hose (*ms*) Metal support. (*rb*) Reservoir bottle. (*sb*) Sample bottle. (*sc*) Spring clamp. (*sh*) Suction hose.

selected position for the predetermined filling interval plus a margin of safety. In case of doubt it is better to double the filling interval and thus be assured that the sample is safe.

IRWIN'S SAMPLING DEVICE

The usual kinds of equipment cannot be used for securing water samples from very small, shallow pools. Irwin (1942) devised means of securing samples from such pools, especially those containing large quantities of debris. Fig. 61 indicates the principal features of its construction. It employs the main features of Hale's sampling apparatus (p. 202) but is adapted for securing samples of small size as is demanded in situations where most or all of the exposed water must be taken in order to secure a usable sample. An intake hose is fitted to the smaller bottle and a suction hose to the larger one. When ready for use, the free end of the intake hose is placed in the water; then by taking the free end of the suction hose in the mouth and withdrawing air both bottles are filled. The sampling bottle (smaller one) is overflowed at least six times its own volume.

In order to make it possible to obtain usable samples from water containing much trash, Irwin devised a special strainer shown in Fig. 62. This device consists of a copper cylinder 5 cm. deep and 13 cm. in diameter. It is open at the top and covered on the bottom with fine-mesh copper screen except for the central region which is occupied by a shallow, cylindrical cup 8 mm. deep and 5 cm. in diameter. In the vertical sides of the main cylinder several circular windows are cut out and covered with fine-mesh

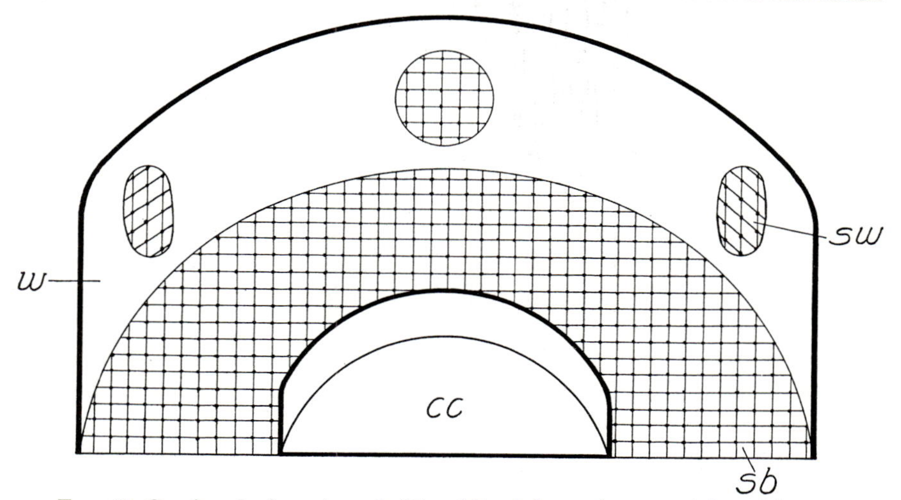

Fig. 62. Sectional view (one-half) of Irwin's strainer used in taking water samples from small shallow pools containing debris. (*cc*) Central cup. (*sb*) Screen bottom. (*sw*) Screen window in side wall. (*w*) Side wall of copper cylinder.

copper screen. When in use, the strainer is placed carefully into the water to be sampled, screen side down, and submerged until outside water surface is above the level of the windows. The water enters through the screened bottom and windows and slowly fills, with minimal agitation, the central cup. The end of the intake hose is held directly above the central cup while taking a sample. The size of the mesh used in the screen is determined by the character of the debris to be excluded and the operator must modify the apparatus to meet his particular needs.

Dissolved Oxygen

SAMPLING

The mode of collection of samples for dissolved oxygen analyses must be chosen in advance on the basis of the necessities of the situation. All subsurface samples should be taken with a sampler which will meet the requirements stated on p. 199. The sampler used must be of such construction that its capacity exceeds several times that of each of the individual sampling bottles.

In multidepression lakes it is evident that the dissolved oxygen picture on any one date cannot be determined from a single vertical series of samples taken only in one of the submerged depressions.

Sampling Bottles. For ordinary field purposes, a good grade, white-glass, narrow-neck, ground-glass stoppered bottle having a capacity of about 250 cc. is recommended. Each bottle should be numbered in some permanent and easily readable way and in some convenient numerical sequence. Each bottle should bear its number on two different places to insure against loss of identity in case one number is accidentally destroyed. A number on the side of the bottle and also on the top of the glass stopper provides satisfactory identification. Glass stoppers must be ground to fit the bottles; hence the necessity of guarding against the mixing of stoppers and bottles. Numbers may be pasted on the glass and then sealed over with two or three coats of shellac. Numbers may also be made as follows: cut out squares of heavy canvas of uniform, convenient size; write numbers on same using Higgins Waterproof India ink or Higgins Eternal ink; through small hole in each square fasten tie-on string of strong cord; put squares and tie-on strings into melted paraffin and keep there until thoroughly permeated; attach label to sampling bottles by tying to neck of bottle. Metal tags should be avoided.

For field transportation and for storage, sampling bottles should be kept numerically arranged in field kits made for that purpose. At the end of each analysis all bottles must be thoroughly rinsed and made ready for the next series.

Field Procedure. When a sample is delivered at the surface by a

Kemmerer sampler, the lower delivery tube of the latter is inserted to the bottom of the sampling bottle and the water allowed to flow continuously until the sampling bottle has been overflowed at least three times its total capacity and all bubbles are eliminated. This threefold overflow insures that the first water entering the bottle and making contact with the air originally contained in it has been removed, all of the air has been displaced, and the final bottle of water is truly an unmodified sample. Satisfactory control of the overflowing may be provided as follows: Determine the length of time required to fill the empty bottle the first time by counting (ordinarily a count of 10 at the rate of 1 per second of time will suffice); then continue overflowing for a count three times the number so determined. The delivery tube of the sampler is gently withdrawn at the end of sampling so that before the flow ceases, the bottle is filled to the top of the neck. The glass stopper is immediately dropped into the neck of the bottle in such a way that it displaces just enough water to enable the stopper to seat itself completely but leaving no air bubbles. With a little practice this trick of dropping the stopper into place without the inclusion of bubbles can be easily acquired.

WINKLER METHOD

The Winkler method, or some modification of it, has become the standard means of determining dissolved oxygen and is more widely used in limnology than any of the other methods. For most purposes it is convenient and sufficiently accurate. Various other methods have been devised some of which are by their very nature strictly laboratory procedures and scarcely adaptable to field operations.

The unmodified Winkler method, to be described first, may be used for waters known in advance to have a very low content of iron, nitrites, or organic matter. For the purer waters (cleaner unmodified water of lakes and streams) it is usually adequate and is simpler and shorter than the Rideal-Stewart modification described later.

Reagents

1. Manganous sulfate: 480 g. $MnSO_4 \cdot 4H_2O$, or 400 g. $MnSO_4 \cdot 2H_2O$ dissolved in distilled water; make up to 1 l.

2. Alkaline-iodide: 500 g. NaOH (or 700 g. KOH) and 135 g. NaI (or 150 g. KI) in distilled water; make up to 1 l.

3. Concentrated sulfuric acid: sp. gr. 1.83–1.84.

4. Sodium thiosulfate: 6.205 $Na_2S_2O_3$ in freshly boiled distilled water; make up to 1 l. This produces a $N/40$ solution. Add 5 cc. chloroform. Make up fresh solution every 2–3 weeks. Standardize occasionally according to directions given below.

5. Starch solution: 5 g. of potato starch, ground in mortar with small

amount of distilled water: pour into 1 l. of boiling distilled water, stir and let settle for several hours; use clear supernatant liquid; add 1.25 g. salicylic acid per liter as preservative.

DIRECTIONS FOR STANDARDIZING SODIUM THIOSULFATE SOLUTION. Dissolve 0.2454 g. potassium dichromate ($K_2Cr_2O_7$), which has been dried at a temperature of 130° C. for one-half hour and cooled in a desiccator, in distilled water; make up to 200 cc. This gives $N/40$ $K_2Cr_2O_7$. Put 10 cc. of this solution into a beaker surrounded by an ice bath; add 1 cc. KOH-KI solution (2) drop by drop, rotating beaker at same time. Add 1 cc. concentrated H_2SO_4 in same way. If odor of iodine is detectable, iodine is being lost and the process should be repeated. Iodine will be released by the KI giving solution a deep color. Now titrate with sodium thiosulfate solution (4) until iodine color begins to disappear; then add starch solution as an indicator and continue titration until blue color just disappears. Record number of cc. thiosulfate used; 10 cc. of sodium thiosulfate should exactly neutralize 10 cc. of potassium dichromate. When this condition is met and 200 cc. of water is titrated for dissolved oxygen the quantity of O_2 in cc. per liter is calculated by multiplying number of cc. of sodium thiosulfate used for titration by 0.698. However, if condition described above is not met, then sodium thiosulfate solution may be adjusted to meet condition, or another constant may be calculated from results, for example, if 11 cc. of sodium thiosulfate is required to neutralize 10 cc. of dichromate solution, then new constant will be 11/10 times 0.698.

Procedure

The first three steps described below must be performed in the field immediately after the sample has been secured. Any delay may result in error.

1. Remove stopper from 250-cc. sample bottle containing sample to be analyzed and, by means of long narrow volumetric pipet, add 1 cc. of manganous sulfate solution (1) well below surface of water. In like manner, add 1 cc. of KOH-KI solution (2).

2. Replace stopper and mix sample by inverting bottle several times. Allow precipitate to settle for a few minutes. Thorough mixing is required; repeat inversion as safety measure or if doubt exists. After final settling of precipitate, a clear fluid should occupy upper portion of bottle.

3. Using slender volumetric pipet, add 1 cc. concentrated sulfuric acid by permitting it to run down neck of bottle; mix well by inverting bottle several times. Allow sample to stand for at least 5–10 min.

After the acid has been added and the stopper of the sample bottle replaced, concluding steps of the analysis may be delayed, if necessary, for several hours without appreciable change. This makes it possible to avoid titrations on board a boat.

4. Transfer 200 cc. of sample to porcelain evaporating dish and titrate rapidly with $N/40$ sodium thiosulfate solution until iodine color in sample has been reduced to pale straw color; then add a few cc. of starch solution and continue titration rapidly but cautiously until blue color *first disappears*. Discontinue titration at this end point. Ignore any return of blue color.

If very precise results are desired, a correction for the loss due to displacement by the reagents may be made as follows: When 2 cc. of reagents (manganous sulfate and KOH-KI) are added to a 250-cc. sample the amount transferred for titration should be $(200 \times 250) \div (250 - 2) = 201.6$ cc. if the amount of the original sample titrated is to be 200 cc. However, this correction is not justified in most kinds of limnological work.

CALCULATION OF RESULTS. If a 200-cc. sample is titrated, the number of cc. of the sodium thiosulfate solution used is numerically equal to the dissolved oxygen content in *parts per million* and no additional calculation is necessary. This value, multiplied by 0.698 (or some corresponding standardization value; see p. 208) will yield the value in terms of *cubic centimeters per liter*. Results in terms of *percentage of saturation* are secured by dividing the titration value in cc. by the solubility value as determined by the temperature of the sample (see Table 8, Appendix).

For altitudes other than sea level, correction for barometric pressure must be made if precise results are desired. This is done by determining the oxygen solubility value at the pressure involved by the use of the following formula:

$$s_1 = s\frac{p}{760} = s\frac{p_1}{29.92}$$

in which
s_1 = solubility at p or p_1
s = solubility at 760 mm. or 29.92 in.
p = barometric pressure in millimeters
p_1 = barometric pressure in inches

For quick reference, Rawson's nomogram (1944) for obtaining oxygen saturation values at different pressures is useful (see Appendix, p. 366).

RIDEAL-STEWART MODIFICATION OF WINKLER METHOD

The Winkler method just described will often yield satisfactory results, but significant errors may occur if the water contains appreciable quantities of nitrites, iron salts, or certain organic compounds. Under these circumstances the Rideal-Stewart modification of the Winkler method should be used. Since in work on unknown waters it may be uncertain whether the Rideal-Stewart modification is necessary, it is common practice to use the latter as a regular procedure on all waters, the only disadvantage being the use of the additional steps in the analyses.

Reagents

(Reagents 1–5 are given under Winkler Method, p. 207.)

6. Potassium permanganate solution: 6.32 g. $KMnO_4$ in distilled water; make up to 1 l.

7. Potassium oxalate: 2 g. in 100 cc. distilled water. (Since this solution may deteriorate rapidly, it should be made up frequently.)

Procedure

1. Remove stopper from the 250-cc. sample bottle containing the sample to be analyzed and by means of a long narrow volumetric pipet add exactly 0.7 cc. of concentrated sulfuric acid (reagent 3, p. 207) just below surface of water.

2. Then with a similar pipet add at once, and well below the surface, enough (usually 1 cc.) of the potassium permanganate solution (reagent 6) to yield a typical permanganate violet color which will remain permanent for at least 20 min. after the bottle has been mixed thoroughly by repeated inverting. If the permanganate color does not persist for 20 min., add a small amount of the permanganate solution; mix by inversions and let stand for suitable interval; continue the same procedure until the violet color persists for the 20-min. interval.

3. Add 0.5 cc. of the potassium oxalate solution (reagent 7); replace the stopper and invert bottle several times; allow to stand for 5 min.; if after this interval the color of permanganate persists, add another 0.5 cc. of the potassium oxalate solution and mix as before.

4. After the permanganate color has disappeared completely, add 1 cc. of manganous sulfate solution (reagent 1, p. 207) and 3 cc. of the alkaline-iodide solution (reagent 2, p. 207); insert stopper and mix by repeated inversion of bottle. Allow the precipitate to settle about halfway in bottle; then mix again and let settle; allow sample to stand for at least 5 min.

5. Then proceed as in the regular Winkler method using steps 3 and 4. The results are calculated as already described (p. 209).

General Considerations

a. In field work, the analyses of the dissolved oxygen should be made immediately after the sample is secured and the reagents added (Winkler: reagents 1–3; Rideal-Stewart modification: reagents 3, 6, 7, 1–3) up to the point where storage and transportation to the laboratory is safe.

b. For dependable work it is necessary that the temperature of the water at each sampling position be measured.

c. If appreciable amounts of iron salts are present, the reduction of the color of the permanganate, even when the proper amounts of oxalate solution are added, may be very slow. To remedy this condition, add 2 cc. of potassium fluoride (40% solution) along with the permanganate and decolorize in the dark.

d. If it is necessary to determine the dissolved oxygen content of waters containing various kinds of industrial wastes, such as sugars, starches, paper-mill wastes, chlorinated wastes, and mine wastes, special correctives and procedures must be added to the methods described, directions for which are available in "Standard Methods for the Examination of Water and Sewage," 9th ed., 1946.

MICRO-WINKLER METHODS

The limnologist is sometimes confronted with situations in which only small samples can be secured, or in which it may be preferable to use only small samples. The standard Winkler method, or the Rideal-Stewart modification, can be adapted for the analyses of samples having a volume of less than the usual 250 cc. by using a sample having a volume representing some convenient divisor of 250 and by the use of the same divisor applied to the quantities of reagents used. For example, if it is desired to determine the dissolved oxygen in a 25-cc. sample, then the quantities of reagents used should be one-tenth of those of the standard method, the only exception being that of the sodium thiosulfate in which instance convenience and accuracy demand that it be considerably diluted to a known degree. Samples as small as 10 cc. and even 5 cc. may be so analyzed. Obviously, the pipets and burets used must be graduated finely enough to make possible the precision necessary to maintain the original accuracy of the standard Winkler method. Addition of reagents may be satisfactorily accomplished by the use of microburets drawn to long capillary points. In the final computation of results, proper allowance must be made for the dilution of the sodium thiosulfate. A dilution of 10 times the original volume is convenient and usually adequate although greater dilution can be used if necessary. Because of the small size of the sample it is advisable to make the final titration in a tall dish and by means of a buret graduated to 0.05 cc. With adequately graduated glassware and with proper care of operation, results are obtainable which are as dependable as are those of the standard method.

Of recent years various mechanical devices have been developed by means of which the micro-Winkler method is rendered more convenient and precise (Thompson and Miller, 1928; Fox and Wingfield, 1938; and others). With certain of these devices it is claimed that samples as small as 0.2 cc. can be analyzed. Analysis of such small samples requires great care in all steps of the procedure.

COLORIMETRIC METHODS

Colorimetric methods for determining oxygen dissolved in water were devised more than 40 years ago. Recently, improved methods have been developed some of which may come to have limnological value. Among them should be mentioned those of Johnson and Whitney (1939) and Isaacs (1935). While these methods appear to have some promise for certain purposes, they cannot be regarded at this time as a substitute for the established ones described on earlier pages.

OXYGEN DEFICIT

By oxygen deficit is meant that amount of dissolved oxygen necessary to raise the quantity already present at a selected depth to some chosen standard. Two methods of computing oxygen deficits have come into use. *Actual deficit* is the difference between the quantity of oxygen present at the selected depth and the amount necessary to saturate the water at that temperature which existed in the sample at the time when it was collected. *Absolute deficit* is the difference between the amount of oxygen present at the selected depth and that necessary to saturate the water at a temperature of 4° C. (9.26 cc. per l., minus a correction for the elevation of the lake surface above sea level).

Actual deficit may be computed by means of the following formula:

$$\frac{(O^{as} - O^{a}) + (O^{bs} - O^{b})}{2} = d^{ab}$$

$$d^{ab}V = D^{ab}$$

$$D^{ab} + D^{bc} + D^{cd} + D^{de} \text{---------} D^{xy} = \text{Total actual deficit}$$

in which

O^{as} = amount of dissolved oxygen which surface water of lake would have if saturated with oxygen at temperature prevailing at time of sampling

O^{a} = amount of dissolved oxygen actually present in sample of surface water

O^{bs} = amount of dissolved oxygen which lake water would have at first depth level in vertical sampling series if saturated with oxygen at temperature observed in sample

O^{b} = amount of dissolved oxygen actually present in sample of water from first depth level

d^{ab} = mean oxygen deficit in uppermost stratum of lake

V = volume of uppermost stratum of lake

D^{ab} = total actual oxygen deficit in uppermost stratum of lake

D^{bc}; D^{cd}; D^{de} = actual oxygen deficits in successive strata of lake

D^{xy} = actual oxygen deficit for final and lowermost stratum of lake

If O_2 is expressed in cc. per liter or in mg. per liter, then V must be in terms of liters. Since results so computed may involve very large numbers, it will be more convenient to express the final results in larger units, as for example, in kilograms or kiloliters.

The quantity of oxygen necessary to saturate the water at the different temperatures can be taken from a table of saturation values. Correction for barometric pressure may be made by the use of the formula

$$S^1 = S \frac{B}{760}$$

in which

S^1 = solubility of O_2 at the sampling station
S = solubility of O_2 at sea level (760 mm.)
B = barometric pressure in millimeters

The data referred to in the general formula given above make possible the computation of (a) the total amount of oxygen present in the lake at the time of sampling, and (b) the actual total oxygen deficit.

If the volume of the lake and that of its various strata are not known, less comprehensive measure of the actual O_2 deficit can be secured by merely using the vertical series of O_2 analyses made in the deepest region of the lake.

Absolute deficit may be determined by the use of the general formula given above for *actual deficit* but modified by substituting the amount of O_2 necessary to saturate the water at a temperature of 4° C. instead of that for the observed temperature at the level involved. Correction for elevation is computed as described above, using 4° C. as the standard. Certain other methods of computing oxygen deficits have been proposed (Hutchinson, 1938). *Oxygen gradient* (Maucha, 1931) may be computed but the method is more complicated and adds little to the information secured by means of oxygen deficit determinations.

Free Carbon Dioxide

Samples for analysis of free CO_2 in water should be collected in the same way and with all of the same precautions as those for dissolved oxygen (p. 206). Since free CO_2 escapes from water so readily, it is highly desirable that the analysis be made immediately after the sample is secured.

REAGENTS

1. Phenolphthalein indicator: Dissolve 5 g. of high quality phenolphthalein in 1 l. of 50 per cent alcohol; neutralize with $N/50$ sodium hydroxide. (Prepare the 1 l. of 50 per cent alcohol by diluting 526 cc. of 95 per cent grain alcohol to 1 l. with boiled distilled water.)

2. Sodium hydroxide: *N*/44 solution. (Limnologists will probably prefer to have this standard solution made up in some properly equipped chemical laboratory; if not, detailed directions can be found in treatises on chemistry or on water analysis.)

Procedure

1. Pour sample into a Nessler tube until 100-cc. mark is reached. In so doing, use every care that sample is not agitated. If a Kemmerer sampler is used, insert delivery tube of sampler deep into Nessler tube and admit water quietly at bottom of tube.
2. Add 10 drops of phenolphthalein indicator.
3. Add *N*/44 sodium hydroxide from buret, stirring gently during titration, until faint permanent pink color appears.

The amount of free carbon dioxide, expressed in parts per million, is calculated by multiplying by 10 the number of cc. of *N*/44 sodium hydroxide used in the titration.

General Considerations

Recently, Peters, Williams and Mitchell (1940) claim to have shown that in a titration method using phenolphthalein and 0.1 *N* sodium hydroxide a sizable error results if the end point used is the "permanent" pink, an error which is eliminated by using as the end point a "lingering pink" which flashes all through the liquid and lasts for but one or two seconds.

ALKALINITY

Alkalinity in most natural waters is due to (a) normal carbonate, (b) bicarbonate, and (c) hydroxides. By the use of two indicators and a standard solution of a strong acid these three sources of alkalinity can be distinguished and their quantities measured.

REAGENTS

1. Sulfuric acid, *N*/50 solution. (This solution must be made up according to standard specifications. Limnologists will probably prefer to secure this solution from some chemical laboratory; if not, directions for the proper preparation of this solution will be found in standard works on chemistry or on water analysis.)
2. Phenolphthalein. Prepare this solution according to same directions as given on p. 213 for phenolphthalein indicator in determination of free carbon dioxide.
3. Methyl orange indicator. Dissolve 0.5 g. of high quality methyl orange in 1 l. of distilled water. Keep in dark glass bottles.

PROCEDURE FOR DETERMINING PHENOLPHTHALEIN ALKALINITY

1. Deliver 100 cc. of sample into white porcelain casserole or into white porcelain evaporating dish, or into Erlenmeyer flask set on white background.

2. Add 4 drops of phenolphthalein indicator. If color appears in solution, hydroxide or normal carbonate is present.

3. If sample becomes pink, add *N*/50 sulfuric acid from buret until pink color just disappears; record number of cubic centimeters of acid used.

The phenolphthalein alkalinity, expressed in parts per million of calcium carbonate, is equal to the number of cc. of the *N*/50 sulfuric acid used multiplied by 10.

PROCEDURE FOR DETERMINING METHYL ORANGE ALKALINITY

1. Prepare sample as described above in directions for determining phenolphthalein alkalinity; or, if desired, use sample to which phenolphthalein has been added.

2. Add 2 drops of methyl orange indicator to sample. If yellow color is produced, hydroxide, normal carbonate, or bicarbonate is present.

3. Add *N*/50 sulfuric acid from buret until color shows first change from pure yellow; end point is a faint orange tint; record number of cc. of acid used.

The *methyl orange alkalinity*, expressed in parts per million of calcium carbonate, is equal to the total number of cc. of *N*/50 sulfuric acid used multiplied by 10.

INTERPRETATION AND CALCULATION OF RESULTS

It has become common practice to record results of alkalinity measurements as *p. p. m. phenolphthalein alkalinity* or as *p. p. m. methyl orange alkalinity*, since from these concise statements various interpretations and

Result of Titration	*Alkalinities Expressed as P. P. M. of Calcium Carbonate*		
	Hydroxide	*Carbonate*	*Bicarbonate*
$P = 0$	0	0	$T \times 10$
$P < \frac{1}{2}T$	0	$2P \times 10$	$(T - 2P) \times 10$
$P = \frac{1}{2}T$	0	$2P \times 10$	0
$P > \frac{1}{2}T$	$(2P - T) \times 10$	$2(T - P) \times 10$	0
$P = T$	$T \times 10$	0	0

Modified from "Standard Methods for the Examination of Water and Sewage." Reprinted with permission of the American Public Health Association.

calculations may be made when desired. Further differentiations of alkalinities due to hydroxide (OH), normal carbonate (CO_3) and bicarbonate (HCO_3) as made possible by the titrations indicated above are shown in the table on page 215 of the five possible conditions which may be met in results of measurements of phenolphthalein and methyl orange alkalinities. P = number of cc. *N*/50 sulfuric acid used in titration with phenolphthalein; T = number of cc. *N*/50 sulfuric acid used in total titration (phenolphthalein plus methyl orange).

GENERAL CONSIDERATIONS

a. Greater precision in the use of the method is attained if the operator uses some means of judging the end point more accurately. Standards may be provided by using two standard buffer solutions one having a pH of 8.0 and another one having a pH value of 4.0. Phenolphthalein indicator is added to the first; methyl orange indicator to the other. Each solution so prepared will give for the indicator used, the exact color tint to be watched for in the titration.

b. If defective color vision interferes with the ability of an operator to use the method as outlined above, certain other indicators may be used as substitutes for phenolphthalein and methyl orange provided the substitute indicators have other colors and have turning points at pH 8.0 and pH 4.0 respectively. If such a substitution is made the results are expressed in parts per million of calcium carbonate and as alkalinities using the substitute names.

Hydrogen-ion Concentration

A general discussion of fundamental aspects of hydrogen-ion concentration may be found in various treatises on that subject. A brief account of the essential features occurs in the author's book, *Limnology* (1935, pp. 108–116). Two methods of expressing hydrogen-ion concentration are in use, namely, (a) the number of moles of ionized hydrogen per liter, and (b) the pH scale, which may be defined as the "logarithm of the reciprocal of the normality of free hydrogen ions." Because of its convenience, the pH scale is now very generally used. In this book hydrogen-ion concentration is expressed in the pH scale.

The wide use of hydrogen-ion concentration measurements in various fields of science has led to the development of many different methods for making such determinations. A great array of materials, indicators, standard colors, and instruments is now available on the market accompanied by detailed descriptions of construction and directions for use. Consequently, it is scarcely worth while for the limnologist to go through the elaborate and tedious processes of making up his own standards or constructing his own apparatus, even if his laboratory is equipped for such work. Costs are such that ordinarily it is preferable to purchase suitable outfits from reliable manufacturers. Because of the highly satisfactory directions which accompany modern hydrogen-ion concentration equipment, only a general discussion will be given here.

Methods of measuring hydrogen-ion concentration are, in general, of two kinds: (a) colorimetric and (b) electrometric. The procedures and equipment for the operation of both are numerous and varied. Of these two general types, the electrometric method has greater precision.

SELECTION OF A METHOD

The selection of a method for measurement of hydrogen-ion concentration depends upon various circumstances prominent among which are (a) physical features of the water, (b) chemical features of the water, and (c) degree of accuracy required. Owing to the fact that all methods have their own sources of error and their own limitations it is highly desirable that, if possible, every working laboratory should possess more than one kind of outfit in order that one may be checked against the other. The choice of a method for a particular job will depend upon the nature of the waters involved. Likewise, the degree of accuracy required will depend upon the use to which the data are to be put. There is no profit in working for a much higher degree of precision than the occasion warrants and the intelligent worker will make the proper discrimination. However, once the necessary degree of accuracy has been correctly ascertained, it is imperative that the method selected be conditioned and operated so that this accuracy is assured in the results. In the final expression of results, all pertinent facts as to method, operation, and accuracy attained, should be recorded for future reference.

COLORIMETRIC METHODS

Colorimetric methods depend upon the addition of the proper sensitive indicator solution to the sample and a comparison of the color so produced with graded, colored standards the pH values of which are known.

Color Charts

Sets of graded colors representing ranges of pH values are sometimes used as standards. Such colors are usually printed on paper in some convenient form. When known to be in dependable condition, fair results may be obtained if all of the limitations and weaknesses of the method are known and the necessary precautions are observed. However, about the only advantages such standards have over better ones are general convenience and low cost. On most occasions more exact and more dependable standards, such as those mentioned later, will be chosen. The following are among the principal weaknesses: (a) instability of colors; (b) flat surface; (c) danger of damage by wetting.

General Considerations

a. Color charts must be protected from exposure to light except when in actual use.

b. Color charts must be checked frequently against some dependable set of standards.

c. High precision cannot be expected from this method.

d. Color charts may be reëvaluated, if changes in color quality have occurred, by checking them against a suitable standard. However, it is usually better to discard color charts which show appreciable change.

e. Since colors on charts cannot be tested directly by potentiometrical methods, they are of necessity only secondary standards made by matching them as closely as possible against some basic standards.

f. One prominent source of error arises from the fact that when a liquid sample is compared with a color standard of an entirely different material the difference in color transmission may be considerable.

Solution Standards

One of the most widely used forms of colorimeter, and in many respects the most satisfactory, is that in which the standards are contained in a series of uniform, hermetically sealed tubes of special nonsoluble clear glass, uniform in bore and thickness. Each tube contains a definite amount of a standardized buffer solution having the desired pH value, to which is added an amount of indicator solution suitable to portray the pH value of the buffer. A preservative is added to prevent growth of mold. All operations are conducted under sterile conditions. The solution is then checked electrometrically, and the tube sealed and labeled with the name of the indicator and the pH value. Such tubes present a graded series of colors each of which represents a particular pH value. Such a series may extend throughout the whole pH range or any part of it and may have whatever pH intervals between tubes the operator desires. For ordinary purposes the interval used is 0.2 pH. These solution standards have two important virtues, namely, (a) they can be checked electrometrically and are thus basic colorimetric standards, and (b) the standard is in all respects similar to the unknown to be tested since the latter is contained in the same kind of a tube, has the same quantity of liquid, and has received the same indicator solution.

General Considerations

a. Solution standards are not permanent. Certain commercial companies guarantee colorimeters for one year against color change. Some colorimeters may remain in good condition much longer than one year.

However, they should be examined frequently, even from the date of original construction, and checked against other standards. Such checks must include every tube in the set since some tubes may change faster than others.

b. Tubes which have undergone change in color may be reëvaluated by careful comparison with other standards, if most of the color is still present. However, reëvaluated tubes must be checked very frequently. If much color change has occurred, a new set is recommended.

c. When not in use, color tubes should be kept in the dark.

d. Color tubes should not be exposed to excessive heat.

e. In the past, some workers have made use of temporary solution color tubes made up rapidly with buffer solution and indicators in test tubes closed with stoppers. Such colorimeters are less expensive but have very limited life, color changes occur very quickly, and a new colorimeter must be made up at least every few days. Ordinarily, such colorimeters are not to be recommended.

f. Solution color tubes should be dated when first received. Some companies date their tubes when they are made.

Colored-glass Standards

Colorimeters are now available in which the standards are in the form of glass units whose colors represent the various pH values in a graded series. The glass standards may be of various forms. A common form is one in which colored glass disks are mounted in circular holders of convenient dimensions. Makers usually claim that such color standards are permanent and that each color has been carefully checked electrometrically. It must be understood, however, that such checking is of necessity indirect and can be made only by comparing the color in the glass with a liquid standard which has been checked electrometrically. In this respect these glass standards resemble color charts mentioned previously. Permanence of color and convenience are prominent virtues of glass standards, but like color charts they present the problem of comparing accurately the liquid sample containing an indicator with a colored standard composed of an entirely different material.

General Considerations

a. Glass standards must be thoroughly clean before being put into use. Since some color disks are composed of several layers of glass sealed together, any cleaner which will attack the sealing materials must be avoided.

b. If dropped on a hard surface, damage to glass standards may occur either as breakage or as a shattering of the sealing materials.

c. It is necessary to have definite information from the maker as to the suitability of a glass-standard colorimeter for use in either daylight or in artificial light; also concerning any accessories necessary to insure as accurate results as possible.

Comparators

Devices known as comparators are commonly used in connection with colorimeters as means of increasing the accuracy and dependability of readings. These aids to operation vary from the simple "block" comparators to elaborate electrical devices equipped with artificial lights and glass screens. Certain comparators and lamps make possible dependable readings at night or in places where ordinary daylight is variable and uncertain. Properly chosen comparators are desirable for many kinds of colorimetric work and, in some instances, higher precision and constancy of reading values can be attained only through them. Ordinarily the worker will do well to follow the advice of the manufacturers in the selection of a comparator for his particular kinds of work.

pH Test Papers

Test papers, prepared from standardized pH indicators, are now available. These sensitive test papers are useful only in the *approximate* estimation of the pH of solutions and for such purposes may sometimes be useful. However, their limitations must be known and their use governed accordingly.

Indicators

Colorimetric measures of pH depend upon the use of dyes which manifest a definite but different color for each particular pH value of the medium into which they are introduced. These differences in color of the indicators is an expression of the differences in hydrogen-ion concentration. Color change by an indicator extends over a certain range of change of pH in the medium; therefore, each indicator has a certain definite pH range beyond which, in either direction, color change is no longer usable. This means that indicators must be chosen which are suitable for the pH range of the samples. There are many different indicators on the market from which the worker will choose those most appropriate for his needs. The following selected list includes indicators serviceable for ordinary purposes:

Indicator	pH Range	Color Change
Bromphenol blue	3.0–4.6	yellow-blue
Bromcresol green	4.0–5.6	yellow-blue
Chlorphenol red	5.2–6.8	yellow-red
Bromthymol blue	6.0–7.6	yellow-blue
Phenol red	6.8–8.4	yellow-red
Thymol blue	8.0–9.6	yellow-blue

Wide-range indicators, sometimes called universal indicators, have a range of color change which is spread over most of the pH scale. For example, certain wide range indicators are claimed to have a pH range of 1.2 to 11.0. These indicators are most serviceable in the preliminary testing of an unknown where the approximate results secured are useful in the subsequent selection of a proper specific indicator for determining the pH value with greater accuracy. Ordinarily these indicators are not used for final results unless only a very rough estimation is sufficient.

Indicators may be secured either in crystalline form or as indicator solutions so prepared that they are available for use at once. Ordinarily, the limnologist will find it convenient and an economy of time to use the indicator solutions already prepared by a reliable supply company. If made up from the crystalline form, great care must be taken to follow rigidly the instructions for the preparation of solutions as given by the firm supplying them.

Indicator solutions should be stored in the dark and securely protected from dust or any form of contamination.

Suitability of Colorimetric Methods

Under ordinary conditions it may be expected that the better colorimetric methods will give reliable results when used properly on the clearer, cleaner, unmodified inland waters. However, they are uncertain for highly colored bog waters, highly turbid waters, and contaminated waters. Any materials in the sample which obscure the color reaction of indicators or which by chemical action alter the nature of the indicator make a colorimetric method subject to errors. In general the colorimetric method is known to be subject to the following kinds: salt errors, protein errors, acid errors, temperature errors, off-color errors, oxidation/reduction errors and colloidal errors (Perley, 1937). While in many of the unmodified inland waters of the cleaner, purer types some of these errors may not be particularly significant, it behooves the limnologist to be on his guard in applying colorimetric methods, especially to bog waters, salt waters, contaminated waters, and various other less common types having unusual chemical and thermal features. It must be remembered that rise in temperature alone generally produces an increase in ionization. Far too often measurements are made without any effort or provision for checking the reliability of the colorimeter for the particular water concerned. This is unsafe practice and it does not clear the case to assume that the error will necessarily remain the same in the same water. Many waters are subject to wide and significant variations.

ELECTROMETRIC METHODS

Electrometric outfits are of several types of which the following are prominent at present: those using a hydrogen gas electrode, a quinhydrone electrode, an antimony electrode, or a glass electrode. Portable outfits are now available and usually meet the needs of limnologists in a satisfactory way. Such outfits are very diverse in construction although the fundamental features of operation may be similar. In addition to the advantage of possessing a greater precision than colorimetric methods, the electrometric methods are suitable for a much wider range of different types of water, functioning well in colored waters, highly turbid waters, and contaminated waters. Makers of electrometric outfits supply full information concerning the construction and operation of their instruments. Since the various outfits often require different procedures, no attempt will be made here to outline them. For work in the field, outfits with special glass parts are subject to breakage and in that respect may be less preferable than others which use little glass and that easily replaceable.

The table on page 223 indicates some of the features of electrometric outfits which employ one of the four principal types of electrodes.

RESERVE pH

Reserve pH, indicated by the symbol RpH, is that value secured by a second pH determination made after the sample has been thoroughly aerated. Aeration of the sample may or may not result in a value (RpH) different from the initial pH; if the total pH results from the presence of substances in the water not releasable by aeration, then the initial pH value and the RpH value will be the same; if, on the other hand, the total pH is in part due to substances in the water releasable by aeration, then the RpH value will differ from the initial pH. Free CO_2 in the sample will be partly responsible for the initial pH reading; aeration will cause the release of the free CO_2 and the postaeration measurement will show an RpH value which is higher on the alkaline side of the scale.

ELECTROLYTES

Pure water is a poor conductor of electricity. Acids, bases, and salts in solution in water produce solutions which are relatively good conductors of electricity. Such substances are called *electrolytes*. Since acids, bases, and salts differ in their solubility in water, they differ in their conductance of electricity, those with slight solubility being *weak electrolytes*, while those with high solubility are *strong electrolytes*. Therefore, a measure of the total electrolytes in natural waters can be secured by measuring the electrical conductance of a sample of the water. Such a measurement provides a means for studying the nature of a solution. Since it is claimed that, other things being equal, the richer a body of water in electrolytes, the greater the biological productivity, any measurement of the electrolyte content of water is of considerable importance in limnological practice.

SHOWING SELECTED SERIES OF CHARACTERISTICS OF DIFFERENT TYPES OF ELECTRODES USED IN pH MEASUREMENTS*

	Hydrogen Gas Electrode	*Quinhydrone Electrode*	*Antimony Electrode*	*Glass Electrode*
Salt Errors	None	In salt concentration above 1 molar	Considerable magnitude	In presence of sodium salts at ranges above pH 9.5
Limit of Error	pH 0.01	0.02 below pH 8.5	pH 0.1	pH 0.03
Range	pH 0.0–14.0	pH 0.0–8.5	pH 0.0–14.0	pH 0.0–14.0
For Turbid or Colored Waters	Suitable	Suitable	Suitable	Suitable
Special Electrode Surface	Required	Not required	Not required	Not required
Addition of Auxiliary Material	Not required	Required	Not required	Not required
Affected by Presence of Dissolved Gases	Yes	No	Yes	No
Oxidation/Reduction Systems	Cause errors	No errors in solutions of mild oxidizing or reducing intensity	Cause errors	Independent of oxidizing and reducing solutions
For Continuous Measurements	Not easily adaptable	Not adaptable	Usable	Usable
Resistance System	Low	Low	Low	High
Other Characteristics	Not suitable for unbuffered solutions over range pH 5.0–8.5	Not dependable in low buffered solutions above pH 5.0	Calibration different for still and moving solutions	Suitable for unbuffered solutions

*Based upon a summary by Perley, 1937.

In methods for measuring electrolytic conductivity, *resistance* is determined by an appropriate instrument. *Conductance* is the reciprocal of the resistance involved. The unit of measure of conductance is the *reciprocal ohm* to which has been given the name *mho*. Some instruments are also so constructed as to read directly in *reciprocal megohms*. Both resistance and conductance depend upon the size and shape of the sample and are therefore not specific properties of a substance. When a resistance value is made to correspond to a sample of centimeter cube dimensions it is designated as *specific resistance;* likewise, when a conductance value is so treated it is designated as *specific conductance* or as *conductivity*.

EQUIPMENT

Rarely if ever will the limnologist find it practicable to construct the necessary equipment for conductivity measurement. For work in a laboratory supplied with electrical current, the operator will find on the market various outfits suitable for his purposes and it will be decidedly to his advantage to purchase such equipment from some reliable manufacturer who will supply full information as to the construction and operation of the outfit. Since it is often necessary to make conductivity tests in the field or

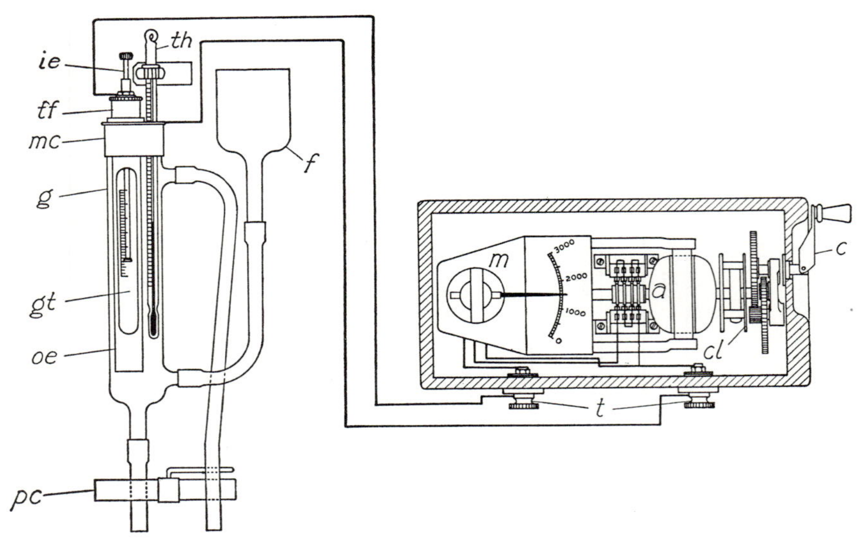

FIG. 63. Diagram of construction of Dionic Water Tester. Unit on the left is the conductivity tube; that on the right is the conductivity meter. (*a*) Armature. (*c*) Crank. (*cl*) Constant speed clutch. (*f*) Funnel. (*g*) Glass container. (*gt*) Glass tube. (*ie*) Inner electrode. (*m*) Conductivity meter and scale. (*mc*) Metal cap. (*oe*) Outer electrode. (*pc*) Pinch cock. (*t*) Terminals for connection to meter. (*tf*) Top fitting. (*th*) Thermometer. (Redrawn from Evershed and Vignoles, London, England; with permission.)

in temporary field quarters not equipped with electric current, a self-contained portable outfit is a necessity. An excellent instrument of this type is the Dionic Water Tester made by Evershed & Vignoles, London, in which a complete generator-indicator unit in very compact form is a special feature. Fig. 63 shows the principal features of construction and assembly. Since it is probably the most satisfactory self-contained portable outfit yet devised the following procedure is based upon its use.

Procedure for Dionic Water Tester

1. Set up instrument in some protected place and on very stable table; make certain that conductivity tube is firmly mounted on some fixed vertical support near edge of table; arrange overflow and outlet tubes to discharge into suitable container below; place conductivity meter alongside conductivity tube and connect them together, attaching insulated wire leads to proper binding posts.
2. Make certain that conductivity tube is scrupulously clean; open outlet tube temporarily to allow escape of any liquid from previous test; close pinch cock on outlet tube and fill glass tube through funnel with liquid to be tested; drain out and repeat to make certain that all traces of any previous test have been removed.
3. Remove thermometer from storage case and pass it through hole in top fitting so that most of its length is within glass test chamber; immobilize it in that position.
4. Fill glass tube with liquid to be tested until excess runs out through overflow tube, insuring that proper quantity is provided.
5. Record temperature of liquid as indicated by thermometer; adjust inner electrode so that its lower edge is opposite mark on scale corresponding to recorded temperature.
6. Turn crank of generator clockwise at a speed above that at which clutch is felt to slip (usually about 120 r. p. m.); as soon as pointer of conductivity meter is at rest, take reading. Reading is in units desired; no computation necessary. Record temperature and conductivity reading.
7. If desirable, repeat test on same sample, making certain that any necessary readjustment for temperature is made.
8. Drain out sample and thoroughly flush out all parts with distilled water; if further testing is postponed, put large cotton plug in funnel and close entire conductivity tube in its box.

General Considerations

a. The temperature compensating scale in the conductivity tube has a range of 10° to 40° C.; therefore water to be tested must be warmed or cooled if outside these limits.

b. Occasionally the operator should test his apparatus for possible current leakage between the lead wires connecting to the conductivity tube, or across the tube itself, by proceeding as if making a conductivity test but with no water in the tube. Under these conditions the meter reading should be zero; if not, there is current leakage which must be removed before further tests are made.

c. Pure rubber tubing only should be used for connections since vulcanized rubber contaminates the sample.

d. Under ordinary conditions the effects of barometric or hydrostatic pressure are negligible.

e. In making a test, care must be taken to avoid trapping air under the inner electrode when filling the tube with liquid. Escape of such air is facilitated by four small holes which pierce the electrode disk. However, if air still tends to become pocketed there, it can be removed by manipulating the electrode stem.

f. Scales of different ranges may be secured for the conductivity meter but for natural uncontaminated inland waters a scale of 0 to 5000 will be preferable.

g. After the apparatus is set up, the time of making a test need not exceed two or three minutes.

h. Since the apparatus is very sensitive to exceedingly small amounts of electrolytes in solution, every care must be taken to keep the inside surfaces of all parts concerned scrupulously clean at all times.

Selected References

(For Chapter 14)

Birge, E. A.: A second report on limnological apparatus, *Trans. Wisconsin Acad. Sci.*, **20:** 533–552, 1922.

Ellis, M. M., B. A. Westfall, and Marion D. Ellis: "Determination of Water Quality." 122 pp. Research Report 9, Fish and Wildlife Service, U.S. Dept. of Interior, 1946.

Fox, H. M., and C. A. Wingfield: A portable apparatus for the determination of oxygen dissolved in a small volume of water, *J. Exp. Biol.*, **15:** 437–445, 1938.

Hutchinson, G. E.: On the relation between oxygen deficit and the productivity and typology of lakes, *Intern. Rev. ges. Hydrobiol. Hydrog.*, **36:** 336–355, 1938.

Irwin, W. H.: The role of certain northern Michigan bog mats in mosquito production, *Ecology*, **23:** 466–477, 1942.

Isaacs, M. L.: A colorimetric method for the determination of dissolved oxygen, *Sewage Works J.*, **7:** 435, 1935.

Johnson, M. L., and R. J. Whitney: Colorimetric method for estimation of dissolved oxygen in the field, *J. Exp. Biol.*, **16:** 56–59, 1939.

Juday, C.: Limnological methods, *Arch. Hydrobiol.*, **20:** 517–524, 1929.

———, and E. A. Birge: Dissolved oxygen and oxygen consumed in the lake waters of northeastern Wisconsin, *Trans. Wisconsin Acad. Sci.*, **27:** 415–486, 1932.

———, ———, and V. W. Meloche: Mineral content of the lake waters of northeastern Wisconsin, *Trans. Wisconsin Acad. Sci.*, **31:** 223–276, 1938.

Maucha, R.: Sauerstoffschichtung und seetypenlehre, *Verh. Internat. Ver. theoret. und angew. Limnologie,* **5:** 75–102, 1931.

Maucha, R.: "Hydrochemische Methoden in der Limnologie." 173 pp. Vol. 12, in Thienemann's Die Binnengewässer. Stuttgart, 1932.

Meloche, V. W., G. Leader, L. Safranski, and C. Juday: The silica and diatom content of Lake Mendota water, *Trans. Wisconsin Acad. Sci.,* 31: 363–376, 1938.

Perley, G. A.: "Modern views of pH measurement." 5 pp. Reprinted for Leeds & Northrup from *Am. Dyestuff Reptr.,* Dec. 27, 1937.

Peters, C. A., S. Williams, and P. C. Mitchell: A simplification of Powers' method for the determination of carbon dioxide in natural waters and comparison with the titration method, *Ecology,* **21:** 107–109, 1940.

Rawson, D. S.: The calculation of oxygen saturation values and their correction for altitude. 4 pp. Spec. Pub. No. 15, Limn. Soc. Am., 1944.

Ricker, W. E.: A critical discussion of various measures of oxygen saturation in lakes, *Ecology,* **15:** 348–363, 1934.

Snoke, A. W.: The determination of dissolved oxygen with the micro-Winkler apparatus of Thompson and Miller, *Ecology,* **10:** 163–164, 1929.

"Standard Methods for the Examination of Water and Sewage," 9th ed., 286 pp. New York, Amer. Pub. Health Assoc., 1946.

"The Dionic Water Tester." 23 pp. List No. 116 h, Evershed & Vignoles, London, 1929.

Theroux, F. R., E. F. Eldridge, and W. L. Mallmann: "Laboratory Manual for Chemical and Bacterial Analysis of Water and Sewage," 2d ed., 228 pp. New York, McGraw-Hill Book Co., 1936.

Thompson, T. G., and R. C. Miller: Apparatus for the microdetermination of dissolved oxygen, *Ind. Eng. Chem.,* **20:** 774, 1928.

Titus, L. and V. W. Meloche: Note on the determination of total phosphorus in lake water residues, *Trans. Wisconsin Acad. Sci.,* **26:** 441–444, 1931.

Welch, Paul S.: "Limnology." 471 pp. New York, McGraw-Hill Book Co., 1935.

Part IV

Biological Methods

PLANKTON METHODS

Collecting Apparatus

In many plankton methods two processes are involved: (a) collection of a sample of some natural, plankton-producing water and its transportation to the surface; and (b) removal of plankton from this water. Both may be performed separately or simultaneously, depending upon the kind of equipment used.

SAMPLERS AND SAMPLING DEVICES

Kemmerer Water Sampler. The modified Kemmerer sampler, described on pp. 199–201, provides trustworthy plankton samples. With it water can be taken from a positively known depth and transported to the surface unaltered. Small capacity of the sampler may be regarded as a disadvantage since the required number of trips to the same depth in order to secure a usable volume of water slows up the work, but, on the other hand, the increased number of trips may add to the efficiency of sampling as a whole. So far as is known, motile plankters do not escape from this sampler. Large-size samplers (3-l. capacity or more) are often convenient in plankton sampling.

Ricker (1938) and others claim that certain plankters, notably *Daphnia*, are able, in the upper water, to see a collecting device (plankton trap; plankton net) and avoid it by day, thus causing a certain loss. To what degree such loss occurs when a Kemmerer sampler is used is yet unknown, but it would seem that because of its shape, size and method of use, error from this source, if present at all, should be minimal.

Pump and Hose Method. The pump and hose method, already discussed on pp. 201–202, has all the disadvantages described there in sampling for plankton. The uncertainty as to the exact level from which water comes is in itself serious enough to ban its use for many purposes. However, another serious defect arises when sampling for plankton, namely, the loss of a certain portion of the plankton Crustacea due to current reaction. Some Copepoda (*Cyclops*, *Diaptomus*, and others) manifest avoiding reactions in water moving toward the intake end of the hose and, as a consequence, 10 per cent or more of such plankters may be lost in the process of sampling. It has been claimed that this loss of plankton Crustacea, due to rheotactic reaction, may be reduced by installing a funnel, with sides making an angle of 30°–40° with the vertical, in the intake end of the hose, the effect being to enlarge the intake area and reduce the velocity of flowage. Because of its various disadvantages, the pump and hose method is now largely abandoned, particularly for quantitative work. For occasional, and perhaps some special kinds of qualitative work it may have some value, particularly because of its delivery of large quantities of water in a relatively short time. Gear pumps produce fragmentation of plankters and must be avoided if the organisms are to be secured intact.

Water Bottles and Similar Devices. Certain sampling devices, often referred to as "water bottles," are sometimes used in plankton work. Hale's water bottle, described on pp. 202–204, belongs in this class; also various types of large bottles so harnessed that they can be sent to various depths on a rope, the stopper jerked out and the bottle allowed to fill. Various accessory devices attached to such bottles improve their performance and speed up sampling. Such bottles fall short, in one way or another, in meeting the requirements of good plankton sampling. They fill by inflowing water thus setting up streaming and automatically leading to losses of certain motile plankters by avoiding reaction, as in the instance of the pump and hose method. For plankton work such water bottles should be looked upon largely as emergency or improvised devices to be used only for certain rough operations.

Other Samplers. Innumerable samplers have been designed, some of which are merely modifications of certain fundamental types. Many samplers, constructed for other purposes, are of only incidental or highly specialized use in plankton work. Certainty in collection of an unmodified sample at a definitely known position is a consideration of prime importance in the choice of a sampler and any device which even in the slightest way falls short of this test should be avoided. Convenience may well be considered but must not be put first in the qualifications of a sampler.

PLANKTON NETS

Some portion of the plankton may be collected by any fine-mesh cloth which will allow water to pass through it. However, for various reasons, certain materials are superior and, except in emergencies, those known to be particularly suitable should be used. It must be understood, however, that irrespective of fineness of mesh or its construction, no material which by any reasonable interpretation might be called a *net* will remove all plankton from a sample. Plankton separated from water by means of a net is generally referred to as *net plankton* and represents only a fraction of the total population.

Silk Bolting Cloth. Silk bolting cloth is generally regarded as the best material for plankton nets. It is commonly manufactured in six different grades and in a large range of sizes of mesh (see Appendix, Tables 9 and 12). It is sometimes possible to secure, in the *standard* grade, a size even finer than No. 25 but it is doubtful whether in most kinds of net-plankton work the additional cost and the slower operation are justified. In the grades listed in Table 9 the *standard* is the lightest grade and the thickness and strength of threads increase in the order of mention, XXX grit gauze being the heavy-duty grade suitable for the severe kinds of work. Some manufacturers stamp the name or the grade and size of mesh on the silk prominently. Unfortunately these marks soon become obliterated on a plankton net. Other makers weave colored threads into the selvages according to some code which expresses the grade, but not the size of mesh, of the material, but since the selvages are not always included on a plankton net they are of little service. The careful limnologist will enter on the cloth where it will not be lost a record indicating the grade, size of mesh, date of purchase, and the maker. These entries can be made with Higgins Waterproof ink or Higgins Eternal ink in some abbreviated way or in some code form which will occupy little space and in no way interfere with the functioning of the silk.

Shrinking of Silk Bolting Cloth. When new silk bolting cloth is first put into water a certain shrinkage occurs, the extent of which depends upon circumstances. Subsequent submergences in cool water may result in further shrinkage although eventually an approximate stability will be reached. Total expected shrinkage cannot be stated in fixed amounts. Published figures indicate a change of as much as 20 per cent in the size of apertures. Therefore it is commonly recommended that new silk bolting cloth be thoroughly shrunk before it is used for precise quantitative work. Shrinking may be done before the silk is cut out and installed in the metal parts of the net, by repeatedly wetting it with a wet sponge and pressing it with an iron just warm enough to accomplish the smoothing. Care should be taken to avoid the use of irons which are *hot* since high temperatures are detrimental to silk. Certain older, published methods which involve boiling are too severe. New unshrunken silk may be built into a net and the shrinking done by repeated submersion and drying if care is taken that the proper allowance is made in advance for subsequent change. If such allowance is not made the silk will become strained, especially in the windows of the bucket, if a bucket is used, and about the fittings on the metal supports; also in a single-seam net the shrinkage is likely to be less on one side thus resulting in a net which does not hang straight.

Linen. For some purposes certain grades of linen may be used for making plankton nets of the general utility type in which uniformity and fine size of mesh are not major considerations. A material sometimes known as India linen is often thus used. Any good grade of thin linen, not too tightly woven, may be suitable. It must be understood, however, that such materials are in no sense a substitute for silk bolting cloth for precise work.

WIRE CLOTH. Wire cloths, made in very precise form and in various sizes of mesh and diameter of wire, are available on the market. Some of them have meshes which approximate those of the fine-mesh silk bolting cloth. These wire cloths are made primarily for certain commercial purposes and manufacturers supply detailed information concerning them. Usually it is not practicable to use them in plankton nets of the ordinary type. However, when made into forms convenient for plankton work, they may be used to collect the larger plankters from water, especially if means are devised for removing the catch from the screen.

Installing Silk Bolting Cloth in Plankton Nets

Since the form of most plankton nets is that of a truncated cone, the preparation of silk bolting cloth for installation in them is facilitated by the use of patterns. In replacement of worn-out silk, new material may be cut to the proper shape by using the old silk, carefully removed from the metal parts, as a pattern, provided that the new silk has been shrunk in advance and that the old silk was originally in satisfactory form. The use of the old silk as a pattern will often be made easier if it is first dampened and then smoothed out carefully with a hot iron. However, it is often preferable, and sometimes necessary, to work with a paper pattern prepared as follows: Measure radius (R) of brass ring at top of net (Fig. 64);

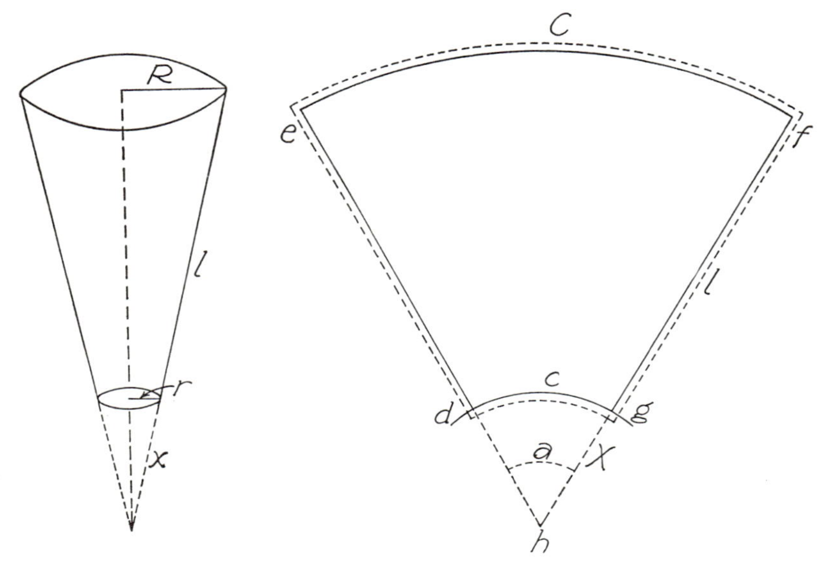

FIG. 64. Diagrams illustrating method of preparing pattern for cutting out new silk bolting cloth for plankton nets.

measure radius r of brass headpiece of bucket; also determine desired length l of side of net. Using these dimensions, convert truncated cone into complete cone by prolonging sides to intersection; determine length of x by use of the formula

$$x = \frac{rl}{R - r}$$

Then x plus l gives length of radius of circle a certain arc of which represents shape and length of upper part of net C. Determine angle a by means of formula $a = \frac{360\ r}{x}$.

Using l plus x as radius, lay off arc C of larger circle on smooth piece of paper. At center h of arc so constructed, lay off angle a by means of protractor and draw lines he and hf. Then with x as radius, draw arc c of smaller circle so that it cuts across lines he and hf.

The pattern is now complete. It consists of the area $gdef$ and represents the surface of the desired truncated cone unrolled and spread out in one plane. However, allowance must be made for seams, the character of which depends upon the structural detail of the net being made. For standard plankton nets with detachable buckets, an allowance of about 1 cm. all around the pattern will usually be adequate (see dotted outline in figure to right, Fig. 64).

The pattern should now be cut out carefully, fastened together temporarily and fitted onto the metal parts as a check on its correctness. If found correct, the pattern is then superimposed upon the silk bolting cloth on a smooth surface, and the margins are carefully marked on the silk with a soft pencil. Under no circumstance should the pattern be pinned to the silk. The silk is then cut out with sharp scissors.

The two long edges of the bolting cloth are sewed together, using a French seam. Silk thread and as small a needle as possible should be employed. A light coat of rubber cement applied along the line of stitches will effectively close all needle holes. If the net to be filled is equipped with an upper canvas section, supported by the brass wire frame (truncated cone form), the silk bolting cloth should be sewed to the canvas as follows: Turn bolting cloth inside out and pass up through larger end of canvas, small end first, until lower edge (larger end of sleeve) of bolting cloth is even with lower edge of canvas; sew two edges (bolting cloth and canvas) together; then turn bolting cloth right side out. The French seam should be on the outside of the bolting-cloth sleeve when the latter is in correct position.

The lower end of the bolting-cloth sleeve must fit smoothly onto the headpiece of the bucket. If the silk sleeve is oversize, wrinkles will develop when the net clamp is screwed on, resulting in leaks. If slightly undersize, the sleeve may be made to fit by longitudinally slitting the edge of the silk for a very short distance. If a bucket is not used on the free end of the net, the form of termination will depend upon the kind of receptacle employed.

The method of installation of bolting cloth in the detachable bucket which is commonly a part of standard plankton nets will depend somewhat upon the construction of the metal parts. In case of replacement, the old bolting cloth may be carefully removed and used as a pattern, proper allowance being made at all points for subsequent shrinkage of the new silk. The new silk, when installed, should show considerable slack in all directions in each of the windows, otherwise use may bring about further shrinkage, causing it to become taut and strained. Slack in the windows, unless excessive, is a matter of no concern; usually it is evidence that the bolting cloth has not been strained by improper installation. New cloth should be cut a little wider than the height of the bucket in order that a small amount of free edge project beyond the upper and lower clamps. It must be long enough to provide the slack mentioned above; also to allow an overlap on one of the strips between the windows. Since screws holding binding strips between windows must pass through the bolting cloth, it is necessary to provide the holes by burning them at proper places with a hot wire.

Birge Cone Net

The Birge cone net (Fig. 65) has three main parts, a cone-shaped top piece, a middle straining section, and a bottom piece.

The top piece consists of a base *tp* of sheet copper 2.5 cm. wide in the form of a cylinder having a diameter of 8 cm., the lowermost edge of which is bent outward and fits closely around a brass wire ring *r;* a cone of brass wire netting *nc* with 3-mm. mesh, soldered to the upper margin of the base and having a side length of about 9 cm.; a piece of brass wire *bw*, diameter about 2 mm., in the form of the letter V, fitted closely to the inside surface of the cone, the two ends soldered firmly to the inside surface of the copper base, and the opposite, apex end projecting beyond the cone and twisted into a loop for attachment of the main draw cord; and two small loops *l*, 180° apart, soldered to the outside surface of the base.

The middle straining section is composed of a sleeve *s* of silk bolting cloth, India linen, or any other material, in the form of a truncated cone fastened to the top piece by being tied or sewed above the ring; also fastened to the upper part of the bottom piece in the same manner. A strong cord *c*, which extends from each eye on the top piece to a similar eye on the side of the bottom part, is slightly shorter than the straining sleeve in order to take the weight or pull off the latter. A band of heavier cloth at the top and at the bottom of the straining sleeve provides a stronger attachment at those points. Supporting cords may hang freely on the outside as shown in Fig. 65, or they may be enclosed in tunnels formed by two extra seams along the side of the net. If the straining sleeve is of some very strong material, the side supporting cords may be omitted, provided the tie-on at top and bottom are secure.

The bottom piece is composed of a brass or copper tube *bp* about 4 cm. long and 3 cm. in diameter; a ring of wire *wr* soldered on the outside surface at the top edge; two loops or eyes, one opposite the other, soldered to the side of the tube for the attachment of supporting cords; and a threaded region on the bottom of the copper tube for the attachment of the receiving cup.

The bottom receiving cup *b* may be in any convenient form which serves the purpose, but should be made of metal and the lowermost piece should screw on snugly to avoid loss. If a screw top from a kerosene can or some similar article of about the same weight is used, it will be necessary to solder a lead weight of appropriate size to the screw cap to insure prompt sinking of the net, otherwise the lower portion when wet may pocket air and tend to float.

All metal parts may be made of galvanized iron, although copper or brass is preferable.

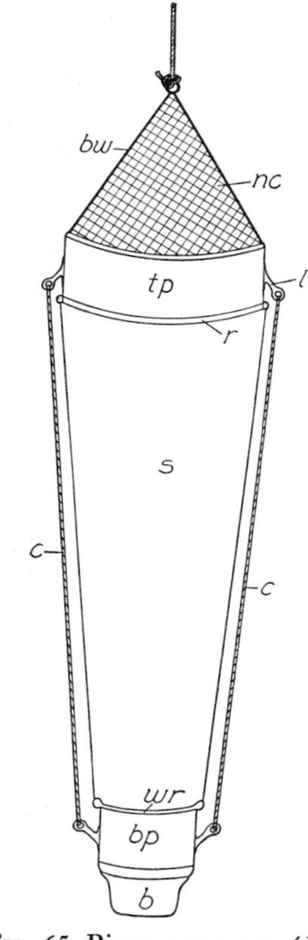

Fig. 65. Birge cone net. (*b*) Detachable bucket or cup. (*bp*) Bottom piece. (*bw*) Brass wire support. (*c*) Supporting cord. (*l*) Loop for attachment of cord. (*nc*) Wire-mesh cone. (*r*) Metal ring on top piece. (*s*) Sleeve of bolting cloth, grit cloth, India linen, or other material. (*tp*) Top piece. (*wr*) Wire ring on top of bottom piece.

Other Tow Nets

Tow nets can be made in many different forms. The following are useful.

A simplified and less expensive tow net may be made, using the same form and general dimensions as those given above for the Birge cone net.

The top piece consists of a brass ring about 1 cm. in diameter around which is sewed a complete, close fitting cover of strong braid. The cone, composed of some good grade, twisted-cord netting, is sewed firmly into braid around the ring and converges to the apex, ending in a brass ring which serves as attachment for the draw cord. Three equidistant strips of braid, sewed firmly to the braid about the large ring, extend to the draw-cord ring; also a similar strip of braid extends from one ring to the other, interwoven along the longitudinal seam of the cone. The other parts of the net are essentially the same as the Birge cone net, except that the bottom receiving cup may be in any convenient form from the screw-cup type to as simple an arrangement as a homeopathic vial or a test tube tied temporarily into place in the end of the net sleeve.

A *larger type* of tow net also resembles the Birge cone net in general form. The top piece is composed of a tube of sheet brass or galvanized sheet metal, 12 cm. long and 11 cm. in diameter, one edge of which is bent around a wire ring. A second ring is soldered around the outside surface of the tube about 1 cm. above the ring in the edge, thus providing a channel for the attachment of the straining net. At the other edge of the tube there is firmly soldered to the inside surface a heavy metal-screen cone, 15 cm. long and with 7-mm. mesh, in the apex of which is firmly soldered a ring or loop for the draw cord. In this top piece the heavy materials used make unnecessary any wire supports. The net is about 35 cm. long. Its larger end is double thickness of material to provide strength and is tied into place with strong cord in the channel between the two parallel rings. The bottom receiving cup may be any convenient type. Lateral supporting cords for the net may or may not be needed depending upon the character and strength of the net material.

A very simple tow net may be made in any convenient size by hanging a cone of India linen or of silk bolting cloth to a brass ring supported by three cords, equidistantly spaced and tied to the ring, which converge and join at point of meeting with the draw cord. To the top of the net is sewed a double strip of stout linen or some similar material, cut bias, and with a heavy cord sewed into the tunnel formed in its upper edge. This net, so prepared, is sewed to the ring by overcast stitches which pass over the ring and below the cord in the edge of the net. Since the bottom of the net is closed, the catch must be removed by turning the net inside out and washing, with an up-and-down motion, the lowermost part in a bottle of water. If desired, the tip may be cut off and a bottle or similar container tied into place.

All tow nets must be weighted in order to insure sinking. Those using some form of metal collecting cup at the lower end can be weighted by the attachment of metal to it. Tow nets which end blindly must be weighted by tying a sinker to the ring at the open end.

The pattern for cutting out the material used in the straining sleeve can be made in accordance with directions given on p. 234. A simpler, but less exact, method of cutting out materials for the straining sleeve in plankton nets, particularly tow nets, is as follows: On a large sheet of plain smooth paper, measure off along one edge a distance equal to circumference of top (largest) ring to which net material is to be attached. Add about 1.5 cm. to provide for seam. Lay off from same corner, along edge at right angle to first, the distance equal to desired length of straining sleeve. Through each of two distance points so located, draw line at right angle to edge and parallel to other edge. Cut smoothly along these lines, thus eliminating excess paper. Fold paper lengthwise from right to left, bringing corners and edges exactly together, and press down fold. Then draw straight line from lower right corner to upper left corner (hypotenuse). With radius equal to length of paper and with center at lower right corner, draw an arc from upper right corner to intersection with diagonal line (hypotenuse). With scissors, make cut full length of hypotenuse and also cut around arc. The pattern so produced will represent the form of the desired net. With the paper pattern make a trial wrapping about the supporting ring to check for accuracy. If the trial shows the pattern to be correct, lay pattern out smoothly upon the fabric to be used and mark margins in pencil; remove pattern and cut out carefully. Under no circumstance pin pattern to fabric. Sew long edges together using French seam. A net so made is closed at the apex. If, however, some form of receiving cup is desired, the apex may be cut off as follows: By rolling upper edge of receiving cup across pattern above apex, locate position where width of pattern equals one-half circumference of cup; mark this level on folded edge of pattern; with distance from apex to mark as a radius, strike off arc from one edge to other; cut off by following arc.

TYPES OF PLANKTON NETS. Plankton nets may be of various forms, depending upon their purposes and the preferences of the operator. Certain types of construction have general advantages.

THE WISCONSIN PLANKTON NET. Many years ago Birge and Juday devised a net, sometimes referred to as the "Wisconsin net," which is now widely used. There are two sizes: the "small net" and the "large net."

The *small net* (Fig. 66) consists essentially of three parts, an upper section in the form of a short inverted truncated cone, a middle section in form of an elongated truncated cone, and a detachable plankton bucket.

The upper section is composed of previously shrunken, light- or medium-weight canvas, supported on a framework of 4-mm. brass spring wire. The upper ring of this framework has a diameter of 12 cm., the lower ring a diameter of 18 cm. These rings are connected to each other by three pieces of straight brass wire of smaller diameter. Loops at the ends of the connecting wires make union with figure-eight pieces of

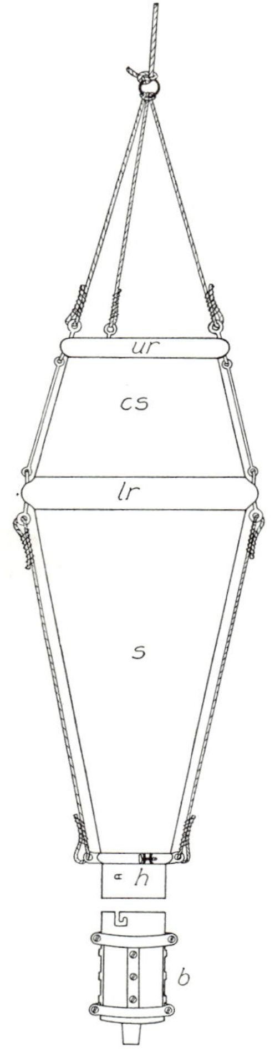

FIG. 66. The Wisconsin plankton net. (*b*) Detachable bucket, shown detached from headpiece. (*cs*) Canvas sleeve between upper and lower rings. (*lr*) Lower ring. (*h*) Headpiece. (*s*) Sleeve of silk bolting cloth between lower ring and headpiece. (*ur*) Upper ring. The third supporting cord, extending from lower ring to headpiece, and the third supporting wire connecting upper and lower ring, are not shown in the figure.

similar wire, 3 strung on the upper ring and 3 on the lower, the pieces of both sets being equally spaced at intervals of one-third the circumference and held loosely in place by bits of small brass wire soldered to the ring on either side. On the upper ring, one end of the wire holding each figure-eight piece in place is soldered around the ring, then bent upward into a loop and brought down onto the other side of the figure-eight piece and again soldered around the ring. These three loops serve as means of attachment for the draw lines. All loops and eyes are oversize, thus forming loose joints, which provision makes this part of the net collapsible. The canvas filling this upper section should be shrunk; then cut out, using a pattern made by the use of the formula given on p. 235. Proper allowance must be made for extra length both at the top and bottom to provide for attachment to the rings. The two cut edges of the canvas should be sewed together with strong thread to form a French seam; then the upper end is put through the inside of the upper ring, the edge turned outward over the ring and sewed below it. The larger end of the canvas is put inside of and against the lower ring with the edge projecting past the ring for a distance of about 1.5 cm.; a piece of strong braid, cut to fit the outside of the lower ring, fastens the canvas to the ring when stitched to it both above and below the ring. The three draw cords converge at a convenient distance above the upper ring and are tied into a loop or a ring to which is fastened the main pull line.

The middle section is an elongated truncated cone of silk bolting cloth installed according to directions given on p. 234. Bolting cloth may be of whatever grade and size of mesh the operator desires. However, for most plankton purposes involving both qualitative and quantitative work, *No. 25 standard* is the grade usually used. From each of the figure-eight connecting loops on the upper ring there extends a piece of strong, nonelastic cord to a corresponding loop on the *headpiece* at the lower end of the silk sleeve. The length of these three cords is adjusted to carry all of the weight of the headpiece and the attached plankton bucket. Consequently, the silk sleeve always shows a certain amount of slack. The brass headpiece, constructed from telescope tubing, is 3.5 cm. long and has an outside diameter of 5.2 cm. A small peg on the outside surface serves as a part of the safety lock when the plankton bucket is attached. Two similar pegs, near the upper rim and 180° apart, fit into holes in the clamp and hold it in position.

Details of construction of the plankton bucket are presented in Fig. 66. The body of the bucket is of brass telescope tubing. It has a length of 9 cm. and an inside diameter of about 5.2 cm. The upper portion, about 2.8 cm. long, serves as a sleeve which fits over the headpiece attached to the bolting cloth. From the upper edge of this sleeve there extends downward an L-shaped cut so constructed that when the plankton bucket is fitted over the lower part of the headpiece the peg on the latter fits snugly into the cut forming a safety lock, sometimes referred to as the bayonet clutch. Two such locks oppositely placed are sometimes provided. Most of the lower part of the bucket is occupied by four windows each about 3 cm. wide and 5 cm. long. The windows are separated from each other by a narrow band about 1.1 cm. wide, on the inside of which is soldered in a vertical position a semicylindrical piece of brass having a diameter of about 6 mm. These semicylindrical pieces strengthen the strips of tubing. The bottom of the bucket is a conical piece of cast bronze soldered into the end of the tube flush with the lower edges of the windows. This bottom slopes toward the middle, ending in a tapering outlet tube about 2.3 cm. long. This outlet tube has a diameter at the upper end of about 2.3 cm.; at the lower end, about 1 cm. Its walls are about 1 mm. thick. A tapering removable plug, exactly fitting the outlet tube, has a handle which extends slightly above the upper rim of the bucket and bears on its free end a milled head.

The *large net* has the same shape and general construction as the *small net*. Its upper ring has a diameter of 25 cm.; the lower ring, 30 cm. The canvas sleeve and the bolting-cloth sleeve have lengths of 33 cm. and 70 cm., respectively.

For most purposes the *small net* is preferable to the *large net*.

The loops, eyes, and connecting wires of the upper section allow it to collapse when not in use. This is convenient in packing and transportation. However, if a collapsible upper section is not needed, it may, with advantage for some purposes, be made of sheet copper or brass.

COMBINATION SAMPLER AND COLLECTION DEVICES

Various types of equipment in which the two operations of sampling and separation of plankton from the water are simultaneous have been devised. The three described here are probably the most useful for ordinary purposes. Only net plankton is secured by these instruments.

Juday Plankton Trap

The general construction of the Juday plankton trap is indicated in Fig. 67. Form and size may differ depending upon the desires of the operator, but a good, general utility type is described below. All parts are made of brass or other nonrust metal.

The trap consists primarily of (a) a frame, 46 × 22.5 × 25.5 cm., supported at the four corners by pieces of cylindrical brass tubing; (b) a box made of thin sheet brass, the four vertical sides of which are immovable; the top and bottom are open but may be closed by lids moving horizontally in close fitting grooves; (c) an open part of the frame, 43 × 24 cm. in horizontal dimensions, having a fixed floor of thin sheet brass; (d) a spring mechanism which operates the sliding doors; (e) a device for releasing the doors and allowing them simultaneously to close the brass box when tripped by a messenger; (f) a short-form plankton net with straining sleeve of No. 25 silk bolting cloth and a detachable bucket of the usual construction, attached to the lower door and opening into the interior of the box through a large circular hole; (g) four supporting chains radiating from a central attachment disk to the tops of the four corner posts and forming suspension for the entire trap; and (h) accessory items as follows: an opening with short piece of brass tubing on the upper surface of the lower fixed floor, covered by a small, fine-mesh cloth bag, permits air to escape from the plankton net when the trap is lowered into the water; a similar hole and tube in the middle of the top door, covered by same kind of cloth bag, permits air to enter the closed box when it is lifted out of the water after a catch; a small rectangular window in the upper part of the side toward the open section of the frame closed by very fine brass gauze also serves a function similar to that of the opening in the upper lid.

Operation of Plankton Trap

1. Provide suitable hoist with which to operate trap. Ordinarily it is not practicable to use trap by hand. Hoist should be provided with graduated rope or wire cable. A hoist with swinging arm (Fig. 51) which makes it possible to deliver trap and enclosed catch onto deck of boat is preferable. Graduation of rope or cable may be eliminated if hoist is equipped with rope meter.

2. Thread free end of hoisting line through mid-channel in brass messenger. Be sure that messenger slides down line with minimum friction. A split messenger, if available, can be put on line at any time or place.

3. Pass free end of hoisting line through hole in upper stirrup of trip mechanism and on through hole in middle plate of suspension chains below; knot or otherwise fasten securely.

4. Adjust hoist and hoist arm so that when in operation the trap will

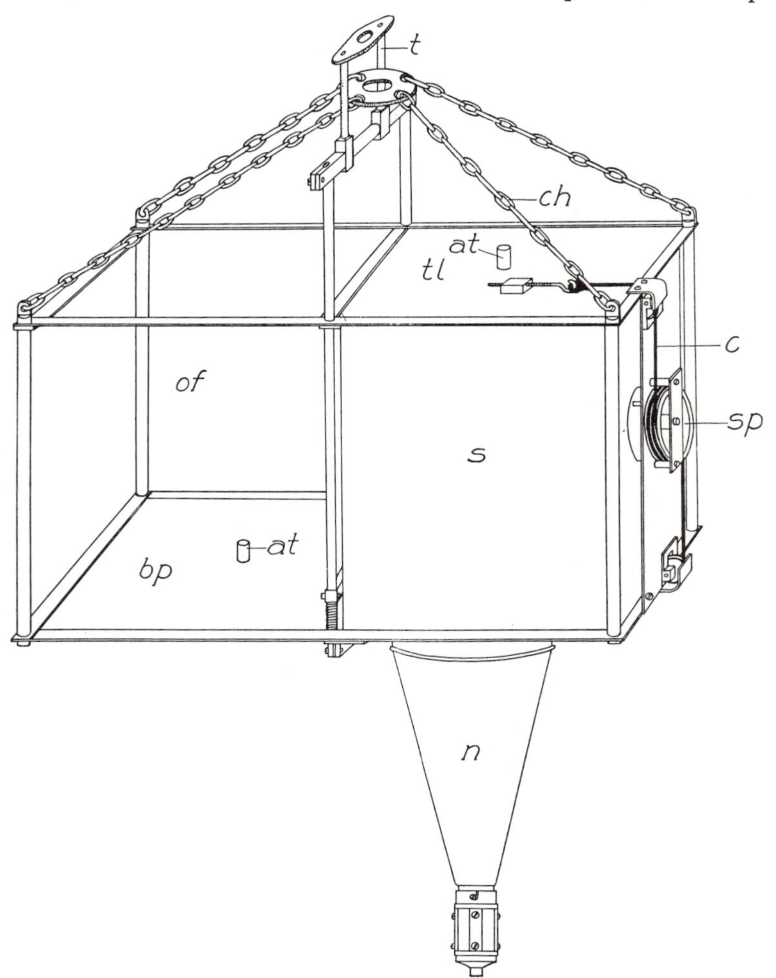

Fig. 67. Side view of Juday plankton trap; shown in closed position. (*at*) Air tube. (*bp*) Bottom plate. (*c*) Operating cord. (*ch*) One of four supporting chains. (*n*) Plankton net. (*of*) Open frame. (*s*) Side plate. (*sp*) Operating spring. (*t*) Top part of tripping device. (*tl*) Top lid.

be lifted out of the water to proper height and will make no contacts with hull of boat.

5. With trap suspended in air, pull upper and lower lids over open part of frame until trip mechanism engages and holds them in new position. Inspect all parts of apparatus to insure good working condition; for trial purposes, trip instrument to see if lids close freely, quickly and completely.

6. With trap set in open position, lower slowly and carefully to desired depth; release messenger; bring trap slowly to surface, lifting it entirely out of water and allowing trapped sample to drain out through plankton net until desired concentration is effected in plankton bucket; transfer catch to proper containers for further use.

7. If succeeding sample is to be from a new depth, wash interior of collection chamber, plankton net and plankton bucket with tap water.

Advantages

The plankton trap has the following advantages:

a. Collection of sample and removal of net plankton are accomplished simultaneously in the field.

b. The sample is taken from the desired depth position, with negligible mixing of water from other levels.

c. Loss of plankters due to reactions to water current is avoided.

d. Uniform quantities of water are taken in each sample.

e. Plankton sampling is accomplished with considerable speed.

Limitations

a. The plankton trap is bulky, a little awkward to handle, and heavy when partly or wholly filled with water. Some kind of hoist is usually a necessity.

b. Only net plankton samples are taken, and it is not practicable to attempt to recover quantitatively the filtered water.

c. The plankton trap can be used only in relatively calm water.

d. The plankton trap can be used only in situations free from water plants and other objects which would interfere with the operation of the instrument.

e. Use of this instrument in very shallow water is likely to result in damage to the suspended plankton net.

General Considerations

a. The plankton net must receive the same care as that described for other nets on p. 253. In preparation for long-distance transportation, the net should be removed from the bottom lid and packed separately. In transportation to and from the field it should be carefully turned partly inside out through the large opening in the bottom lid into the collection chamber and all metal parts wrapped in soft packing to prevent injury to the bolting cloth.

b. A specially designed carrying case or kit is a necessity.

c. The grooves in which the lids operate must be kept free from grease or gummy substances. They should be cleaned carefully from time to time.

d. Operation of the plankton trap depends upon the proper alignment of lids and their grooves, proper tension of the operating spring, proper functioning of the tripping device, and good condition of the plankton net.

e. It has been claimed (Ricker, 1938, and others) that certain plankters, notably *Daphnia*, are able to "see" a plankton trap during day sampling in the uppermost waters and avoid it, thus leading to the loss of a certain fraction of the reacting plankters. To what extent this is an error of consequence remains to be determined.

f. Clarke (1942) has modified the plankton trap as originally described by Juday by (a) having the sliding doors, in open position, on opposite

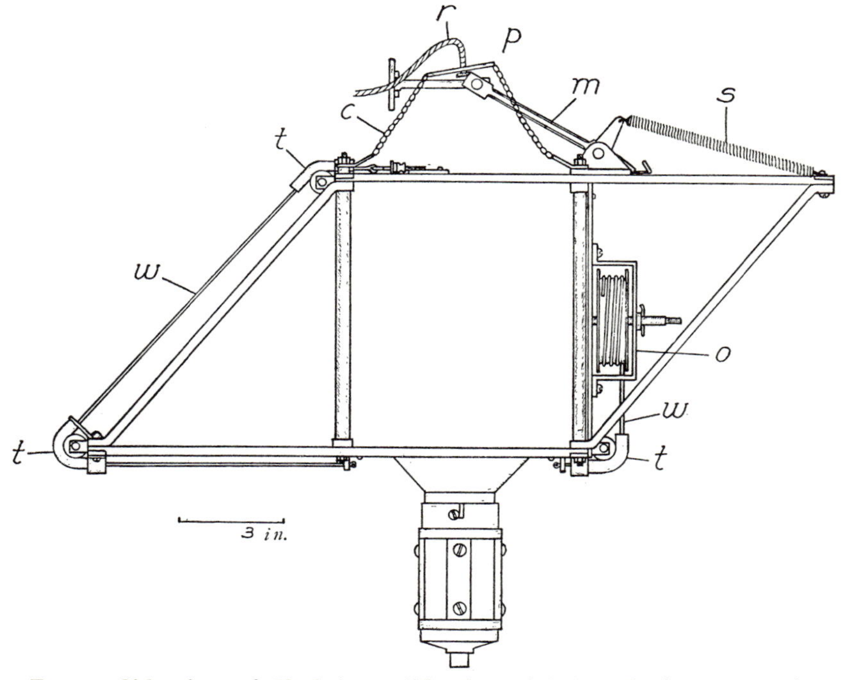

FIG. 68. Side view of Clarke's modification of Juday plankton trap; shown with doors in closed position. (*c*) Supporting chain. (*m*) Trip mechanism. (*o*) Operating spring. (*p*) Metal plate to which supporting chains are attached. (*r*) Draw rope. (*s*) Coiled spring. (*t*) Metal tube to keep wire on pulleys. (*w*) Wire for operating doors of trap. (From Clarke, 1942.)

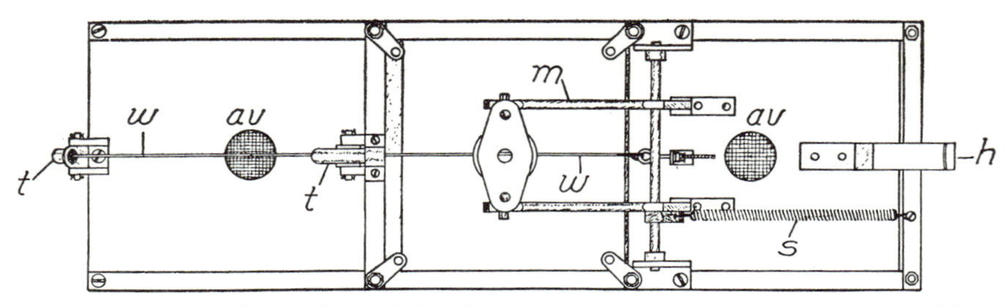

Fig. 69. Top view of Clarke's plankton trap with doors in open position. (*av*) Air vent covered with fine wire screen. (*h*) Handle. (*m*) Trip mechanism. (*t*) Metal tube to keep wire on pulleys. (*w*) Wire for operating doors of trap. (From Clarke, 1942.)

sides thus providing an even distribution of weight and a better balance; (b) reducing the capacity to 5 l.; (c) providing a different mechanism for controlling the action of the sliding doors, making their performance more certain; (d) omitting the plankton net (sleeve) and straining all of the trapped water through the plankton bucket; and (e) passing the operating wire through sections of metal tube so located and constructed that the wire cannot leave the pulley. Figs. 68 and 69 indicate the essential features of the revised sampler.

Closing Nets

A closing net is so constructed that it can be lowered into the water to a selected position, made to function from that point by drawing it through the water, then closed at the conclusion of the haul, and the catch brought to the surface without further additions to the sample. A closing net now in common use consists essentially of a long truncated cone composed of light canvas or heavy muslin; a long straining sleeve of silk bolting cloth or other material also in the form of a truncated cone; a removable plankton bucket of the form described on p. 241; and a mechanical release operated by a messenger. Fig. 70 indicates the principal features of construction. At the upper end of the canvas sleeve a brass ring, 12 cm. in diameter, is sewed firmly into place. From this ring radiate three supporting cords, equidistantly placed, which converge to attachment to a small brass ring. At the lower end of the canvas sleeve, another brass ring, 17 cm. in diameter, is also sewed firmly into place slightly above the edge of the sleeve so that a margin of the latter, about 2 cm. wide, extends below the ring. The canvas sleeve is 40 cm. long.

The straining sleeve is about 47 cm. long, is sewed to the projecting margin of the canvas sleeve below the larger supporting ring, and is attached at the lower end to the brass headpiece which receives the plankton bucket. Three cords, equidistantly placed and extending from the ring at the top of the straining sleeve to the headpiece, are so adjusted in length that they carry all of the weight of the headpiece and bucket thus allowing the straining sleeve to hang loosely.

The main draw line is attached to the brass ring at the bottom of the canvas sleeve. At a point on this draw line equal in distance to the combined length of the canvas sleeve and the three supporting cords and their ring, a knot is tied, determining the lowermost position of the tripping mechanism. The trip is threaded onto the main draw line and rests upon this knot. The construction of a suitable trip *r* is indicated in Fig. 70; when a messenger of suitable shape and weight is sent down the draw line it strikes the top of the trip, opening the release mechanism.

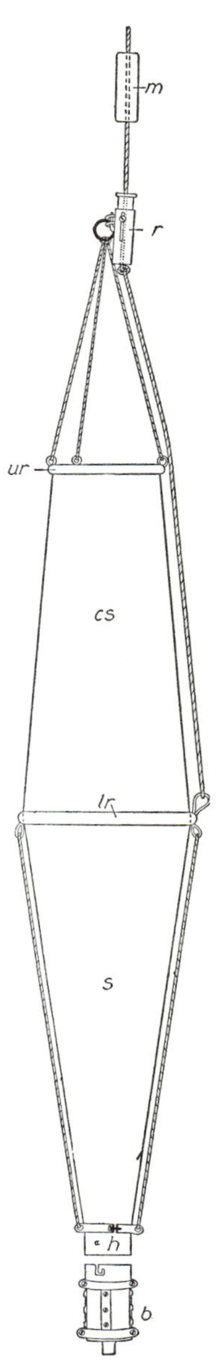

Operation of Closing Net

1. Attach plankton bucket securely on headpiece at lower end of straining sleeve. Hold main draw line high enough in air so that net is suspended vertically. Lift canvas sleeve into vertical position, push down plunger of trip and hook ring at upper end of suspension cords over upturned hook of trip between two guard wires; release pressure on plunger and allow ring to be locked into place. The whole net is now suspended in a vertical position.

2. By means of graduated main draw line, or by use of recording meter and line, lower net promptly until top of canvas sleeve is at lowermost depth level of horizontal stratum through which sample is to be taken; then draw net slowly but steadily upward (rate about 0.5 m. per sec.) until top of stratum is reached. Send down messenger which, on striking plunger of trip, releases suspension ring in top of net, allowing entire canvas sleeve to fall over and assume a reverse position, transferring weight of whole net to lowermost end of main draw line, and causing supporting ring at top of straining sleeve to take an edgewise position. Thus the net is effectually closed.

Fig. 70. Side view of closing net of usual type. (*b*) Detachable bucket, shown detached from headpiece. (*cs*) Canvas sleeve. (*h*) Headpiece. (*lr*) Lower ring. (*m*) Messenger. (*r*) Release. (*s*) Sleeve of silk bolting cloth, or other suitable material. (*ur*) Upper ring. Third supporting cord, extending from lower ring to headpiece, not shown in figure.

3. Bring net to surface by steady lifting motion and secure plankton catch by procedures described on p. 276; also see same page for directions for washing net and bucket with tap water.

Limitations

1. The closing net takes a sample only when drawn through a stratum of water having considerable thickness. It is not suitable for securing samples from one restricted region.

2. It is not suitable for work in very shallow water.

3. It must be used in calm weather since rough water may cause contamination of the sample.

4. Theoretically, and under ideal conditions of operation, a closing net strains a column of water with a diameter of that of the upper ring of the canvas sleeve and a length equal to the distance through which the net was pulled prior to closing. Actually, this result is seldom completely attained. Much depends upon the speed of pull. Excessive speed results in overspill at the receiving end of the net; undue slowness also gives uncertain results. An approach to the optimum speed depends upon a combination of good judgment by the operator and some test trials made before the work is begun. Suitable speed of pull also depends much upon the size of the mesh in the straining sleeve; also upon abundance of plankton and the rate at which the meshes of the net become more and more clogged. It is likely that under most circumstances the operator should not expect an efficiency of more than about 80 per cent, often less, in quantitative work.

5. When plankton is scanty, and the thickness of the selected water stratum is small, workable concentrates can be secured only by repeated draws through the same stratum. However, in several hauls in succession the efficiency of the net diminishes due to clogging.

6. Closing nets of the type described here are designed primarily for vertical hauls. With special technic and accessory equipment, oblique and horizontal hauls may be made but ordinarily must be regarded as impracticable.

7. When used in daytime it is claimed (Ricker, 1938, and others) that an error arises because certain plankters, notably *Daphnia*, are thought to be able to see the plankton net in the upper waters during the day and avoid it, thus leading to a certain loss. Whether such a loss occurs generally, and to what extent, remains to be determined.

Advantages

1. Collection of sample and plankton in the field is simultaneous.

2. Loss of plankters because of negative reaction to current is minimal.

3. A transverse section of the plankton of a body of water or any stratum of such water is secured.

4. Collections may be made easily and rapidly by one person.

5. For qualitative work of a reconnaissance type, the closing net is rapid and effective.

General Considerations

a. If properly done, the necessary lowering of the net *open* involves no entry of plankters from strata other than the selected one.

b. The straining sleeve may be made of any grade or mesh of silk bolting cloth, grit cloth, or other appropriate fabric, depending upon the needs of the work in progress.

c. Because of its large size and its general delicacy, a closing net should always be transported to and from the field in a specially designed case.

d. On return from the field, closing nets should be suspended full length in the air, in subdued light, and allowed to dry at once.

e. When wet, the canvas sleeve becomes watertight and allows no net plankters to enter through its sides.

Clarke's Plankton Sampler

Recently a sampler (Figs. 71 and 72) has been devised by Clarke and Bumpus (1940) which overcomes some of the more serious limitations of the ordinary closing net. This is the most satisfactory closing-net type of quantitative plankton sampler that has been produced. The principal features are: (a) a brass tube about 6 in. long and 5 in. in diameter; (b) a straining sleeve of bolting cloth of some desired grade attached to the rear of the brass tube by means of a ring with a bayonet lock; (c) a brass frame which carries the brass tube on pivots and which provides attachment to the draw cable; (d) two vanes, one on each side of the brass tube, so set that they assist in holding the tube in a horizontal position; (e) a disk-shaped controlled shutter, mounted on vertical pivots, in the front end of the tube; (f) a large propeller mounted within rear half of tube and geared to a counter which registers the number of revolutions and consequently the volume of water which passes through the tube and the net; (g) messengers, triggers, lugs, draw-line weights, and other accessories necessary to the functioning of the apparatus. The draw cable is attached by a device which allows the supporting frame to swivel freely about the cable, thus insuring that the open end of the tube is always forward when the net is being towed.

CALIBRATION. Before this instrument can be used for quantitative work, the propeller-counter mechanism must be calibrated in terms of liters of water filtered per revolution of the propeller. Such calibration must be done in some properly equipped laboratory by passing known volumes of water through the tube at various speeds. It is said that, in instruments made according to the specifications of the inventors, the rating would be

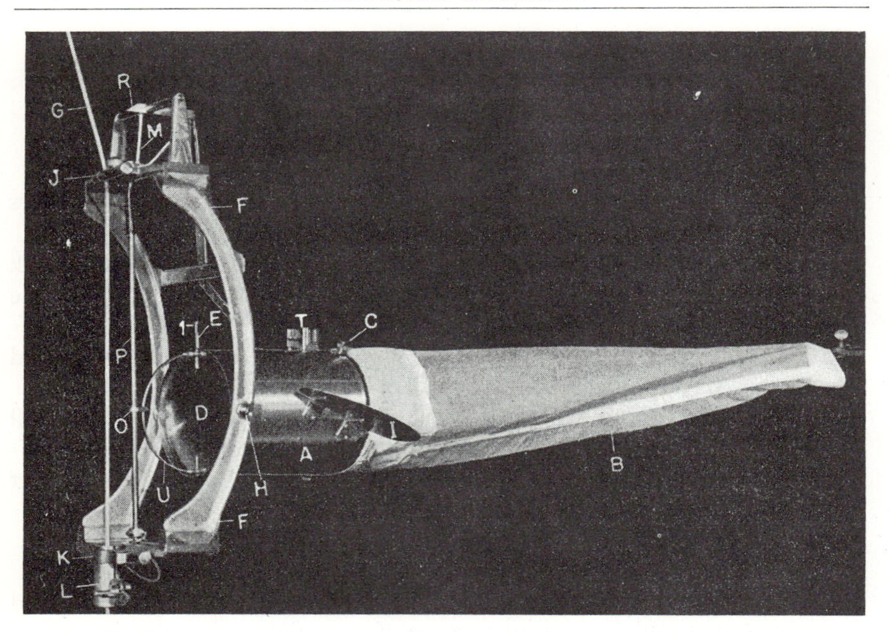

Fig. 71. Side view of Clarke and Bumpus plankton sampler attached to supporting cable and with shutter in initial closed position. The receptacle at the free end of the net and the drag which holds the net in a straight position are not shown in this figure. (*A*) Tube. (*B*) Net. (*C*) Bayonet lock. (*D*) Shutter. (*E*) Pivot for shutter. (*F*) Frame. (*G*) Cable. (*H*) Pivot for tube. (*I*) Plane. (*J*) Spring pin. (*K*) Gate lock. (*L*) Supporting clamp. (*M*) Rod fixed to trigger. (*O*) Long finger lug. (*P*) Rod. (*R*) Trigger. (*T*) Counter. (*U*) Semicircular bar. (*1*) Spring. (From Clarke and Bumpus, 1940.)

found to vary less than 5 per cent when operated under boat speeds of about one-half to 4 nautical m. p. h. Such a constancy makes it unnecessary to know precisely the boat speed as long as it is within these limits.

Operation of Net

1. Before instrument is attached to cable, set shutter in tube as follows: Turn shutter to final closed position with semicircular bar inside tube; rotate shutter 90° counterclockwise against spring, at same time rotate lug rod clockwise against its spring until edge of shutter engages longer lug; make necessary adjustments of trigger mechanism; continue rotation of shutter counterclockwise for another 90° and rotate lug rod clockwise until semicircular bar of shutter engages shorter lug and trigger rod engages proper horizontal arm. Make certain that proper spring tension is present in this position to insure prompt return of shutter to former positions when released; if not, remove stops and give shafts additional turns

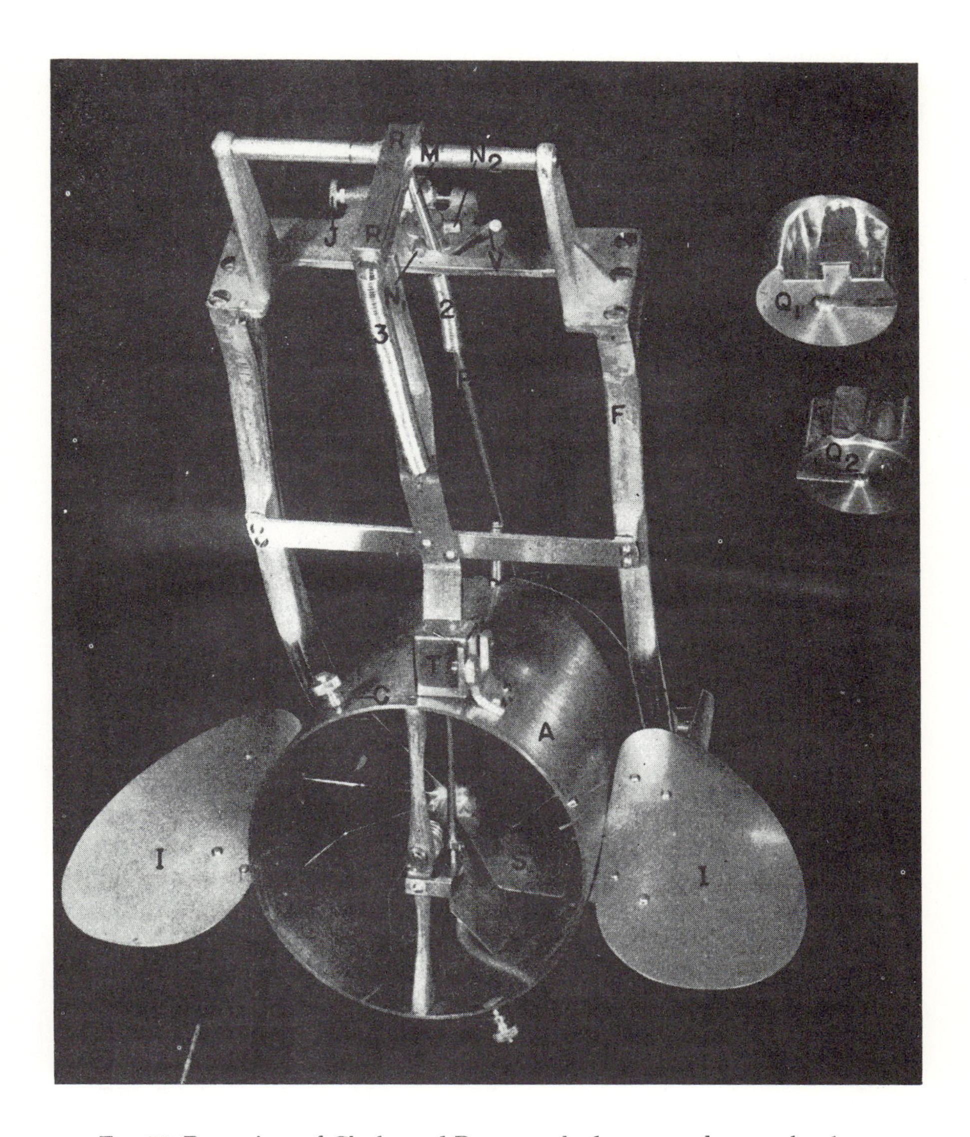

Fig. 72. Rear view of Clarke and Bumpus plankton sampler; net has been removed. (*A*) Tube. (*C*) Bayonet lock. (*F*) Frame. (*I*) Planes. (*J*) Spring pin. (*M*) Rod fixed to trigger. (N_1, N_2) Arms in cap. (*P*) Rod. (Q_1, Q_2) Messengers. (*R*) Trigger. (*S*) Propeller. (*T*) Counter. (*V*) Escapement rod. (*2, 3*) Springs. (From Clarke and Bumpus, 1940.)

in direction opposite their direction of operation; shutter is now in operating position.

2. Fasten instrument to cable, carefully following directions supplied by maker.

3. Record position of dial on counter.

4. Attach weight to lower end of cable.

5. Start boat forward; lower instrument slowly into water; if tail of net shows tendency to whip forward, attach to it a drag (tin funnel drawn mouth forward by three equidistantly spaced converging pull cords, or similar device) to insure that it pulls in proper position.

6. When ready to begin making plankton catch, slip messenger onto cable; drop messenger causing shutter to rotate 90° into *open* position; water now entering mouth of tube passes through propeller and is measured quantitatively, then on into net through which it flows, leaving net plankton behind in terminal collecting cup.

7. At end of selected towing period, send down second messenger which causes shutter to rotate another 90° and close aperture of tube, thus ending collection.

8. Bring instrument to surface and secure catch in usual way.

Advantages

In addition to possessing some of the same advantages presented in the ordinary closing net (p. 248), this instrument has the further ones listed below:

a. The meter records exactly the amount of water which passed into the net.

b. Any clogging of the net as the tow proceeds is no source of error.

c. The instrument is particularly suited to horizontal tows. In addition, with proper arrangements, it can be used for vertical and diagonal hauls.

d. Overspill of water at the mouth of the net due to excess speed of towing is of no consequence.

e. Calm water is not necessarily required.

f. In waters of large size and depths, two or more of these samplers may be operated simultaneously on the same cable, each at a depth different from that of the others.

g. It is usable from boats of any ordinary size, even rowboats operated by oars.

h. The plankton may be caught over a considerable distance, making the catch more representative because of irregularities in horizontal distribution.

Limitations

1. Clarke's sampler is not suitable for very small or very shallow waters.

2. It does not give a sample from one very restricted position.

3. It gives no measure of nannoplankton; nor does it make possible the securing by some other method of the nannoplankton from the column of water strained.

General Considerations

a. If the closing shutter is not required, a greatly simplified and less expensive form of the instrument may be provided in which the tube, meter, and net are combined in the same way but mounted on a simple, light, rectangular frame.

b. All working parts should be kept well lubricated. If long intervals are to exist between times of use, the meter should either be taken apart and dried, or it should be flushed with kerosene and filled with motor oil.

c. If sampler is to be used in sea water or in inland saline waters, it is recommended that all parts of the instrument be nickel plated in order to avoid corrosion. Skunk oil is recommended as the lubricant in salt water. After each use the salty water must be washed from all parts of the instrument and in particular the meter. The meter may be kept full of some heavy grade of motor oil.

d. Special attention must be given to the strength of the cable and its attachment to insure against loss of the instrument.

e. This instrument must not be operated through beds of aquatic vegetation, in waters containing submerged objects, or close to the bottom.

f. A modified form of oblique haul may be made by starting the haul at the deepest selected depth and towing the instrument horizontally, but periodically shortening the cable until all selected levels have been sampled.

CARE AND REPAIR OF PLANKTON NETS

a. In storage, plankton nets should be hung from hooks, in normal position, away from contacts with other objects, and in subdued light. Silk deteriorates rapidly when exposed to direct sunlight.

b. After a net has been used in water, it should be dried promptly by hanging it in open air away from direct sunlight.

c. A plankton net is very delicate and must at all times be carefully protected. During transportation to and from the field, it should be stored, preferably hung, in some suitable container. The small Wisconsin net may be easily suspended within the neck of a large milk can.

d. Holes and slits, due to age, wear, or accidents, may be repaired temporarily by the use of soft paraffin, beeswax, rubber cement, or some similar waterproof material.

Settling Method

The settling method has a limited but sometimes important use. It is based upon gravity settling of plankters in plankton-bearing water after the addition of a killing and preserving agent. It may be operated as follows:

1. Provide several duplicate glass tubes having diameter of any convenient size (not less than 1.5 cm., preferably larger) and length of not less than 80 cm.; also provide suitable rack for storage of such tubes in vertical position. Clean tubes thoroughly. Attach to each tube a permanent serial identification number.
2. Ascertain capacity of each tube up to graduation mark located near top of tube; record this value in some permanent form on outer surface of tube.
3. Install tubes, and supporting rack in small darkened room which can be locked to insure against disturbance.
4. Secure desired samples of plankton-bearing water by means of sampler and deliver into sample bottles; bring to laboratory immediately. Invert sample bottles several times to offset possible settling effects during transportation; then pour samples into tubes mentioned above, filling them to graduation mark. Divide one sample among several tubes, or use only one tube per sample if capacity is adequate.
5. To each tube add sufficient 10 to 15 per cent formalin (or other suitable killing and preserving fluid) to produce quick killing and preservation of all plankters.
6. Close tops of tubes to prevent evaporation and entry of dust. Allow to stand in closed, darkened room for about two weeks. At end of that period most of total plankton will have settled to bottom of tubes. Draw off very carefully, by siphon or pipet, all of upper five-sixths or more of water column, avoiding removal of settled material.
7. Stir remaining portion of water column, preferably by gentle swirling, until all settled materials are again in suspension; then transfer to another long tube, but much smaller in diameter, which will hold residual portion from first settling. Again place in isolation room and allow to settle for another period of about two weeks. Then draw off uppermost part of water column. For some purposes, sufficient settling and concentration may have been accomplished at this stage; if not, the processes of settling and transferring should continue until it reaches satisfactory degree. Usually three to four such settlings and transfers will be adequate for many purposes.

GENERAL CONSIDERATIONS

a. The operator must use his judgment as to how much of the settled water column can be siphoned off without significant loss of plankters. In quantitative work, a certain margin of safety must be allowed since some settled or almost settled materials may not be easily visible.

b. In the instance of division of an original water sample among several duplicate tubes, results of the first settling may be combined into one tube of the original lot for the second settling period. However, final steps must usually be accomplished in smaller tubes.

c. The number of settling periods and the allowable dilution of the final plankton concentrate must be determined by the operator on the bases of practicability and the use to which the concentrate is to be put.

d. No fixed rule can be stated concerning the length of settling periods. For qualitative work, the periods may be much reduced without serious disadvantage; for quantitative work, preliminary tests on the plankton of the selected waters may be of service in arriving at a dependable settling period.

ADVANTAGES

a. This method involves a minimum of equipment and can probably be operated with materials regularly found in any laboratory.

b. For qualitative work, this method has the virtue of providing plankters which are free from breakage, crushing, or other forms of general damage inherent in net or centrifuge treatment.

DISADVANTAGES

a. The long periods of time necessary for settling are sometimes a serious disadvantage.

b. In plankton-poor water, the required large size of the sample demands the use of many duplicate tubes if an adequate amount of concentrate is to be secured.

c. Determination of the required settling period for different waters may require preliminary tests.

d. A measure of the possible loss of plankton in the removal of overlying water is almost impossible to determine in any precise way.

e. Living plankton concentrates cannot be secured by this method.

Plankton Centrifuges

Of recent years centrifuge methods have become firmly established in plankton work, particularly when the nannoplankton is of major interest, or when total plankton is to be measured volumetrically or gravimetrically.

Various types of centrifuge are in use but the one described below comes more nearly qualifying as a standard instrument.

FOERST ELECTRIC CENTRIFUGE

The principal features of the Foerst centrifuge appear in Fig. 73. The form commonly used has a maximum speed of 20,000 r. p. m. It is of the revolving-bowl type in which the organisms are deposited in the angle formed by the junction of the side wall of the bowl with the bottom. The plankton-bearing water enters at the top and the centrifuged water is thrown over the rim of the bowl. A rheostat control permits adjustment of speeds within the ranges inherent in the instrument.

Operation of Foerst Centrifuge

1. Oil principal shaft bearings before starting run. As commonly made, centrifuge has three places for oil, two on upper section of main shaft, and one at lowermost end of shaft below motor, all closed by screws with "oil" stamped on head. Remove "oil" screws and put into each hole about 3–4 drops of some good grade light oil, such as sewing-machine oil, then replace screws. This quantity of oil will usually serve for at least 1–2 hrs. of steady running. Carefully avoid adding more oil than necessary.

2. Set up centrifuge as indicated in Fig. 73. Use some type of support or frame for firm installation of all items of equipment. The support indicated in Fig. 73, designed by the writer, has been very satisfactory. Put base of centrifuge on a rubber mat or thick felt pad to reduce noise and vibration. In this figure r^1 and r^3 are iron rings of the ordinary ring-stand variety, diameters of which are suitable for the selected funnel and sample bottle; r^2 is a similar ring, of smaller diameter and with a section removed to facilitate introduction of neck of sample bottle.

3. Adjust glass funnel with its terminal rubber tubing and pinch cock so that water is delivered into centrifuge without danger of loss. Start centrifuge, building up speed to operating level, and pour tap water into funnel; control delivery of water into centrifuge by pinch cock; use this step as trial to test complete working order of apparatus.

4. Set centrifuge at selected operating speed by means of graduated rheostat knob, or with cap of centrifuge chamber removed, determine speed by applying speed indicator to end of revolving shaft, replacing cap when proper speed is determined.

5. With upper ring (r^1) lifted higher than its working position and with the middle ring (r^2) fixed at proper level, remove stopper from sample bottle, and, with a quick motion, invert bottle containing sample into funnel with mouth about halfway into it; slip bottle in place in middle ring and bring upper ring down over bottom of bottle in supporting position. Water flowing from sample bottle will rise in funnel to certain level

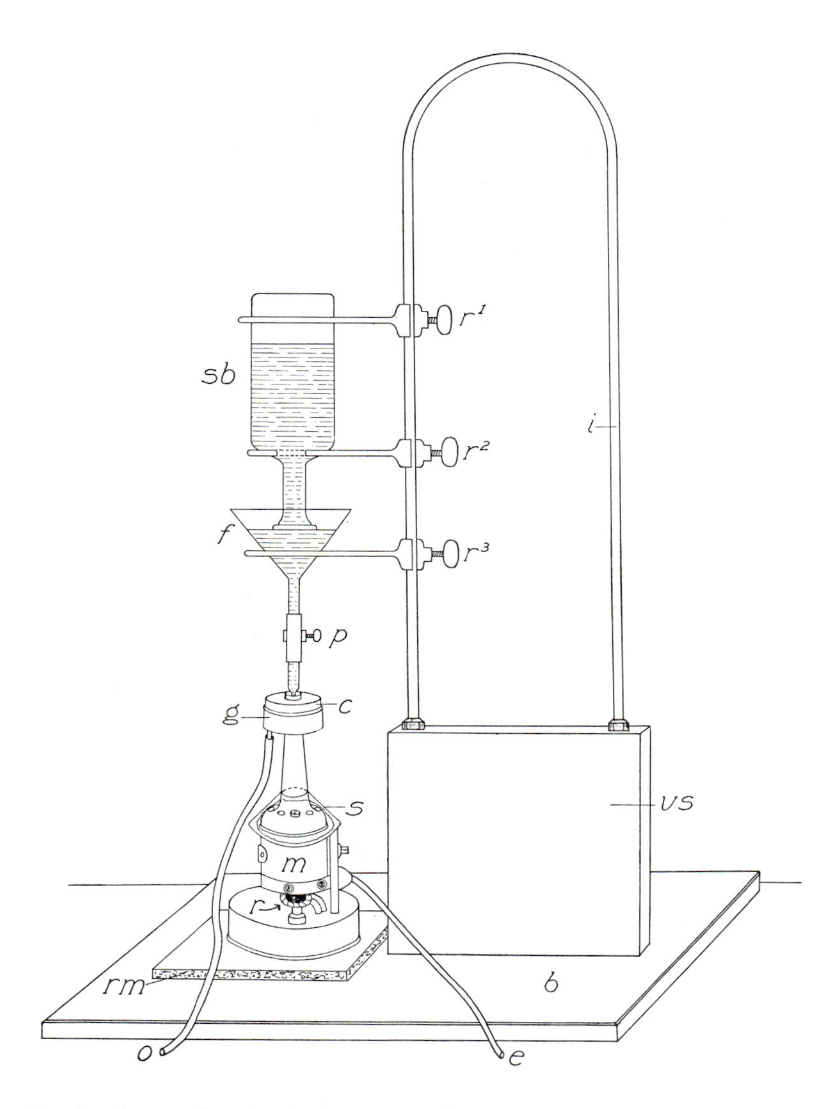

FIG. 73. Foerst electric plankton centrifuge set up in operating position. (*b*) Base of support. (*c*) Cap of centrifuge chamber. (*e*) Electrical connection to motor. (*f*) Funnel. (*g*) Housing of revolving bowl. (*i*) Iron rod support. (*m*) Motor. (*o*) Outflow tube. (*p*) Pinch cock. (*r*) Rheostat. (r^1, r^2, r^3) Iron rings from ring stand. (*rm*) Rubber mat. (*s*) Metal shield (front cut away to show top of motor beneath) to protect motor from water dripping from bowl. (*sb*) Sample bottle. (*vs*) Vertical wooden support for iron rod.

but will not overflow; further inflowage of the sample into funnel will depend upon outflow past pinch cock into centrifuge. When proper adjustment of outflow is made by pinch cock, delivery to centrifuge becomes automatic and should continue until sample bottle is emptied.

6. Adjust screw pinch cock so that proper rate of flow into centrifuge is provided. If centrifuge tends to slow down in speed, shows signs of laboring, or shows a tendency for water to be forced to outside at edge of cup, delivery of sample water is too rapid. Crowding of centrifuge leads to imperfect results; too slow delivery of sample water does no harm but unduly prolongs time of running sample. Operator will soon learn to detect correct adjustment. The small Foerst centrifuge will usually require about 7–10 min. per l. of sample.

7. Maintain centrifuging speed for short time after water disappears from stem of funnel to insure that all of sample has passed through instrument and only a small quantity of water remains in bowl. Turn off current and stop centrifuge. Remove tip of funnel delivery tube from centrifuge and with back-and-forth rotary motion remove cap from top of centrifuge, exposing interior of bowl.

8. Slip specially made wrench into narrow space between lower end of motor and upper surface of centrifuge base and engage the lower projecting end of main shaft, thus immobilizing it. Grasp upper rim of rotating cup in bowl and turn *clockwise*, unscrewing cup and removing it from machine. Cup will contain small amount of water. Plankton accumulation tends to be compacted into angle between floor and sides of cup. By means of piece of soft rubber tubing, cut diagonally at one end to form elongated tip, and other end slipped on a glass rod for support, very gently brush loose compacted plankton from bowl. Pour plankton concentrate into selected receptacle; add small amount of distilled water (about 1 cc.) to bowl and wash carefully to secure plankters possibly still remaining; add residue so collected to concentrate in receptable; repeat washing at least once.

9. Wash bowl and inside of removable cup with distilled water; dry with suitable absorbent cloth or sponge; hold main shaft by means of wrench at lower end below motor and replace bowl by screwing it counterclockwise onto shaft; replace top of cup. Centrifuge is now ready for another run or for storage.

10. On completion of use, store centrifuge by laying it on its side, thus safeguarding against possibility of oil running down into motor from upper bearing.

General Considerations

a. Size of sample centrifuged will depend upon the wishes of the operator and upon the richness or paucity of the plankton. For ordinary purposes, one liter is sufficient when plankton is at least moderately abundant; for plankton-poor waters, two or more liters may be required.

b. For removal of all kinds of plankton, the centrifuge should be operated at about the maximum speed (20,000 r. p. m.). For separation of certain groups of plankters, the centrifuge may be made to exercise a limited differential sorting effect by the use of lower speeds. Differential effects of different reduced speeds must be determined by preliminary experimental tests.

c. Protective shield (Fig. 73, *s*) installed just above the motor will protect against water from the bowl finding its way into the motor.

d. When operated at about 20,000 r. p. m., the Foerst centrifuge, in the first run, will remove about 98 per cent of the plankton, exclusive of bacteria. Some bacteria, perhaps 25 to 50 per cent, are also included. A second run of the water from the first run will remove most of the remaining 2 per cent. Certain Algae are particularly resistant to centrifuging, as for example, *Aphanizomenon*, of which only about 50 per cent may be secured in the first run. In ordinary kinds of plankton work the first run will often suffice, the remaining error falling within the general error of other operations involved.

e. Disposal and use of material secured with the centrifuge depends upon the aim of the work. If it is not to be worked immediately, preservative should be added as soon as material is collected.

f. On all occasions when the bowl is replaced in the instrument, it should be screwed snugly into place, otherwise it may become loose during a run and wobble. A loose bowl may be detected during a run by the changing sound of the machine.

g. If for any reason, the form of set-up indicated in Fig. 73 is not convenient or desirable, a large aspirator bottle can be used for a feed bottle and if it is fitted with a rubber stopper, a very satisfactory control of rate of flow into the centrifuge can be provided by the use of a glass stopcock.

Advantages

a. This instrument combines the advantages of portability, compactness, small weight, ease of use, efficiency and minimal accessory equipment for operation.

b. All organisms commonly included in plankton work are removed; only some residual bacteria will usually remain behind.

c. Secures material suitable for numerical, gravimetric, volumetric, and microchemical studies.

Disadvantages

a. Includes all detritus as well as plankton.

b. Requires electric current for operation.

c. Produces some crushing effect on plankton catch when operated at higher speeds.

d. Limited capacity results in a certain slowness in handling samples.

e. Rather high initial cost.

OTHER CENTRIFUGES

Various centrifuges, with speeds which develop about 50,000 r. p. m. have been tried in plankton work. It appears that for many kinds of plankton work the added cost and other disadvantages are not justified. Larger centrifuges with greater capacity are useful in shortening the time of running a sample, and higher speeds may effect a more complete removal of the plankton during the first run. Large, high-speed centrifuges usually require a permanent installation, thus losing the advantage of portability. Initial costs are also much greater. Power centrifuges designed for other purposes may be adapted for plankton removal but are likely to be inconvenient and relatively inefficient.

The common hand centrifuge is convenient for some of the grosser aspects of plankton collection.

Glass-slide Racks

Glass slides suspended in open water accumulate various microscopic and semimicroscopic organisms common to such situations. For certain kinds of plankton work effective use can be made of this simple technique. A convenient form of the method is the suspension from anchored floats

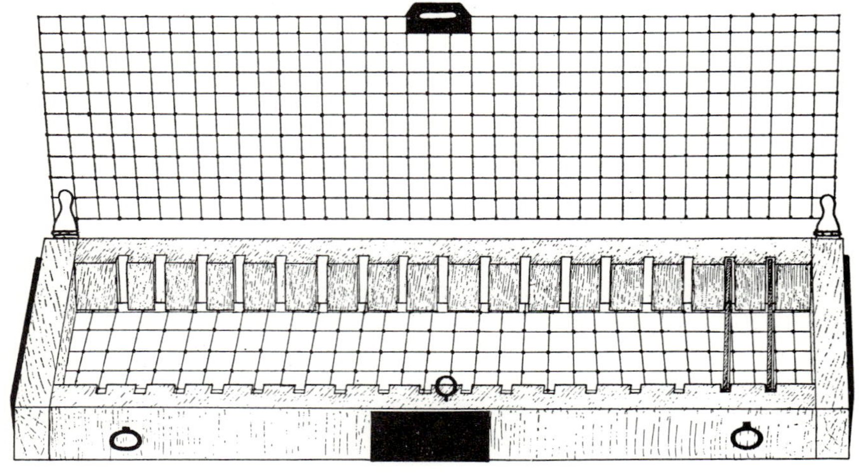

Fig. 74. Rack for suspending glass slides in water, shown with lid open and two ordinary 1 × 3-in. microscope slides in place in right end. Black areas on each end and on middle of top surface of frame represent pieces of sheet lead which load rack so that it will sink.

of clean, ordinary 1 × 3-in. microscope slides in a specially designed rack (Fig. 74) constructed as follows: Fasten two strips of cypress wood, 34.5 × 2.7 × 1.5 cm., and two strips of same material, 7 × 2.7 × 1.5 cm., together to form rectangular frame so that inside transverse dimension will be 7 cm. and longitudinal dimension 31.5 cm. On inner surface of each of longer wooden strips, make 18 saw-cuts so located that those of one side are exactly opposite those on the other, thus forming pairs. Make each saw-cut of such depth that a 1 × 3-in microscope slide can easily be inserted with extra allowance for swelling of wood and variation in slides. Cover one side of frame with zinc wire screen having mesh of about 0.8 cm. and fasten permanently. To other side of frame, hinge similar piece of zinc screen so that access to interior of cage so formed is provided. On free edge of lid, near middle of its length, solder a piece of sheet brass in which is cut a slender slit so made as to fit over a small brass eye on frame forming a locking device. To outside of wooden frame, attach strips of lead sufficient to insure sinking of rack. Attach one or more brass eyes in outside of frame from which to suspend rack in desired position. All nails, screws, and metal parts other than screen and lead weights should be brass to avoid rusting. Fill rack with clean microscope slides. Number each rack with serial number for individual identification. Such numbering may be done by attaching to frame a brass plate on which number is stamped; or if preferred, number may be stamped plainly on lead plates.

OPERATION OF SLIDE RACKS

1. Fill racks with slides which have been thoroughly cleaned and rinsed.

2. Suspend racks from buoys, piers, or other supports so that they do not make contact with each other or with other objects. Make record of rack numbers, depths, and other pertinent data.

3. Leave racks in suspended position until attachment of organisms occurs. This will require at least several days; in some waters even longer. No fixed directions can be set for period of submergence. In some instances duplicate sets will be advantageous; one set can thus be used as trial set to determine if submergence period has been long enough.

4. At end of submergence period, lift racks slowly, cautiously and only in calm water. Remove from water as follows: (a) if submergence was in the uppermost level of water, completely submerge a large pail in surface water, slip beneath rack when it approaches surface, then lift out so that rack is brought out in pail; (b) if submergence is in a deep stratum in which character of water differs distinctly from surface water, fill pail on deck by securing water from submergence level with a sampler, then bring rack to surface, lift from water very gently and transfer immediately to pail. Transport to laboratory with minimum of agitation.

5. If only a few slides in a rack are to be examined at one time, transfer slides from rack to Petri dishes containing same kind of water from which rack was taken and cover. Return rack immediately to original level for continued submergence.

LIMITATIONS

a. The slide-rack method is a rough but useful method which merely supplements qualitative work done with more precise methods. In no direct way does it supply any information of a quantitative sort.

b. In plankton studies it is usually most useful for studies made in the uppermost strata. When brought from deeper strata there is always the possibility that plankters may be added from the upper water when the rack is pulled through it, although the additions may be few in number.

c. Only those plankters which adhere to objects, or which become associated one way or another with the various other organisms which eventually form a growth on the slide, are collected in this way.

d. Slides may collect certain organisms which at no stage could be classified as plankton.

ADVANTAGES

a. This method is an excellent means of securing undamaged specimens.

b. Plankters on slides are more easily kept alive in the laboratory than in concentrates taken with nets.

c. Plankters on slides may be kept alive for some days by storage in refrigerators or ice boxes whose temperatures do not fall to the freezing point.

d. Information concerning plankters having attached stages may be secured.

e. Under some circumstances, this method may aid in the collection of some of the rarer plankters.

f. Slide racks may be used to collect plankters in situations in which nets and samplers can be used only with great difficulty.

g. Information concerning plankton succession may be secured.

h. Plankters may be conveniently killed, fixed and preserved *in situ* on the slide.

Plankton Filters

Plankton filters have been largely superseded by other devices. However, for some purposes they may be convenient and a brief account is presented here. For the most part, they are more efficient than silk-bolting-cloth nets in retaining very small plankters. They all have certain disadvantages in common, namely, (a) the filtering process is so slow that it is difficult to pass through the filter an adequate quantity of the plankton-bearing water; and (b) there is difficulty in separating all of the collected plankton from the filter.

FILTER-PAPER METHOD

Samples of water containing plankton may be filtered through hard-surface filter paper fitted in a glass funnel or some other appropriate supporting device after which the collected plankton is washed off the paper into a receptacle. Such filtration is very slow unless speeded up somewhat by means of a filter pump. Some of the nannoplankters are likely to adhere very closely to the paper and others become caught in its meshes thus making their removal difficult and in some instances almost impossible. This method is suitable for qualitative work and with the proper care some kinds of quantitative work are possible, especially if a direct count is made of all of the more numerous plankters. For the rarer plankters the method is impracticable since the catch is likely to contain too few of the individuals to make possible a dependable enumeration. Any grade of hard-surface filter paper may be used but it must not be assumed that all such papers give the same results. No. 575 Schleicher and Schüll or No. 50 Whatman are suitable. In records made by the filter-paper method the specifications of the paper used should be recorded in full.

SAND-FILTER METHOD

The degree of completeness of plankton removal by sand filters depends upon the condition and quality of the sand used and upon the manner of its use. Much detailed information concerning sand filters is available in reference works on sanitary engineering and elsewhere. If ordinary sand is used it must first be thoroughly washed and ignited. However, for most purposes it will be desirable to follow the specification of "Standard Methods for the Examination of Water and Sewage" (9th ed., 1946) which is as follows: "White sand is required as a filtering medium. This may be Berkshire or Ottawa sand, ground quartz or white beach sand. It should be washed and screened, only that portion being used which passes U.S. Series No. 60 screen and is retained on a No. 120 screen." Such sand is commonly referred to as *60 to 120 sand*. Such sand is satisfactory for ordinary samples. However, if it is necessary that very minute organisms be retained, a smaller size of sand grain must be used as for example, 60 to 140 sand. In order to function properly, a sand filter must be installed in some specially designed holder.

Whipple Sling Filter

The Whipple sling filter (Fig. 75) is a simple device convenient for some purposes. The cup at the bottom, telescoped onto the funnel and locked in place by a bayonet clutch, carries a column of sand about 0.5 in. deep supported on fine wire gauze. The water to be filtered is poured into the funnel, passes through the sand and on out through the open bottom. This filtration is speeded up by the centrifugal force produced by whirling the funnel around the handle.

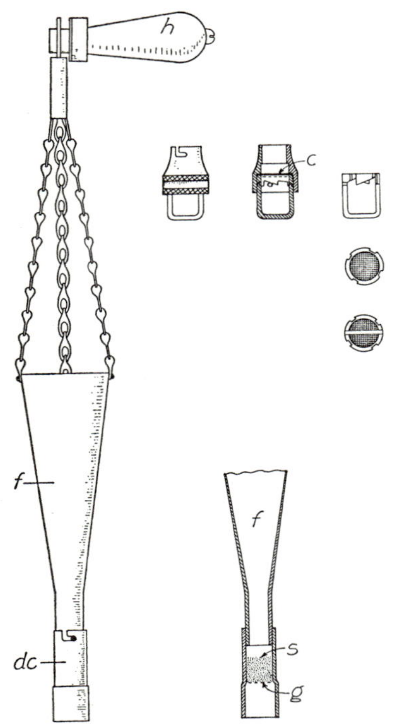

FIG. 75. Whipple sling filter and attachments. (*Left*) Sling filter assembled and ready for use. (*Bottom right*) Sectional view of lower portion of sling with filtering material in place. (*Top right*) Attachment for use of cotton disk, in which the cotton rests upon a wire screen and is held in position by a clamp; top and bottom view of supporting screen shown just below figure. (*c*) Cotton disk. (*dc*) Detachable cup. (*f*) Funnel. (*g*) Fine wire gauze. (*h*) Handle. (*s*) Sand. (Reprinted by permission from "The Microscopy of Drinking Water," by Whipple, Fair, and Whipple, published by John Wiley and Sons, Inc.) (Redrawn.)

Cotton-disk Filter

This method consists in passing a sample of water through a thin layer of cotton. Whipple devised for this sling filter an attachment (Fig. 75, *top right*) whereby such a sheet of cotton can be substituted for the sand layer. He also developed the use of cotton-pad collection by the use of a mechanism for filtering a sample of water from one bottle to another (Wizard Sediment Tester). After the filtration is completed the cotton disk is removed, dried, bottom side down, on blotting paper, and the disk mounted on a card and labeled. The result of such filtration of a representative sample is the production of a visual, approximate record of the suspended materials in the water produced by discoloration of the disk. This method is not designed to do more than produce such a visual record. It is not practicable to try to remove plankton from the disk. However, for some purposes this method may be of value in a comparative study in which the mounted stained disks become permanent records. Commonly, in making such disk records it will be necessary to ascertain in advance whether the amount of water filtered will produce disks sufficiently colored to make them usable.

Sedgwick-Rafter Method

The time-honored Sedgwick-Rafter method is still the type of sand filter recommended by "Standard Methods for the Examination of Water and Sewage" for sanitary work. Equipment for operation of this method consists of the following:

1. Funnel: Cylindrical glass funnel; upper portion about 23 cm. long, diameter 5 cm., and sides parallel; lower portion, diameter diminishing gradually over length of about 7.5 cm. until diameter becomes 12 mm.; terminal part about 6.5 cm. long and 12 mm. in diameter. Capacity of funnel, 500 cc. Funnel sometimes with series of graduation marks.

2. Stopper: Perforated rubber stopper (Fig. 76) fitted tightly into bottom of funnel; capped on inner end with circular piece of silk bolting cloth having about 200 meshes per inch (linen cloth may be used); cap of silk or linen should be cut with a wad cutter and to a dimension of slightly less than 12-mm. diameter; small glass U-tube in stopper, outer end of which extends about 2.5 above inner end of stopper (this device suspends filtration when water in funnel drops to level of its outer end).

3. Filter: Sand which meets specifications stated on p. 263; must be fine enough to accomplish desired filtration, yet not so fine as to interfere with prompt settling in decantation process; not less than 12 mm. deep on top of inner end of rubber stopper; supported by bolting-cloth disk.

4. Supports: Appropriate supports for funnel; multiple supports in instances of need for several funnels operating simultaneously.

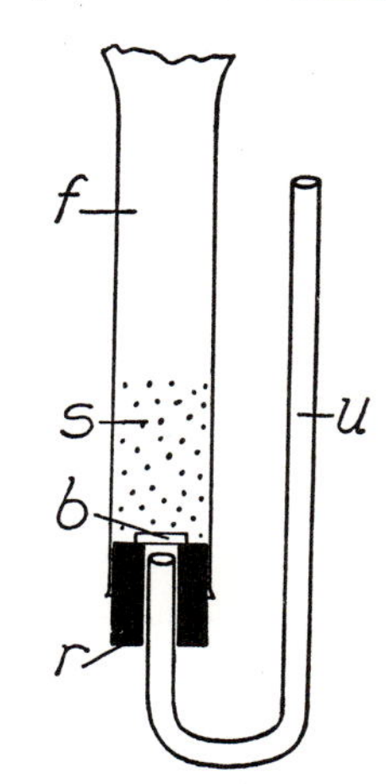

FIG. 76. Sectional view of lower end of Sedgwick-Rafter tube with attached U-tube. (*b*) Silk bolting cloth. (*f*) Lower end of funnel. (*r*) Rubber stopper. (*s*) Sand. (*u*) U-tube which controls gravity filtration so that water level in funnel cannot drop lower than the outlet level.

Operation

1. Set up funnel in operating position; pour 5–10 cc. distilled water into funnel; then introduce sand to form layer, not less than 12 mm. thick, on top of stopper; tilt funnel from side to side to permit escape of air included in sand; flush sides of funnel with stream of distilled water to wash down sand adhering to walls of funnel.

2. Fill funnel to top of neck with distilled water and allow drainage

to begin. Mix original sample of plankton-bearing water gently but thoroughly; immediately measure in a graduate that portion of original sample to be filtered (250–1000 cc. depending upon abundance of plankters in sample); add measured portion of original sample to funnel, taking every precaution not to disturb sand of filter bed; allow water to filter out; during period of filtration, flush sides of funnel with stream of distilled water from wash bottle to wash down any plankters adhering to sides of glass.

3. When water level in funnel reaches level of outlet and of U-tube, carefully remove U-tube and allow all remaining water to drain through filter.

4. Turn funnel into horizontal position and slip small beaker under lower end; with gentle twisting motion, remove stopper slowly; raise funnel to vertical position with lower end inside beaker, allowing sand filter bed to fall out; flush inside walls of funnel with 5–10 cc. of distilled water delivered from pipet, collecting water in beaker.

5. Mix contents of beaker gently but thoroughly to free plankters from sand and to bring them into suspension; allow sand to settle; then decant immediately into second beaker; add 5 cc. distilled water to sand in first beaker, wash and mix thoroughly for second time, and decant again into second beaker. Second beaker now contains plankton concentrate ready for analysis. With graduate of appropriate precision measure carefully total volume of concentrate so secured.

General Considerations

a. Gravity filtration commonly requires about 30 min. but this time can be much shortened by the use of suction. When suction is employed, it should not be too strong and should be discontinued before water level reaches the constricted portion of the funnel otherwise plankters may be pulled into the sand and damaged. Suction is applied directly to the stopper after removal of the U-tube.

b. Lodgment of plankters on the glass produces a certain "funnel error." Funnel error may be reduced by the use of (1) cylindrical funnels; (2) suction to speed up filtration; and (3) scrupulously clean funnels.

c. The sand error—failure of the sand filter to retain organisms—depends much upon careful choice of size of sand for the filter bed and proper compacting of filter bed prior to the introduction of the sample.

d. The original sample should be put through the sand filter as soon as possible after collection in the field to minimize errors of disintegration. In some waters certain plankters may be present which disintegrate quickly after introduction into the storage container. Immediate storage in a refrigerator will make it possible to delay filtration for a few hours. It is probable that errors of disintegration cannot be avoided entirely but they may be reduced by prompt and careful filtration.

e. In decantation, some plankters are inevitably left behind; also some of the wash water. However, with repetition of washing and decantation, together with careful drainage and other safeguards, the decantation error should not exceed 5 per cent (Whipple *et al.*, 1927).

f. Filtration outfit may be set up using an ordinary glass funnel, about 8 inches in diameter, within which the sand filter bed is supported on a mass of rolled wire gauze or glass wool. However, the specially designed cylindrical funnel described above is superior.

Qualitative Plankton Methods

Quantitative plankton methods are often of such a nature that material usable for qualitative work is automatically provided. However, it sometimes happens that qualitative plankton studies are in themselves the center of interest, or perhaps a necessary part of some preliminary approach to later quantitative work.

COLLECTING METHODS

Any method of collecting plankters for qualitative work may be employed, provided it secures the organisms in good condition and is otherwise suitable. Because of the crushing effect of high speed centrifuges, plankton so collected is often in poor condition for identification. The pump-and-hose method (p. 231) can be used since here its quantitative errors are of no consequence. Various kinds of tow nets may be employed, provided their limitations are known. Most water samplers are also satisfactory. Slide racks, described on p. 260, are sometimes valuable adjuncts to qualitative collection. In some waters, the stomach contents of plankton-feeding fishes and of tadpoles yield limited plankton collections of interest. Periphyton surrounding submerged parts of aquatic plants acts as a lodgment place for plankters; also some of the components of periphyton itself comprise certain plankters in their sessile stages.

For preliminary purposes, qualitative collections of plankton may be made where and when the operator chooses. However, when the goal is an approach to the completeness of a dependable survey, the whole program of sampling and collection must be organized with considerable care, taking into account the fact that waters differ both within themselves and at different times with respect to the kinds of plankters present. Because of the great diversity of waters and the great variety of changes which occur within them, no fixed rules for the organization of a collection program can be laid down. The limnologist must either feel his way into an adequate program if no previous data are available, or plan a program with the help of such information as may be available from already existing sources. In either instance, proper account must be taken of the following possible differences in the qualitative aspects of plankton populations (a) in different depths in the same body of water; (b) in different seasons; (c) in day and night; and (d) in different areas of the same body of water, such as the limnetic region; various kinds of littoral regions, especially the exposed littoral, as contrasted with the highly protected littoral, regions; regions of extensive inflow of other waters; regions of plankton drift; and plant beds. Still other conditions may need to be taken into account.

IDENTIFICATION OF PLANKTON

Because of the tremendous diversity of organisms composing the plankton of inland waters, the difficulties of identification tax severely the taxonomic skill of even the most experienced worker. Ideally, and often practically, it is just as important to the limnologist as to the taxonomist to identify plankters to *species*. But with the exception of a certain number of well-marked forms, identification to species must be restricted to those particular groups among the plankters in which the limnologist may have taxonomic command through long experience. Beyond this limitation he must count on securing the aid of experts in the other groups involved. Commonly, in general plankton work, strict practicality precludes identification of all plankters to species and in many kinds of work no attempt is made to identify beyond *genus*. There is reason for believing that, in certain types of plankton work, identification to genus is adequate; however, this point must be settled on the basis of the inherent demands of the work involved. In some kinds of plankton work, identifications are made only to family or even order. Clearly such gross identification has only limited value. By force of necessity, limnologists have been compelled thus far to forego even gross attempts to identify the bacteria and the fungi found in the plankton, depending entirely upon the bacteriologist and mycologist for such aid as can be secured.

The beginner cannot be warned too strongly against the risks inherent in plankton identifications. Unfortunately there is a deceiving and non-valid simplicity in the taxonomic treatment of plankters in many works dealing with plankton, both in figures and descriptions, which can lead to grave mistakes by the inexperienced. It is imperative that one fact be kept in the foreground, namely, that the procedures and problems of making valid taxonomic identifications among plankters are just as crucial and exacting as in other organisms; that only by the same meticulous checking of extended descriptions and other sources of information can dependable identifications be made.

CONDITION OF PLANKTON FOR IDENTIFICATION

LIVING MATERIAL. Plankton in living form is, in general, vastly superior to any other condition for identification. Unusually active plankters may be slowed down or immobilized by the use of anesthetics or mechanical

means. Whenever possible, plankton should be studied immediately after collection. Living materials can be kept for some days by storage in an ice box or refrigeration chamber in which the temperature stands at a few degrees above freezing.

PRESERVED MATERIAL. Unfortunately, plankton collections must, more frequently than not, be preserved for study at some later time. No ideal preservative has been found. The best of the various preservatives produce effects which may interfere more or less seriously, depending upon the kind of plankter, with identification. No preservative has been adopted as a standard by limnologists, although formalin (p. 271) has been proposed as such by some workers and it is probably more widely used for general plankton preservation than any other liquid. However, some of the shortcomings of formalin have long been known. Recently, Lackey (1939) made a critical study of changes in river plankton caused by formalin preservation, listed a considerable number of structural changes produced, and pointed out, among other things, that of 234 species studied, 33 per cent could not be identified to species and many of them not even to genus, while a few were totally destroyed, all of which indicates the great advantage of studying unkilled samples. The sudden plunge of a plankton concentrate into killing and preserving fluid often produced severe contraction and distortion of body form of certain rotifers and others, resulting in an object identifiable only, if at all, by the veteran worker. Plasmolysis, loss of colors, loss of parts, and other effects may not only lead to puzzling results but may also involve the actual disappearance of some distinguishing characters. The worker who is obliged to use preserved samples must be on the alert for misleading effects of the preservative.

MOUNTING OF MATERIAL. Directions for permanent mounting of plankters do not come within the province of this book. The reader is referred to various works on microtechnique for instructions for killing, fixing, staining, and mounting of microscopic and semimicroscopic organisms. Temporary water mounts of various living plankters are often usable for several hours. Living plankters may be mounted directly in dilute glycerin and so kept in usable form for some time.

The worker will be able to devise his own methods for slowing down or immobilizing those plankters which are too active for study. A small crystal of chloretone placed at the edge of the cover glass, chloroform on a bit of cotton attached to the cork in the top of the bottle, 1 per cent cocaine solution added to the water, and other similar treatments will be found helpful.

PLANKTON IDENTIFICATION RECORDS

PERMANENT MOUNTS. Good permanent mounts of plankters, made by the use of carefully selected methods of killing, fixing, mounting, and completely labeled, have no peer as records of qualitative work. Various plankters differ in their microtechnique requirements and the methods are frequently time-consuming, tedious and demanding in experience. For details, the reader is referred to standard works on the microtechnique of smaller organisms. The inexperienced worker will soon discover the necessity of practice if acceptable mounts are to be secured. The proper choice of methods is often critical. Careful anesthesia is commonly required in the instance of sensitive, contractile forms. In many kinds of plankton work, time limitations will render permanent mounting impracticable; hence some other form of record will be required.

Drawing and Measurements. For many purposes, rapid but careful drawings accompanied by significant measurements and notes constitute important qualitative records, and often must be used in lieu of permanent mounts. Such drawings should be confined to strictly essential taxonomic details and thus reduce the time involved. Careful measurements are often required in plankton identification and the records of such measurements should be entered on the drawings. Drawings need not be on a large scale, in fact, for many purposes they may be contained in the area of a 3 × 5-in. card. In addition, the record should contain supplementary notes on salient taxonomic features not readily represented in drawings, such as color, behavior, and environmental relations.

FORM 7

PLANKTON IDENTIFICATION RECORD

Genus

Species

Locality	*Date*	*Identified By*	*Remarks*

CATALOG. In addition to the records described above, a qualitative card catalog will be found useful if substantial work is being done. Such a catalog may be of a form somewhat like the one shown in Form 7. The ordinary 3 × 5 in. card is often a suitable size.

PRESERVATION OF PLANKTON

In the past many methods of preserving plankton have been proposed, some of which are elaborate, time-consuming, and not suited to the needs of general plankton work. It now appears that the more complicated methods have little or no advantage over the simpler ones discussed below and they are therefore omitted here. It must be understood that suitability for *general plankton work* is the main concern of the following discussion of preservatives.

FORMALIN. A 5 per cent solution of commercial formalin; a good general killing and preserving fluid which tends to preserve some colors of plankters; in general, produces little shrinkage of tissues; preserves immediately and permanently if strength of solution is kept at proper level; is a powerful antiseptic; and is generally available and inexpensive. It has disagreeable effects on operator; tends to dissolve calcareous materials unless used in *neutralized* forms; action is not uniform, producing in at least a few plankters such different effects as body-form distortion, severe contraction, loss of cilia or flagella, distortion of internal structures, change of color, and change in external structures such as pellicles and shells; and causes complete destruction of a few plankters (see p. 269).

It is imperative that the strength of all formalin solutions be known before they are used; also that proper provision be made to insure against loss of strength by evaporation. Preserved plankton samples should be stored in a darkroom. Confusion in the meaning of the terms *formalin* and *formaldehyde* must be avoided.

ALCOHOL. Pure methyl alcohol; 80 per cent strength. Lacks disagreeable effects of formalin; is an excellent preservative; causes shrinkage of tissues; has a decolorizing effect; is not always a good fixative; plankton must be introduced directly into full strength since transfer through grades of ascending strength can be effected only in the case of net plankton and then only with some difficulty; is more expensive than formalin.

Use of denatured alcohol should be avoided; denaturing chemicals may have deleterious effects upon plankters.

Plankton concentrates should be introduced directly into the storage bottle containing alcohol of such strength that the addition of the concentrate will still leave the preservative strong enough for permanent storage.

FORMOL-ALCOHOL. 5 per cent formalin added to 70 per cent alcohol, equal parts. An excellent fluid for general preservation.

Quantitative Methods

SAMPLING PROGRAMS

Reconnaissance Work. For preliminary or exploratory work, sampling does not necessarily require more than one or two stations, preferably located remote from shore. A generous amount of water should be taken if the sample is to be reasonably representative.

Comprehensive Survey. If a dependable measure of the plankton of any water is desired, the sampling program must be planned with great care, particularly if the lake or stream is of larger size and diversified. No set rules of procedure can be formulated. Each body of water presents its own circumstances on which the details of the sampling program must be based. In small lakes with regular shore lines and uniform bottom configuration, one station near the middle or in the deepest region may yield material representative of the whole lake provided collecting procedures are adequate.

Large, diversified lakes can seldom if ever be adequately sampled from one station and the number and position of required stations must be determined by the features of each lake concerned. Careful consideration must be given to those influences which produce diversity in the plankton distribution. No set rules can be laid down which apply equally in all lakes. However, these suggestions will help in planning such a program:

1. Never depend upon one sampling station.
2. Consider carefully the possible necessity of having at least one sampling station: (a) opposite sizable inflowing and outflowing streams; (b) in each horizontal flowage area; (c) at deepest region, if only one "deep" is present; at each submerged depression if more than one; (d) in partly enclosed bays and coves; (e) over extensive shallow areas; (f) in vicinity of protected shores; (g) in vicinity of exposed shores; (h) behind large islands; (i) behind large peninsulas; (j) in vicinity of marsh or swamp shores.
3. Consider desirability of selecting stations along significantly placed transects.
4. Consider possibility of adequate sampling by strictly random collections in chance positions scattered over all open water.

The number of sampling stations must not only be chosen to meet the demands of the water examined but also must be kept within limits of practicability. An idealized program might involve so many stations that plankton materials so secured would far outrun the ability of the operator to make proper use of them; also it might be made to exceed the general accuracy of the whole method.

SEASONAL SAMPLING. If a true measure of plankton production in any body of water is to be secured, the sampling program must be extended to cover, at regular and significant intervals, all seasons of the year. Only in this way can the well-known and often wide fluctuations of the plankton crop be taken into account. Intervals between sampling must be so chosen that significant variations are not missed. In some programs, practicality may force the use of one-month intervals, but in some waters a truer picture may depend upon the use of two-week intervals. Sampling intervals must be so chosen that particularly significant events are properly provided for, as for example, the seasonal highs and lows of plankton production, the overturns, intervals between broods of plankters, and the changing conditions under permanent ice cover.

DURATION OF SAMPLING. Since plankton production may fluctuate markedly from year to year, no dependable appraisal of the productivity of a body of water can be obtained unless the sampling program is extended over more than one year.

QUANTITY OF WATER PER SAMPLE. In sampling at any one station, sufficient water must be secured to insure a plankton concentrate truly representative of that situation. Here, again, no fixed rule can be followed. Obviously in plankton-poor waters a much larger quantity must be used than in plankton-rich waters; also the same is true for the plankton-poor and plankton-rich seasons of the year in the same water. The operator must use his judgment. Trial collections may be necessary in order to plan dependable sampling. Summer collections of net plankton in the average larger inland lake of the Great Lakes region should be based upon not less than 25 l. per sample, preferably more. Experience has shown that 75 l. usually provides satisfactory concentrates during the seasonal lows of plankton production and 3 l. may provide satisfactory concentrates of nannoplankton secured by the high-speed centrifuge during the same periods. During seasons of plankton maxima, the amounts may be reduced. In lakes of high plankton production, smaller amounts are more satisfactory.

VERTICAL SAMPLING. Much depends upon a proper vertical sampling program. The depth levels at which samples are taken must be chosen in accordance with the circumstances which prevail at the particular locality sampled. Since these circumstances differ greatly in different situations and vary from time to time within the same water, no fixed set of rules for sampling procedures can be laid down. The most satisfactory guide to a vertical sampling program on any particular date of collection is a set of physicochemical records made just prior to the proposed sampling. A vertical series of temperatures, dissolved oxygen, pH, and free CO_2 records will usually give the necessary clues to the requirements for a plankton series. The following recommendations are likely to apply in general and serve as a framework for necessary modifications and extensions.

In temperate lakes of the *first* and *second orders*, samples should be taken at no fewer positions than the following ones during open season: surface, middle of epilimnion; lower region of epilimnion; top, middle, and bottom of thermocline; and top, middle and bottom of hypolimnion. Under permanent ice cover this sampling program must be considerably modified.

In temperate lakes of the *third order* the minimum number of samples in a vertical series should be as follows: surface; midway between surface and bottom; and just above bottom.

The distribution of certain plankters in concentration zones is a possibility which must be taken into account in sampling. Such concentrations may be thick enough to make it reasonably certain that almost any sampling program would include them. However, some concentration zones are so thin that they may be missed entirely if the sampling levels are too far apart. The position of concentration zones differs in various lakes and with the seasons; therefore, no definite directions for locating them can be proposed. However, the possibility of their occurrence must not be overlooked.

Day and Night Sampling. Sampling carried on exclusively during the day may fail to yield a true picture of the plankton for two reasons, namely, the effects of diurnal movements of certain plankters and avoidance reaction.

The various kinds of diurnal movements (Welch, 1935) result in vertical shiftings of populations or parts of populations from one region to another. Surface waters may teem with certain plankters during most of the night but the same plankters may disappear into deeper water or, in some instances, even into the bottom mud with the onset of day. During the day certain plankters may be distributed deeper in the water and in narrow concentration zones which might be missed entirely if sampling levels are too far apart.

Ricker (1938) and others claim that during the day and in the illuminated waters certain Entomostraca have the ability to see and avoid plankton samplers, resulting in a loss of a portion of the population. Sampling at night is recommended by these workers as a means of avoiding this error. Knowledge concerning this behavior is still too meager to warrant definite statements as to its extent, but the possibility of this error should not be overlooked.

SAMPLING PROCEDURES

The following procedures are satisfactory for general purposes and may also serve as a guide to the operation of modified programs.

Sampler-and-Plankton Net Procedure

SUITABILITY. The sampler-and-plankton net procedure is suitable for most quantitative purposes. In addition to collection of net plankton, means are also provided for securing nannoplankton.

EQUIPMENT. Plankton net (p. 239) containing No. 20 or No. 25 silk bolting cloth; 2 or more milk cans (5 or 10 gal.), with known capacity and graduation marks on inside of can; one 5-gal. milk can full of tap water; two 1-pt. dippers; field kit containing set of wide-mouth 250-cc. capacity glass bottles for receiving plankton concentrates, each bottle having a serial identification number and containing 5 cc. of strong formalin; one 3-l. capacity Kemmerer sampler (p. 199) or some similar type of sampler; 1 graduated line, or a line and some form of rope meter; 1 brass messenger; 1 meter stick; tie-on cords; record forms; field kit containing set of 1-l. capacity bottles, each with serial identification number; 3 ordinary buckets.

Procedure

On arrival at selected station, anchor boat, and proceed as follows:

1. Suspend plankton net in neck of milk can with canvas sleeve erect and opening upward; attach one end of 5- to 6-ft. tie-on cord to ring at top of plankton net and other end to some convenient object on hull of boat; place can and included net in convenient place.

2. Put messenger on graduated line; thread zero end of line through sampler and fasten securely. Tie other end of line to convenient object on hull of boat.

3. Set sampler in open form and, with messenger in hand, lower sampler into water to desired depth as measured on graduated line or by rope meter; release messenger and when sampler has closed, bring to surface; lift entirely out of water by use of line; dash a dipper of tap water over outside to wash off any adhering plankters from other levels.

4. Insert lower end of sampler into uppermost sleeve of plankton net and release sampler load into net either by rotating lower valve until it loosens or by judicious endwise pushing of central rod of sampler.

5. Reset sampler and repeat operation. Keep careful record of number of sampler loads put through net; watch water level in milk can until it becomes filled with strained water; if more than one container is needed, transfer plankton net to another can and proceed as before.

6. At conclusion of sampling, read level of strained water in each can by means of graduated marks on inside and put on record; also for purposes of further check, insert meter stick vertically into can and measure depth of strained water, and record result. Record identification numbers of cans used at each station and at each depth level.

7. Lift plankton net vertically in mouth of can and allow contained water to drain into can until water level in net drops below upper rim of plankton bucket. Detach plankton bucket, set lower end into open neck of concentrate bottle, and remove bucket plunger, delivering contents of bucket into bottle. Replace plunger and reattach bucket to net; with dipper, dash wash water all around inside of net to wash down adhering plankters, at same time grasping metal parts of bucket and giving it a rotary motion to wash its inner surface. Drain as before, this time discarding wash water, remove bucket from net, and deliver collected material into concentrate bottle; record number of bottle used.

8. Wash plankton bucket in pail containing tap water; also, with hand over lower open end of headpiece, dash tap water into inside of net and release by removing hand in order to eliminate any plankters which possibly remained after first washing; reattach bucket to net.

9. Stir water in milk cans and fill one or more liter bottles with water from each different station or level sampled for subsequent nannoplankton measurements; stopper carefully and make proper records; remaining strained water may be discarded.

General Considerations

a. Telescoping surfaces of plankton bucket and headpiece should be kept lubricated with a minimal amount of petrolatum or some similar substance. From time to time these surfaces should be cleaned with a rag saturated with gasoline, dried, and relubricated. Excess lubricant may easily find its way onto the silk bolting cloth.

b. In all operations, avoid touching silk bolting cloth with hands, sampler, or other objects.

c. Washing of net and bucket should be done inside the boat. Washing over the side of the boat in lake water is poor technique; also there is the hazard of loss of net or its detachable parts.

d. Before sampling begins, the operator must make certain that the silk of the net is in good condition in every respect. In an older net close scrutiny is required since small breaks may appear under the best of handling.

e. Cans containing water which has been strained through the plankton net should be capped and inverted four or five times before filling liter bottles for nannoplankton measurement.

Plankton Trap Procedure

Suitability. The plankton trap procedure is suitable for most quantitative work, provided net plankton alone is adequate. Nannoplankton from the same sample cannot be secured.

Equipment. Plankton trap, complete with attached plankton net of

No. 20 or No. 25 silk bolting cloth; crane or hoist suitable for attachment to boat to lift plankton trap; large milk can or similar receptacle containing tap water; two 1-pt. dippers; field kit containing set of wide-mouth 250-cc. glass bottles for receiving plankton concentrate, each bottle with serial identification number and containing 5 cc. of strong formalin; 1 graduated line, or plain line or cable if hoist is equipped with rope meter; brass messenger; record forms; 2 pails.

Procedure

On arrival at selected station, anchor boat and proceed as follows:

1. Install hoist in convenient position; thread messenger on line; attach line securely to plankton trap; set lids of trap in open position.

2. Swing arm of hoist over water and in such position that trap will not make contact with boat; lower trap slowly into water; control its sinking so that vertical descent is made certain; lower to desired depth. Drop messenger and permit sufficient time to elapse for messenger to release lids of trap; lift slowly to surface; lift gently out of water and allow trapped water to drain out through net until level is just below upper rim of plankton bucket.

3. Detach plankton bucket from net; set lower end of bucket into open neck of concentrate bottle; remove plunger, delivering contents of bucket into bottle. Replace plunger and reattach bucket to net; pull upper trap lid into open position; with dipper, dash tap water all around inside of both trap box and net to wash down any adhering plankters, at same time grasping bucket by its metal parts and giving it a rotary motion to wash its inner surface more thoroughly. Drain as before, discarding tap water which drains through; remove bucket from net; deliver collected material into same concentrate bottle; record all essential data.

4. Wash bucket in pail containing tap water; also dip net into tap water; reattach bucket to net and proceed to next sample.

General Considerations

a. A plankton trap must be kept in a specially built field case during transportation to avoid damage. On return to the laboratory the trap, net and case should be dried in open air but not in direct sunlight.

b. A plankton trap may be used without a hoist by the hand-over-hand method, but ordinarily this is not a satisfactory procedure because of the unwieldiness of the trap and the risk of damage.

c. Ordinarily the field party should be composed of two persons although, with the right kind of a hoist and accessories, one person can make collections with the plankton trap. In the absence of a suitable hoist, it is ordinarily impracticable for one operator to work alone.

d. Before going into the field the trap should be inspected thoroughly for good condition and operation. The grooves in which the lids slide may, at times, require cleaning.

e. If traps of the larger capacities are used, it may happen that one sample will yield a suitable concentrate. However, if more than one sample from the same place or level is required and the resulting concentrate is too dilute, reconcentration of the accumulated concentrates may be desirable, but this must be done before the formalin is introduced.

Net Coefficients

Formerly, when collection of net plankton by hauling nets through water was a more common practice, much attention was given to the correction of certain errors involved. It was realized that the degree to which a net, drawn through water, filters that entire column of water represented by the size of the net aperture and the distance through which the haul is made, depends upon certain factors such as speed of haul, size of mesh, age of net, and clogging of net. Even under the best operating conditions, a certain amount of overspill occurs at the net aperture with the result that not all of the water passes through it and the plankton catch is less than it should be. The ratio of the actual catch to the catch that would have been taken had all of the water of the column passed through the net is known as the *net coefficient* and must be determined experimentally for each individual net at the various proposed speeds of operation. The ratio so secured is used in applying a correction to actual catches. For example, if for a certain net operated at a given speed the ratio $\frac{\text{total plankton in water}}{\text{total plankton caught}} = 1.3$ (net coefficient), then the plankton actually caught must be multiplied by 1.3 if the final result is to approximate the total plankton in the water column.

Since of recent years plankton nets are commonly used only for straining operations in which definitely measured quantities of plankton-bearing water are delivered through the net, and since nets drawn in the water, such as Clarke's net, may be equipped with meters which measure the amount of water actually entering the net, the determination of net coefficients is a matter of decreasing importance. However, the closing net, described on pp. 246–249, is still a useful device for some purposes and if used for precise quantitative work, the efficiency of the net should be tested and coefficients determined. Various methods of determining net coefficients have been developed but for common purposes the following one will probably be convenient and adequate for the closing net: If a plankton trap (p. 242), having a net of same mesh and age as the closing net is available, make catches at meter or half-meter intervals across the same stratum through which the closing net is to be hauled; from these trap catches compute the total plankton content of the whole water column having transverse dimensions of plankton trap; from this information, compute the plankton content of the column having the transverse area of the closing-net aperture; compare this result with the plankton catch made by drawing the closing net through selected stratum. This procedure will give a fair index of the efficiency of the net. If a plankton trap is not available, the same procedure may be followed, using a sampler, such as a Kemmerer sampler, bringing samples to the surface and passing them through the same closing net as used for the haul.

Analyses of Plankton Concentrates

Choice of a procedure for analyzing plankton concentrates quantitatively must be made in accordance with the use to which the results are to be put. The various methods may be grouped under three main headings, viz., numerical methods, volumetric methods, and gravimetric methods.

NUMERICAL METHODS

Under the designation of Numerical Methods are grouped those quantitative methods which result in measures of plankton expressed in terms of *numbers of plankters* or *numbers of units of plankters*, or both, per unit volume of natural water.

Equipment

All numerical methods require the use of suitable optical equipment.

Compound Microscopes. A good grade compound microscope is a necessity and should be equipped at least with a draw tube, one 10× ocular and one 16-mm. (10×) objective. More extensive accessories may be necessary, such as (a) other oculars and objectives of higher and lower magnification; (b) substage condenser, with iris diaphragm; (c) mechanical stage; and (d) graduated fine-adjustment screw.

Binocular Microscopes. For certain kinds of plankton work a binocular microscope is useful since it provides the most satisfactory means of making total counts of large plankters within a whole counting cell. The optical equipment of such an instrument should include 10× oculars, one 40-mm. objective, and one 25-mm. objective. Other optical combinations may be used but because of the small size of the largest plankters it is likely that ordinary needs will fall near the combinations indicated above. A mechanical stage is often desirable.

Ocular Micrometer. Some type of ocular micrometer adapted for plankton counting is required. Such a device may be of various forms but the most satisfactory one is the Whipple micrometer (Fig. 77). Its rulings are engraved on a disk of thin clear glass the diameter of which is such that it will just fit into the interior of an ocular. The dimensions of the large square are such that with a certain combination of objective, ocular, and draw-tube length the area enclosed on the microscope stage is exactly 1 sq. mm. (1 mm. on each side). Such micrometers may be in the form of a simple glass disk, or they may be rimmed with metal for greater protection against breakage.

COUNTING CELL. A counting cell suitable for ordinary purposes is designed for the examination of plankton concentrates on the stage of a microscope. It may be constructed in many forms depending upon the character of work to be performed. The type commonly used is the Sedgwick-Rafter cell consisting of a brass rectangle, 50 × 20 × 1 mm.

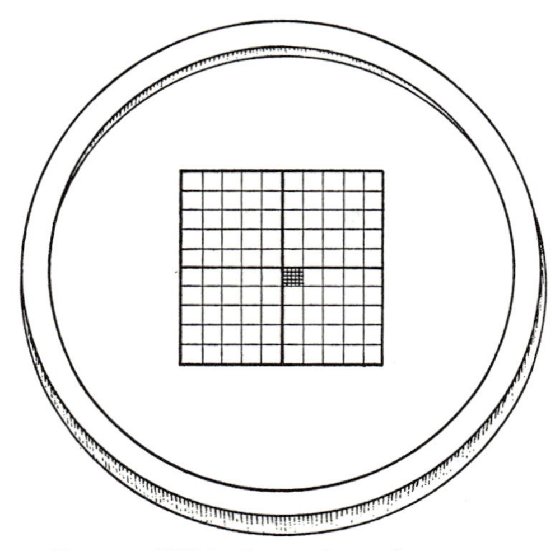

FIG. 77. Whipple ocular micrometer.

sealed to an ordinary 1 × 3-in. glass microscope slide (Fig. 78). It encloses an area of 1000 sq. mm., has a depth of 1 mm., and has a capacity of exactly 1 cc. A rectangular cover glass, large enough to cover the whole cell, is required.

The principal objection to the Sedgwick-Rafter counting cell arises from the fact that its thickness prevents the use of higher magnifications of the microscope. Very shallow counting cells which permit the use of high-power objectives may be made by carefully constructing from very thin sheet brass a rectangle similar in length and width to the Sedgwick-Rafter cell and sealing it to a glass slide. Similar counting cells may be made by cutting four strips of very thin glass and sealing them on a glass slide in the form of a rectangle. Canada balsam is a suitable sealing material.

The form of a counting cell may be immaterial. A circular cell is easily made and quite satisfactory. The depth and diameter may be made to suit the desires of the operator. For some purposes a long narrow counting cell has advantages, especially when the larger plankters are counted under a binocular microscope and the entire width of the cell is within the diameter of the field of view.

Pipets. Samples of plankton concentrates are usually transferred from the concentrate bottle to counting cells or to other receptacles by means of pipets. These pipets must be of such form that the samples taken are truly representative of the concentrate as a whole. If ordinary laboratory pipets are to be used, the operator must make certain that the intake apertures are several times greater in diameter than the longest dimension of the largest plankter thus insuring against the possibility that the pipet

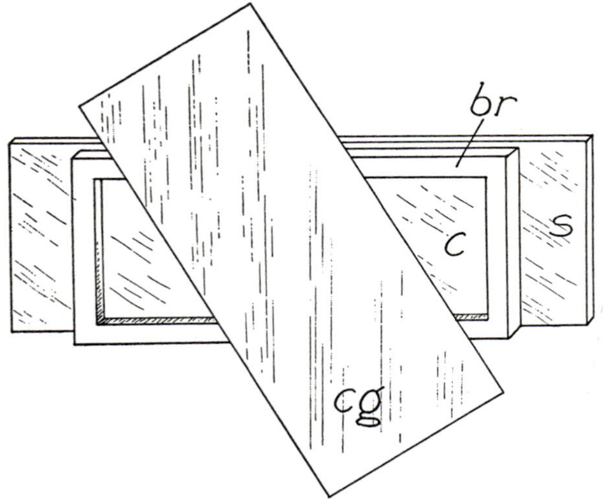

Fig. 78. Sedgwick-Rafter counting cell. (*br*) Brass rectangle. (*c*) Cell enclosed by brass rectangle. (*cg*) Cover glass. (*s*) Glass slide to which rectangle is sealed.

is exercising any selection of organisms when concentrate is drawn into it. Usually the safest procedure is to cut off the tapered tips of pipets, leaving a large opening. Such a pipet should be tied with a string to the concentrate bottle in which it is to be used, thus safeguarding against the possibility of mixture of pipets among the different concentrate bottles.

Special devices for accurate quantitative sampling of plankton concentrates have been devised. The well-known Hensen-Stempel pipet (Fig. 79) consists of a piston, composed of alternate layers of metal and cork held tightly together, fitted into a strong glass tube. To the lower end of the piston is attached a spool-shaped piece of metal, the flanges of which fit closely the inside surface of the glass tube. When pulled into the tube, the spool forms with its walls a closed space of known volume. The metal spool is removable and others of different size may be substituted. If the cork in the piston has become dry from long storage, it should be soaked in water for several hours before it is put into use. To operate, push piston handle down, exposing spool to outside; thrust lower end of pipet deep into the thoroughly mixed plankton concentrate; pull handle up until spool is entirely within walls of glass tube thus enclosing sample; then lift pipet fròm concentrate; carry to its destination, and discharge by pushing down handle until spool is again exposed. The principal virtue of the Hensen-Stempel pipet is its convenience and its speed in taking samples.

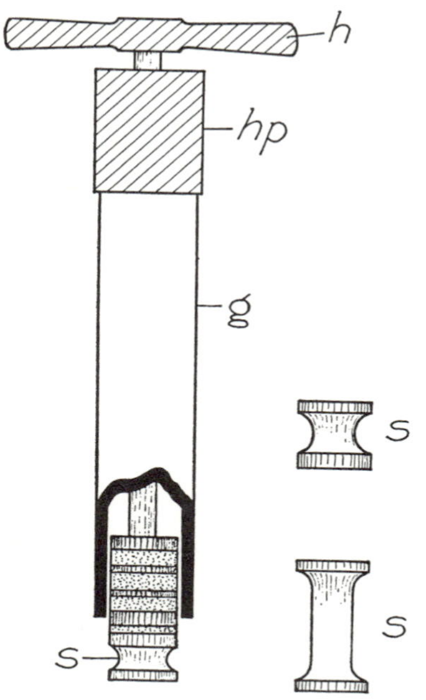

FIG. 79. Hensen-Stempel pipet for quantitative sampling of plankton concentrates. Lower end of pipet shown is cut away to expose piston composed of alternate layers of cork and brass; crosshatched parts are brass, stippled parts are cork disks. (*g*) Heavy glass tube. (*h*) Handle. (*hp*) Brass headpiece. (*s*) Metal spool, detachable, determines size of sample; three sizes are shown in figure.

Calibration of Ocular Micrometer

Irrespective of type of construction, every ocular micrometer must be calibrated. Also it should never be assumed that calibration values worked out for one individual micrometer can, without verification, be applied to another. Details of calibration may differ for micrometers of different forms, but fundamentally the procedures are similar.

WHIPPLE OCULAR MICROMETER. Since the Whipple ocular micrometer is the one in most common use and is recommended for general purposes, its calibration procedure will be described in detail. It will also serve as a basis for the calibration of other micrometers.

Procedure

1. Select compound microscope and accessory equipment; clean thoroughly all optical parts; record identification numbers of microscope, objectives and oculars.

2. Select Whipple ocular micrometer; record its identification number; clean thoroughly.

3. Remove 10× ocular from microscope; unscrew eye lens (lens near-

est eye of observer); insert micrometer into ocular, ruling side downward, and make certain that micrometer comes to rest flat on shelf extending from side of ocular about midway of its length; replace eye lens in ocular; replace ocular in microscope and examine for clarity of micrometer rulings. These rulings must be in clear, clean-cut focus; if not, try different positions of micrometer in ocular until clearly defined image of rulings is secured.

4. Select suitable stage micrometer with rulings of known value (values usually indicated on one end of slide); compare rulings with some dependable standard as check for accuracy; rulings with values of 0.1 and 0.01 mm. are usually suitable; place stage micrometer on stage of microscope, turn on 16-mm. objective, and focus on stage micrometer rulings for as clear an image as possible. Since rulings are on glass, careful adjustment of light is a necessity.

5. Superimpose ocular micrometer rulings upon those of stage micrometer and orient so that one side of former rests across scale of latter, one end coinciding exactly with one line on stage micrometer; pull out draw tube of microscope barrel, observing effect of same, until one side of ocular micrometer square becomes exactly 1 mm. in length; turn ocular 90° and test length of adjacent side of square in same way; repeat similarly for other sides; by means of graduated scale on side of draw tube, ascertain tube length used and make permanent record of same. In similar way check accuracy and determine value of subdivisions within large square of ocular micrometer.

6. If ocular micrometer has been properly constructed, and if the conditions mentioned above have been met, then area enclosed by large square = 1 sq. mm. on stage; that enclosed by each of 100 interior squares = 0.01 sq. mm.; that enclosed by one of smallest squares = 0.0004 sq. mm. One side of smallest square has length of 0.02 mm.

7. Calibrate other combinations of objectives and oculars in similar way.

8. Record all data in table constructed similar to Form 8.

General Considerations

a. Occasionally the operator may find an ocular in which the interior supporting ledge is not properly placed; consequently, there is no position for the ocular micrometer in which the rulings are in clear view. In such instance the substitution of another ocular of the same grade is the quickest solution of the difficulty.

b. Some ocular micrometers are rimmed with brass or other metal as protection against breakage. In such rims the glass may be so mounted that one side is deeper than the other. Commonly the deeper side should be uppermost in the ocular.

c. Close-fitting, metal-rimmed ocular micrometers sometimes are difficult to introduce into, or remove from, the ocular because of a tendency to jam. Gentle shaking, jarring on palm of hand, judicious use of a rubber-tipped probe, or similar treatment will correct the trouble. Use of force must be avoided.

d. In the calibration of the ocular micrometer used with different optical combinations of oculars and objectives, it is possible, and under some circumstances more convenient, to find a draw-tube length for each optical combination which will make some lines on both ocular and stage micrometers coincide exactly. However, this may necessitate a different tube length for each combination, thus leading to errors through the operator's forgetting to change tube length with changes of optical parts. Ordinarily it is safer to calibrate all optical combinations on the basis of one tube length.

e. If the large square of the ocular micrometer is to be used as a whole, the operator must determine at the outset whether the entire square falls within the field of view. If two adjacent corners are slightly cut off, gentle tapping on the side of the ocular may shift the micrometer on its supporting ledge and bring all corners within view; if all four corners are cut off, or if large amounts of two corners are cut off, there is no remedy except to use an ocular of different construction. However, cut-off corners will not interfere with the use of the smaller squares of the micrometer.

f. Dimness of ocular micrometer rulings on one side results from failure of the micrometer to rest flat on the supporting ledge, or from a defective installation of the supporting ledge in the ocular.

FORM 8

Microscope No.	*Ocular Micrometer No.*	*Stage Micrometer No.*	*Ocular*	*Objective*	*Tube Length*	*Value of Counting Unit*

Survey Count

In the *survey count* the large net plankters, such as the Microcrustacea, rotifers, and others of similar size which can be readily seen and identified under very low magnifications of the microscope, are counted in the entire counting cell.

Suitability. The survey count, as commonly used, is suitable (a) when only an enumeration of the largest plankters is needed; (b) when in the interests of economy of time and energy the largest plankters are counted in the entire counting cell, leaving all smaller ones to be counted by the application of another more suitable method, and (c) when some of the largest plankters occur in such sparse numbers that a much larger quantity of concentrate must be examined in order to secure a more adequate measure of them.

Equipment. Compound microscope with low-power objective and ocular, or binocular microscope with assortment of objectives and oculars; Sedgwick-Rafter counting cell; pipets for transferring concentrate; thin sheet of translucent paper, cover glass, or other similar material, ruled in squares. Mechanical stage and hand tally are convenient but not required.

Procedure

1. By means of appropriate type of calipers, measure dimensions of counting cell very accurately and compute its capacity.

2. By use of graduate of suitable size and graduation, measure total volume of original plankton concentrate; enter on record form.

3. Fill counting cell with tap water and test for leakage; empty cell and dry completely.

4. Place counting cell on level surface; lay cover glass diagonally across middle of counting cell (Fig. 78) leaving uncovered area at each end.

5. Make certain that concentrate bottle is securely closed or stoppered; take concentrate bottle in hand, placing thumb over stopper or cover to insure that it stays in place; invert concentrate bottle slowly and gently at least seven times to provide thorough mixing; open concentrate bottle immediately and with pipet secure sample before any settling of plankters occurs; transfer pipet sample immediately to counting cell, introducing into one of open ends a quantity sufficient to slightly more than fill cell; return unused part of pipet sample to concentrate bottle.

6. Keeping cover glass in contact with rim of counting cell, carefully swing it around until whole cell is covered, cutting off excess concentrate and leaving cell completely filled. If air bubbles are enclosed within cell, whole procedure of filling must be repeated.

7. Transfer counting cell to stage of microscope and examine. If, as is likely, only a portion of cell is visible, slip under counting cell a sheet

of translucent paper, a large cover glass, or some similar object, ruled in squares of convenient size, thus dividing whole cell into recognizable areas; attach mechanical stage if available.

8. Determine what plankters may be dependably counted in survey; begin at one end of counting cell and progressively count plankters in each of square areas. If survey includes only one kind of plankter, counting may be speedily recorded by hand tally if available; if not, use ordinary tally method on record sheet. If survey includes several kinds of large plankters, count one kind at a time as just indicated, or if simultaneous counting of all selected plankters is desirable, use tally method on record sheet.

9. Examine as many different samples from concentrate bottle as practicality and desired precision of result require.

10. From averages of individual counts, compute total number of plankters in entire concentrate; then compute number per liter of original water as it occurred in nature.

Total Count

By *total count* is meant an enumeration of all plankters recognizable under the conditions of the examination without any attempt to distinguish between the different kinds.

SUITABILITY. The total count is suitable for quantitative determinations of plankton in which the mere *number of individual* plankters is adequate. It may be used (a) in making a count of all plankters, large or small, in the total concentrate or (b) in combination with the survey count in which instance the smaller plankters are counted. It is commonly used for work with net plankton but if very thin counting cells are employed, counts of nannoplankton are possible.

EQUIPMENT. Compound microscope with suitable series of oculars and objectives; counting cell, either standard Sedgwick-Rafter type or specially made thin cells; pipets for transferring concentrate; Whipple ocular micrometer, or similar type; record forms. A hand tally is a convenience and adds to the speed of counting.

Procedure

1. With appropriate type of calipers measure carefully depth of counting cell.

2. Make careful measurement of volume of original plankton concentrate; enter on record form.

3. Fill counting cell with tap water and test for leakage; empty cell and dry completely.

4. Install micrometer in ocular and calibrate as described on p. 282. After areal value of largest square has been determined, convert into cubic

unit by multiplying it by depth of counting cell. Record value so computed.

5. Mix contents of concentrate bottle and fill counting cell as described on p. 285; allow counting cell to stand for short time until organisms have settled; transfer cell to microscope.

6. Select, at random, one area, wholly within cell and count included plankters according to following directions: (a) count only plankters which were alive at time concentrate was killed; (b) ignore all kinds of debris, cast skins, fragments of plankters, and objects and organisms accidental to the concentrate; (c) count as *one plankter* each filament, colony, egg mass, partial colony or mass irrespective of size; (d) count each separate individual organism as one plankter; (e) in instance of plankter partly inside and partly outside counting square, estimate fractional part within square and so record; (f) counting fields which by chance of random selection contain no plankters must be counted as zero and included in computing averages; (g) any stage of life cycle of plankters must be counted when isolated from parent.

7. Focus microscope through entire thickness of cell to make certain that all plankters are counted.

8. With eye removed from ocular, move counting cell any distance to new field and count as before; repeat until ten fields selected strictly at random have been counted. Use of 2 different counting cells full of concentrate and 10 fields counted in each gives higher degree of accuracy.

9. From results so secured, compute number of plankters per liter of original lake or stream water by use of following formula:

$$n = \frac{(a1000)c}{l}$$

in which n = number of plankters per liter of original water
a = average number plankters in all counts in counting unit of 1 cu. mm. capacity
c = volume of original concentrate in cc.
l = volume of original water expressed in liters

If the survey method has been employed for the larger plankters of the same concentrate, then the total count may be computed from the following formula:

$$n = \frac{wc}{l} + \frac{(a1000)c}{l}$$

in which w = the number of larger plankters enumerated in survey count in whole counting cell (1 cc.), or average of all counts of whole counting cell (1 cc.)

General Considerations

a. Counting cells should be tested at all four corners for uniformity of depth; resealing of cells sometimes results in one side or one end being higher than the opposite one.

b. Distribution of plankters in the counting cell is not, and cannot be, uniform; therefore, it may happen that counts of the various counting units differ considerably; hence the value of making a considerable number of counts.

c. No rule can be laid down concerning the number of random units to be counted in each cell. The ten indicated above is commonly regarded as reasonably good practice; a larger number taken from more than one counting cell will result in increased accuracy.

d. Sometimes it is desirable to make a preliminary examination of the concentrate to discover if the amount of dilution is favorable. If a concentrate is so thick that counting is difficult, the test sample should all be returned to the concentrate bottle and tap or distilled water added until a desirable dilution has been produced; then the newly diluted concentrate must be remeasured for volume. If a concentrate as taken from the field operations is found to be too dilute for satisfactory and dependable counting, it may be necessary to reduce the concentrate to a smaller volume by passing a portion of it through hard surface filter paper, returning the catch on the filter paper carefully to the other portion of the concentrate and remeasuring the new volume of the total concentrate.

e. Mixing of concentrate prior to counting must be done slowly and gently; rough treatment may result in damage to plankters.

f. The operator must not be misled by accidental attachment and other odd forms of relation of diverse plankters to one another in concentrates—results which come about from the very fact of concentration.

g. Ordinarily, it is regarded as impracticable to attempt to count bacteria in concentrates.

h. Difficulties sometimes arise in positive discrimination between remains of plankters which were dead before the sample was collected and plankters which were alive at the moment of killing by preservative in the concentrate bottle. Ordinarily doubtful cases should not be counted.

i. Objects so small as to be near the limit of vision may not be recognizable with certainty as plankters. It is usually necessary to ignore them.

j. When, for any purpose, it is desirable that a measure of the detritus be included, count the *masses* of detritus and list separately under that designation.

k. A hand tally adds to the speed and accuracy of counting; however, in using a tally, the estimation of fractional parts of plankters partly within the field must be done separately and then added into the total.

l. If the microscope is equipped with a mechanical stage, the *zonal* method of counting may be substituted for the random unit-area method. The procedure is as follows: Start at one end of counting cell and move cell by means of mechanical stage toward opposite end, counting plankters as they pass through ruled field of ocular micrometer; repeat until several such zones have been counted. The dimensions of a zone will be the product of the length of the counting cell, its depth, and the length of one side of the ocular micrometer square. That product multiplied by the number of zones will give the amount of concentrate actually examined. Zones may be made across the counting cell if desired.

m. Final computation is facilitated if, by the addition of wash water (distilled water, tap water, or filtered water), the volume of the concentrate is made to be some exact multiple of 10, provided, of course, that the addition of more water will not produce too great a dilution.

Differential Count

By *differential count* is meant the enumeration of some or all of the different kinds of plankters, distinguishing them qualitatively, counting and recording the numbers of individuals of each. The extent to which the taxonomic subdivision is carried will depend upon the conditions and purpose of the study, and upon the experience of the operator in the recognition of species, genera, and other groups in the classification. Ideally, such a count should involve identification to species, but depending upon circumstances, this may be impossible, or impracticable, or even unnecessary (p. 268).

SUITABILITY AND EQUIPMENT. Same as in the total count (p. 286).

Procedure

1. Follow procedures outlined in preceding section on *total count*, except that each different kind of plankter is counted separately.

2. Enter counts of each kind of plankter on record form by the tally method. If only certain selected kinds of plankters are considered, counts may be made with a mechanical hand tally.

3. Make calculations for each kind of plankter by means of formulas indicated in preceding section (p. 287) and express results in *number per liter*.

Limitations of Survey, Total, and Differential Counts

Results obtained by survey, total and differential counts are enumerations only in which no account is taken of great differences in size of plankters. Because of the necessary use of counting cells having some depth, high magnification is precluded and the smallest of the nannoplankton can receive at best only very uncertain consideration. Precision of these methods, under the best of conditions, probably cannot be expected to be better than about 90 per cent.

Areal Standard-unit Method

The areal standard-unit method was devised originally by Whipple to provide a quantitative means of measuring plankton in which the size of the individual plankters is considered. Thus one of the obvious limita-

---

FORM 9

PLANKTON RECORDS

Locality	Depth	Records By
Date	Bottle No.	Apparatus
Time	Lake Water	Corrections
Wind	Concentrate	Counting Unit
Sky		

Organisms	1	2	3	4	5	6	7	8	9	10	Total	Av.
Totals												

Number per

tions of the survey, total, and differential counts is largely eliminated. The areal standard-unit method, when properly employed, has commendable features and is widely used. It is based upon the principle that since most organisms in a counting cell rest in such a way that their greatest surface area is in a horizontal position, a measure of that *area* of each plankter visible to the operator gives a close estimate of its volume. The measurement is, then, actually one of area although it is used as an index to volume. In order that such a value may be in usable form, an *areal standard unit* is employed in terms of which all measurements are expressed. This standard unit has an area of 0.0004 sq. mm. (400 square microns) enclosed within a square each side of which has a length of 0.02 mm. (20 microns). The Whipple ocular micrometer was designed specifically to meet the requirements of this method.

Suitability. The areal standard-unit method probably finds its most convenient use when applied to net plankton and when employed with that combination of objective, ocular and tube length on the miscroscope which provides that one side of the large square of the ocular micrometer exactly covers 1 mm. on the stage. However, its use may be extended to higher magnifications and very thin counting cells by reëvaluating the ocular micrometer for each optical combination in terms of the areal standard-unit dimensions as given above.

Equipment. Whipple ocular micrometer; compound microscope with 10$\times$ ocular and 16-mm. objective, other objectives and oculars as desired; Sedgwick-Rafter counting cell, or equivalent, or thinner cells if desired; pipets for transferring plankton concentrate.

Procedure

1. Introduce Whipple micrometer into 10$\times$ ocular of microscope; test for clear visibility of micrometer rulings; set microscope with 16-mm. objective and proper tube length (see p. 282 for directions for calibrating micrometer) so that each side of large square in ocular micrometer exactly covers 1 mm. on stage.

2. Measure dimensions of counting cell as described on p. 285; fill counting cell with concentrate following directions given on p. 285.

3. With microscope set as indicated above, each of series of 25 smallest squares ruled near middle of large square (squares of third order) has area exactly representing one *standard unit* (0.0004 sq. mm.). Since all other squares (squares of first order; squares of second order) are multiples of the standard unit, the eye of operator, with practice, can easily divide sides of squares of second order into fifths.

4. Select one plankter wholly within large squares; determine, by means of squares or partial squares which it crosses, length and breadth of selected plankter in terms of length of one side of standard unit (0.02

mm.); multiply number of units of length by number of units of width and result will be in *standard units*. Proceed in same fashion with all other plankters within large square. In instance of plankters partly within and partly outside square determine number of standard units only in those portions within square.

5. Proceed with nine other counting areas (or some larger number) selected at random within counting cell; on basis of averages of total count and differential counts, compute number of *areal standard units* in 1 l. of original water.

General Considerations

a. It will be convenient to designate 0.02 mm. (length of one side of areal standard unit) as one *linear unit*. Thus, for example, a certain filament can be referred to as 20 linear units long and 0.6 linear unit wide.

b. The efficiency of this method depends much upon careful practice. The eye rapidly gains accuracy and, with practice, speed in estimating linear dimensions and areas in terms of the standard unit, a procedure which at first seems slow, tedious, and uncertain, soon develops into an effective one.

c. It will usually be desirable to combine simultaneously the total count, the differential count, and the areal standard-unit count in the same procedure, thus yielding results which supplement each other.

d. Some plankters, e.g., certain diatoms, are so constant in size that they may be counted individually and then converted to standard units by multiplying the total number of individuals counted by a constant representing the size of the individual. Some filamentous plankters have very constant diameters, hence counting is faciliated by determining the length in linear units of the individual filaments and then multiplying by the number representing the common diameter.

Cubic Standard-unit Method

The method of measuring plankters quantitatively by the use of a *cubic unit* is in some respects the most desirable one of the various enumeration methods due to the fact that the results are actually volumetric. It is essentially an extension of the areal standard-unit method in which the third dimension of plankters is measured and results expressed in terms of a *cubic unit* whose size is 0.000008 cu. mm. (8000 cubic microns) and the length of one side of which is 0.02 mm. This unit, then, is merely the areal unit converted into a cubic unit by multiplying it by the length of one side of the cube, namely, 0.02 mm.

SUITABILITY. The suitability of the cubic standard unit method is very similar to that of the areal standard-unit method (p. 291).

EQUIPMENT. The equipment needed is the same as that required for the areal standard-unit method, except that if the third dimension is to be measured directly, the compound microscope must be equipped with a graduated micrometer head on the fine adjustment so that the third dimension (thickness) may be measured by determining the distance through which the microscope tube moves in focusing from the uppermost to the lowermost surfaces of a plankter. The value of the change in vertical distance produced by the rotation of the fine adjustment screw is commonly supplied by the makers of the microscope. If the value is not known, it will be necessary to calibrate the screw.

Procedure

1. Set up microscope and fill plankton counting cell as described previously for areal standard-unit method (p. 291).

2. Before beginning count, decide which one, or combination, of the following ways of obtaining value of third dimension is most practical and suitable for work in hand:

(a) Focus carefully on topmost and bottommost surfaces of plankters and determine thickness from vertical distance through which objective moved by reading on graduated fine-adjustment screw head. If plankter is too opaque to permit focus on bottom with certainty, focus on top, then on periphery, and regard vertical distance through which objective moved as one-half the thickness of plankter.

(b) Note whether form of plankter approaches that of a sphere, a cylinder, or an ellipsoid; if so, use diameter as third dimension. If form approaches cube, use linear measurement as third dimension. This method requires no direct measurement of thickness.

(c) Measure directly thickness of several individuals of each species and compute average; use this average for all individuals of that species subsequently counted.

3. Proceed to counting of 10 or more fields selected at random. Also simultaneously keep record of number of individuals as in total differential count. From average of all counts, compute total cubic units in total concentrate; then compute number of cubic units of plankton per liter of original water in usual way.

4. Record results in cubic standard units. If for any reason it should be preferable to express results in parts per million by volume, divide number of cubic standard units by 125.

LIMITATIONS. Since the cubic standard-unit method involves measurement or estimation of three dimensions, the work is necessarily slower than in the areal standard-unit method; also much greater experience on the part of the operator is required if results are to be reliable. Even in the hands of an experienced operator, any speeding up of the work may lead to errors.

General Considerations

a. The chart (Fig. 80) facilitates the calculation of cubic units from linear dimensions. In this chart the spherical and cubic volumes are obtained directly by the use of any measured diameter or side. Cylindrical volumes are secured by applying the diameter of the circular cross section to the chart as linear standard units, then reading the corresponding cubic standard-units value, and multiplying it by the length of the cell or filament in linear units. Ellipsoidal volumes are obtained by applying the third diameter of the plankter and multiplying the value so determined by the product of the long and short diameters of the elliptical section.

b. As an aid to the computation of cubic standard units, use may be made of a table commonly included in engineers' handbooks from which may be read directly the area of a circle, the volume of a sphere, and the volume of a cube, from a graded series of diameter values.

c. When in some kinds of work a volumetric measure of the amorphous material is desirable, volumes in terms of cubic standard units

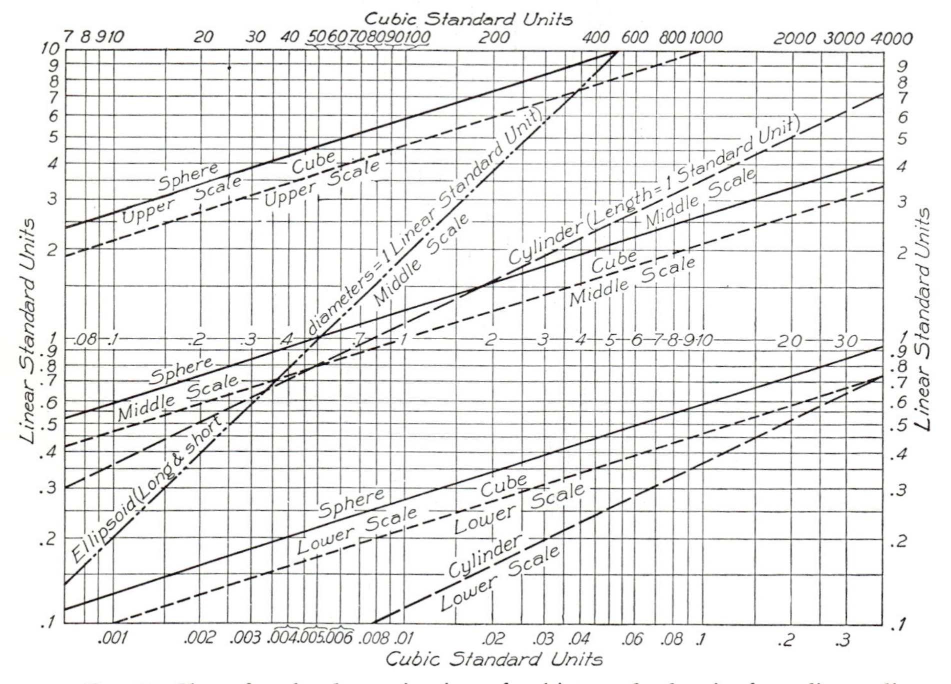

FIG. 80. Chart for the determination of cubic standard units from linear dimensions of organisms. (Reprinted by permission from "The Microscopy of Drinking Water," by Whipple, Fair, and Whipple, published by John Wiley and Sons, Inc.) (Redrawn.)

may be secured by estimation of the number of units in each particle.

d. A convenient form of keeping simultaneous records of both total differential count and volumetric count is the use of a common fraction in which the numerator represents the number of individual plankters while the denominator represents the number of volumetric standard units.

e. For work of this kind a specially prepared plankton-count form with a complete set of headings and ruled columns should be used for all records.

GRAVIMETRIC METHOD

Gravimetric methods are concerned with measurement of plankton in terms of units of weight per selected quantity of plankton-bearing water. The procedures described here are based upon those developed by Birge and Juday.

SUITABILITY. This method may be applied to net plankton, nannoplankton, or both. It is also suitable for plankton concentrates irrespective of the method of collection, provided that all of the necessary collection data are available.

EQUIPMENT. Plankton centrifuge and accessories (p. 256); platinum dish with capacity of 8 cc.; electric constant-temperature oven maintaining temperature at 60° C.; analytical balance sensitive to 0.0001 g.; electric furnace which can maintain a temperature of 600° C.

Procedure

1. Secure centrifuge concentrate as indicated on pp. 258–259. If plankton is abundant, 1-l. samples will be adequate; if plankton is scarce it may be necessary to centrifuge two or more liters of water; measure volume of concentrate (should not exceed 8 cc.).

2. Place concentrate in platinum dish; put in electric oven maintaining controlled temperature at 60° C.; leave in oven until all water is evaporated (about 24 hrs.).

3. Weigh dish and contents; transfer to electric furnace maintaining temperature at about 600° C.; leave in furnace for 30 min.

4. Weigh dish and contents; result of second weighing subtracted from that of first weighing gives *gross loss* due to ignition.

5. To correct for organic matter in solution in that lake water included in the concentrate, run a blank consisting of same amount of centrifuged lake water at same time and in same way as plankton sample. Subtract weight of furnace-incinerated sample of blank from weight of oven-dried condition of same sample; difference represents organic matter in lake water and constitutes correction to be subtracted from *gross loss* mentioned in paragraph 4 above.

6. Compute amount of organic matter in original water, expressing results in milligrams per liter.

General Considerations

a. It must be understood that a centrifuge concentrate includes all of the various materials in suspension in the water, such as silt, wind-blown materials from shore, dead and disintegrating animal and plant remains in finely divided condition, and exuviae. Therefore for this and certain other reasons it is not correct to assume that a centrifuge concentrate is a measure of the living plankton alone. No satisfactory method of separating the living plankton from these other suspended materials has been devised. The magnitude of this error will vary greatly depending upon circumstances; in certain clean waters it may be so low as to be practically negligible for most purposes; in other waters which are heavily laden with suspended matter, it may be so high as to render the method useless. Preliminary examination and the exercise of good judgment will determine whether this method can be expected to provide useful results.

b. The effectiveness of the first run of a sample through the centrifuge has been mentioned on p. 259. If the sample should contain appreciable amounts of *Aphanizomenon*, only about one-half of this plankter will be obtained in the first run, but in the second run practically all is secured.

c. If for any reason it is desirable to express the amount of plankton as *wet weight*, evaporation is stopped at the end of step No. 2 above, care being taken to remove as much free water in the sample as possible without withdrawing water from the plankters. Weight of empty dish subtracted from weight of dish containing evaporated sample will yield the amount of nondesiccated plankton (and possibly other associated materials) and results can be expressed in terms of any convenient unit of weight per liter.

VOLUMETRIC METHOD

By obvious modifications of the gravimetric method just described, samples, wet or desiccated, may be measured volumetrically and results expressed in terms of volume per liter. Ordinarily, however, the gravimetric method will be preferred because of convenience.

Selected References

(For Chapter 15)

Allen, W. E.: Methods in quantitative research on marine microplankton, *Bull. Scripps Inst. Oceanog., Tech. Ser.*, **2**: 319–329, 1930.

Clarke, E. B.: "A Modification of the Juday Plankton Trap." 4 pp. Spec. Pub. No. 8, Limn. Soc. Am., 1942.

Clarke, G. L., and D. F. Bumpus: "The Plankton Sampler—An Instrument for Quantitative Plankton Investigations." 8 pp. Spec. Pub. No. 5, Limn. Soc. Am., 1940.
Juday, C.: Limnological apparatus, *Trans. Wisconsin Acad. Sci.,* **18:** 566–592, 1916.
———: A third report on limnological apparatus, *Trans. Wisconsin Acad. Sci.,* **22:** 299–314, 1926.
Kofoid, C. A.: Plankton studies. I. Methods and apparatus in use in plankton investigations at the biological experiment station of the University of Illinois, *Bull. Ill. St. Lab. Nat. Hist.,* **5:** 1–25, 1897.
Lackey, J. B.: The manipulation and counting of river plankton and changes in some organisms due to formalin preservation. U.S. Public Health Service, *Pub. Health Repts.,* **53:** 2080–2093, 1939.
Littleford, R. A., C. L. Newcombe, and B. B. Shepherd: An experimental study of certain quantitative plankton methods, *Ecology,* **21:** 309–322, 1940.
Ricker, W. E.: An adequate quantitative sampling of the pelagic net plankton of a lake, *J. Fisheries Research Board Can.,* **4:** 19–32, 1938.
Seiwell, H. R.: Patterns for conical silk plankton-nets of one and half-meter diameters, *J. conseil permanent intern. exploration mer,* **4:** 99–103, 1929.
"Standard Methods for the Examination of Water and Sewage," 9th ed., New York, Amer. Pub. Health Assoc., 1946.
Sverdrup, H. U., M. W. Johnson, and R. H. Fleming: "The Oceans, Their Physics, Chemistry and General Biology." 1087 pp. New York, Prentice-Hall, 1942.
Welch, P. S.: "Limnology." 471 pp. New York, McGraw-Hill Book Co., 1935.
Whipple, G. C., G. M. Fair, and M. C. Whipple: "The Microscopy of Drinking Water," 4th ed., 586 pp. New York, John Wiley and Sons, 1927.
Winsor, C. P., and G. L. Clarke: A statistical study of variation in the catch of plankton nets, *J. Marine Research,* **3:** 1–34, 1940.

BOTTOM-FAUNA METHODS

Methods suitable for the study of bottom faunas must differ in order to meet the requirements of various kinds of bottom as they exist in lakes and streams.

METHODS FOR PROFUNDAL BOTTOMS

QUALITATIVE METHODS

Since bottom deposits in profundal regions of lakes are usually of the soft, finely divided type, any bottom sampler such as an Ekman dredge, core samplers, or ooze suckers, which will collect these materials and transport them in unmodified form to the surface, may be used for qualitative work. The construction and use of these samplers have been described on pp. 175–186. The location and number of samples taken must depend upon the judgment of the operator, but clearly there should be a sufficient spread of an adequate number of samples over the area under consideration to insure that the survey is sufficiently complete for the purpose of the work.

The treatment and analysis of samples taken for qualitative work by the Ekman dredge are very similar to those of samples for quantitative work, and since the latter must be described, the reader is referred to p. 300.

Material collected with the ooze sucker (Fig. 53) is particularly favorable for qualitative analysis of microscopic organisms in the uppermost layers of bottom deposits. Detection and isolation of microscopic animals in bottom mud are usually very difficult and examinations of living materials are much easier than those of preserved samples. Such samples should be examined first under a low-power binocular microscope. If necessary the sample should be diluted with filtered or tap water to reduce the concentration of the material. As organisms are found they may be removed by means of a capillary pipet, transferred to a microscope slide, and examined under a compound microscope. The ooze sucker is not suitable for quantitative work since it disturbs the finely divided ooze as it settles on the bottom; it collects a sample only from the uppermost layer of bottom deposits; and its efficiency is probably different on different kinds of bottoms. Collections of profundal mud stored in a refrigerator will usually remain in good condition for several days. Sometimes isolation of the organisms can be facilitated by dividing the sample into fractional portions and putting them through one or more fine-mesh sieves.

QUANTITATIVE METHODS

Ekman Dredge Method

SUITABILITY. Suitable for macroscopic bottom fauna; also for soft, finely divided bottom materials only.

EQUIPMENT. Ekman dredge; strong line for operating dredge; several large-size tubs or similar containers; 2 or 3 ordinary pails; several quart-size, wide-mouth jars with tops and with serial identification numbers, in field kit; several large, strong, circular screens with No. 30 mesh (p. 186); forms for sampling records. If large-size Ekman dredge is used, a suitable hoist is necessary.

Procedure

1. Inspect dredge for good working condition; thread messenger on line; attach line securely to dredge and tie opposite end of line to boat or hoist.

2. Select first sampling position and locate boat above it; set dredge in open form; lower dredge into water so that it will descend vertically; on arrival at bottom, allow dredge to settle; release messenger, holding line with just enough tension to keep it straight. When dredge has closed, lift to surface with moderate, steady speed.

3. On arrival at surface, lift dredge promptly from water, swing over large tub and discharge contents by pulling up each jaw chain two or three times, first on one side, then on the other, to empty dredge completely. Again set dredge in open position, shift position of boat, lower into water and proceed with sampling as before. Keep accurate record of number of samples taken.

4. Spread sampling over selected area according to prearranged plan (scattered at random, or distributed along transects) and accumulate dredged materials until number of samples and quantity of bottom materials meet requirements of work.

5. When sampling is concluded, or when available tubs are filled, transfer dredged materials to shore for screening.

6. Select convenient place at shore in shallow, calm water; deliver small quantities (about 1–2 qt. at a time) into screen; lower loaded screen into water to depth equal to about one-half height of sides; rotate horizontally, alternately clockwise and counterclockwise, and with sufficient vigor to wash finely divided contents through screen, making each swing through about 180°; facilitate screening by occasionally varying motions, such as short, rapid, vertical dips or sharp rocking motions; avoid loss of bottom material over upper edge of screen.

7. After all finer materials have passed through screen, concentrate residue (screenings) at one edge of screen by judicious scooping motion of bottom of screen against surface of outside water; by means of spoon or similar article, carefully transfer screenings to wide-mouth sample bottle; refill screen and continue screening activities until all dredgings are used; careful cleaning of sieve is not necessary until end of screening of sample involved since residues left in sieve during process are not lost.

8. Do not crowd screenings in bottles; include plenty of water in bottles, otherwise certain organisms will die; record serial numbers of bottles; close bottles securely, place in bucket of cool water, and protect against sun.

9. If screenings are analyzed at once, no further directions are necessary. If analyses must be delayed a few hours, store screenings in refrigerator; if delayed for longer time, preservation is necessary.

Limitations

a. This method is limited to bottom materials which are soft and finely divided. Hard bottoms are excluded. Intermixtures of sand interfere with mechanical operation of the dredge.

b. Because of the use of screens, only macroscopic bottom fauna is secured, the size of which depends upon the dimensions of the mesh.

General Considerations

a. Samples may be scattered irregularly and at random over the area concerned, or samples may be made at intervals along transects which cross the area. In either instance, the sampling is random since there can be no actual selection by the operator. Transects have an additional value only when a knowledge of exactly where the samples were taken is important.

b. The number of sampler loads collected must be a matter of judgment by the operator, based upon the size of the area concerned, its degree of uniformity, the density of the population, and perhaps other circumstances. A sufficient number must be taken to insure that the samples are truly representative.

c. When the open Ekman dredge is lowered too rapidly it may tend to "sail" off at some angle from the vertical. The dredge must be lowered as nearly vertically as possible, otherwise it may not meet the bottom in the correct position.

d. If on reaching the surface the dredge shows leakage through the jaws, the difficulty will usually be caused either by grit between the side pieces of the jaws and the wall of the dredge body, or by hard objects caught between the edges of the two opposing jaws. Such a sample must be discarded.

e. Some workers cover the top opening of an Ekman dredge, just under the top lids, with a fine-mesh brass screen to protect against loss of organisms by overspill when the dredge settles into the bottom. Much depends upon the size of mesh of this screen since too coarse a mesh would accomplish little or nothing and too fine a mesh may interfere with the proper passage of water and bottom materials into the dredge. Ekman dredges made in tall form are sometimes used.

f. Field work is facilitated when boats providing high degree of stability are used, particularly on larger bodies of water.

g. Change in position of the boat is imperative if the sampling is to be adequate. Such changes can be made in various ways, such as use of a very long anchor rope, portions of which are pulled inboard from time to time; periodic use of motor; and allowing boat to drift unanchored across the area to be sampled. If transect positions are to be accurately known, the boat must describe definite courses from one set of shore signals to another, or operate under some other kind of guidance.

h. In transporting dredgings to shore, cover tubs with a piece of wet canvas to protect against hot sun and against loss by sloshing caused by motions of the boat.

i. Screening of dredgings may be done on board boats at the time of sampling, but it is usually a laborious and sometimes a less accurate process, especially during times of rough water.

j. The labor of screening may be reduced by the selection of convenient place and certain facilities at shore, as for example, a small dock, protected water, and a firm bottom. However, it must be remembered that large quantities of profundal muds screened into shallow water at shore may produce objectionable effects locally.

k. On occasion it may be more desirable or convenient to wash dredgings on land at some outdoor hydrant. Tap water may be directed gently from a hose into the screens and screening accomplished satisfactorily in this manner.

l. Some workers have used strong bags with large windows of coarse grit cloth as a substitute for brass screens. Dredgings are placed in such a bag, the top closed, and the bag manipulated in the water or trailed behind a boat until the fine materials are removed. For most purposes it appears that screens are preferable.

m. If screenings must remain in storage for an indefinite time after collection, they should be preserved in a sufficient quantity of 10 per cent formalin to insure safekeeping. However, it should be remembered that preserved materials are more difficult to analyze.

n. At the end of sampling activities, wash all equipment, ropes, and boats before adhering bottom muds become dry. Wash screens with particular care.

Analysis of Ekman-dredge Samples

EQUIPMENT. Large white porcelain pans, or 8-in. shallow culture dishes with white backgrounds; finger bowls, watch glasses, and similar glassware; pipets and medicine droppers with large apertures; sorting needles with recurved points; absorbent paper; burets or tall graduates; balances of suitable sensitivity.

Procedure

1. Transfer small portion of screenings to wide, shallow porcelain dish or 8-in. glass culture dish with a white background; add tap or distilled water until conveniently diluted; systematically examine all material, removing macroscopic animals by hand picking and sorting them in smaller dishes into various taxonomic groups.

2. When all animals have apparently been removed, pour residue into special container; take another portion of screenings and continue examination until all screenings have been worked over.

3. Reëxamine in similar way all residues and repeat at least twice. Each succeeding examination of residues will probably yield animals although, if previous work has been well done, in rapidly declining numbers.

4. Sort, classify, and identify animals taken in sample; carry identification to that taxonomic step (family, genus, species) required by purposes of work; then employ one or more of following methods for calculating standing crop of organisms, depending upon use to be made of results.

a. *Survey Count:* Count all individuals in each of various sorting dishes; count only living animals, or if preserved screenings must be used, count only animals which were alive when sample was taken; record totals for each species, genus, or family, in proper space in some record form; compute for each individual group (differential count), or for grand total, the number of macroscopic animals in *1 square meter* by use of formula:

$$n = \frac{o}{as} 10{,}000$$

in which n = number of macroscopic animals in 1 sq. m. of profundal bottom

o = number of animals actually counted

a = transverse area of Ekman dredge in sq. cm.

s = number of samples taken at one sampling station.

If total number of animals in whole profundal area is desired, multiply n in formula given above by total number of square meters in whole profundal area secured by measuring area within bounding submerged contour on hydrographic map; note that results will be in terms of number of profundal macroscopic animals which would occur if bottom within bounding contour were a plane.

b. *Volumetric Method: Procedure 1.* Transfer animals to absorbent paper and allow visible water to drain from them for about 1 min.; if sample contains considerable numbers of organisms, move them gently about over paper, without scattering them individually; if mollusks and case-bearing caddis fly larvae are present, either the shells must be removed or some factor representing shell volume (or shell weight if gravimetric method is used) must be worked out and applied as a correction to the results. *Procedure 2.* Place animals in graduated centrifuge tube previously calibrated against a buret having same graduation values; fix centrifuge tube in support which will hold it at about 45° angle, mouth downward, allowing liquid to drain off for about 3 min. If drainage is done while animals are alive and active, insert wire-gauze plug in mouth of tube to prevent their escape, removing gauze plug at end of drainage period. Pour water into buret and read quantity; place centrifuge tube containing animals under buret outlet and admit water from latter until mass of animals is just covered; subtract amount of water admitted from buret from reading on centrifuge tube and result is volume of contained animals.

Compute for each individual group, or for grand total of all animals, *volume* of animals per square meter of profundal bottom by use of formula:

$$V = \frac{v}{as} 10{,}000$$

in which V = volume of macroscopic animals under 1 sq. m. of bottom surface
v = volume of animals actually measured expressed in cc.
a = transverse area of Ekman dredge expressed in sq. cm.
s = number of samples taken at one sampling station

If total volume of animals in whole profundal area is desired, multiply V in formula by total number of square meters in whole profundal area. Same limitation holds here as described in preceding method.

c. *Gravimetric Method:* Follow procedure outlined under volumetric method, except that animals are weighed on balance of appropriate sensitivity and results expressed in terms of *wet weight.*

General Considerations

a. Instruments appropriate for isolation and sorting of bottom organisms are a necessity. Forceps, if used at all, must be employed sparingly and cautiously when soft-bodied animals are concerned since they are apt to cause fragmentation. A dissecting needle bent at the point into a recurved hook is especially useful in handling annelids and insect larvae; also both ordinary insect pins (size No. 4 or 5) and insect "minuten" pins mounted in small wooden handles and with points bent into sharply recurved hooks are recommended. Medicine droppers and pipets of various types are also needed.

b. Those animals, such as Tubificidae, which fragment easily, present difficulties in counting. Counting all fragments and dividing the sum by 2 is probably as good a solution to the problem as can be expected, although at best only an approximation.

c. In isolating and sorting live animals from the screenings, certain ones, notably Tubificidae, may tend to aggregate into compact masses. Usually such masses may be dispersed by judiciously thrusting a needle through the mass, transferring it to clean water in a large culture dish and then washing the mass vigorously back and forth through the water.

d. If for any reason two sets of screenings have been obtained, as for example, one from a No. 30 screen and one secured by allowing the material passing through the No. 30 screen also to pass through another screen of finer mesh such as a No. 100, the material from the finer screen may be treated as follows: Because of smaller size the organisms cannot be hand picked and sorted as in the coarser screenings; therefore, they must be examined and counted under a binocular microscope. Furthermore, it will probably be impracticable to count the entire screenings; in that instance, an aliquot sample of the screenings (0.1 of screenings may be satisfactory) is selected and placed, wholly or in successive portions, in a shallow glass vessel of convenient size on the bottom of which fine parallel guide lines have been ruled, or to the lower side of which a piece of translucent, ruled and paraffined paper has been attached. Guide lines may be so disposed that the intervening spaces have a width slightly less than the diameter of the field of view of the microscope, thus making it possible to examine the entire field by moving its various subdivisions successively, from end to end, under the microscope. Calculations of number of organisms in entire screening are made by computing from the number found in the aliquot sample the total number in the entire sample.

e. Certain animals common to profundal muds are very transparent and may be overlooked unless a suitable background for the examination dish is provided and unless the observer is on the alert.

f. It seems to have been demonstrated that benthic organisms in general undergo shrinkage in volume after several months of storage in preservative and that the amount of such shrinkage is greater in alcohol than in formalin. It therefore appears that for precise work, a correction factor must be worked out if and when specimens which have been stored for some time in preservative are compared with fresh material.

g. In *Procedure 2*, the excess liquid drains off but specimens do not lose moisture subsequently by evaporation, an important advantage when material preserved in alcohol is handled. By this procedure it is possible to eliminate much of the variation inherent in other methods used to reduce moisture to a near constant.

h. The volumetric method, particularly as outlined in *Procedure 2*, is distinctly advantageous for field use. It does not require special, expensive, or delicate apparatus, and is relatively rapid in operation.

Core Sampler Method

SUITABILITY. Particularly suitable for quantitative study of microscopic organisms at various depth levels in soft and medium hard bottoms.

EQUIPMENT. Vertical core sampler (p. 182); several serially numbered glass tubes to fit interior of sampler; field case for glass tubes; operating line for sampler; corks for ends of glass tubes; collar weight for sampler to insure penetration in harder bottoms.

Procedure

1. Examine sampler for good working condition; unscrew steel nose and insert glass tube; replace steel nose; attach operating line to sampler and tie other end of line to boat.
2. Select first sampling position and locate boat above it; lower sampler rapidly and vertically into water; allow sampler to penetrate into bottom in vertical position with full force of its weight.
3. Haul sampler to surface and keep in vertical position; insert cork in bottom of sampler; unscrew steel nose and remove glass tube containing core of bottom deposit; insert cork in each end of tube, record number of tube, and store in field case *in vertical position.*
4. Insert another glass tube in sampler and repeat procedure until sampling program is completed; transfer collected cores to laboratory. Store cores in refrigerator until ready for analyses. Never allow cores to be put into any position other than vertical.

Analyses of Cores

Analyses of cores depend, in their details, upon the kind of data desired. The following procedure is designed to effect a complete general analysis in which information on both horizontal and vertical distribution is obtainable. Modifications of this procedure may be made in order to adapt the procedure to special or limited purposes.

EQUIPMENT. Piston for removing cores from glass tubes (Fig. 56); rubber cup for top of tubes; rubber base for tubes; bulb syringe; 100-mesh screen; counting cell.

Procedure

1. Remove stoppers at upper and lower ends of glass tube; insert properly fitted piston, made from part of rubber stopper, into lower end

of glass tube; install rubber cup (p. 184) on upper end of tube; siphon off any clear water above core of bottom material in tube. By means of graduated piston rod, push up piston some selected distance, displacing uppermost portion of core and spilling it into rubber cup; scrape off materials at top of tube even with margins; remove materials in cup with small bulb syringe, wash cup with tap water and recover any residual materials. Remove next lower portion of core in same fashion, and continue until all desired segments of core are successively removed.

2. If necessary, take more than one core at same station; treat all cores from same sampling station in same way; mix together into one composite sample all materials representing same stratum in each core; if sampler sinks to different levels because of inequalities of bottom, make certain that *sections* from different cores are from corresponding levels; mix composite sample gently but thoroughly; remove aliquot part of composite sample for direct examination; transfer to counting cell for examination under microscope.

3. Transfer remaining portion of composite sample to 100-mesh screen; wash by passing tap water through sample; transfer screenings to counting cell for identification and count of larger animals held by screen.

4. Repeat procedures just outlined for all other strata of cores.

5. Determine quantity of population by survey count method (p. 303) and express results in number of individual organisms per selected unit area of surface of bottom, e.g., number of organisms per square decimeter of bottom area.

General Considerations

a. Choice of depth (thickness) of sections of cores must be made on bases of aim and desired precision of work.

b. Since it is unlikely that the distribution of microscopic animals in bottom muds is uniform and since the transverse area of the sampler is small, a single core will have little more than exploratory value. For dependable results several cores must be taken at each sampling station.

c. Counting cells convenient for this kind of work are made in various ways. A handy one is constructed by placing strips of adhesive tape on frosted-glass lantern-slide plates and coating the tape with paraffin. The dimensions of the spaces between strips may be of any convenient value. Several parallel chambers on the same plate may be so made. The chambers are placed on the frosted side of the glass. Frosted glass results in more evenly distributed illumination. Chambers having the width of the microscope field of view are convenient.

d. The amount of the composite sample counted directly will be determined by the degree of precision desired; also by the time required since at best this type of counting is very slow.

e. Theoretically, the use of the areal standard unit and the cubic standard unit (pp. 290–295) is desirable, but practical difficulties arising from the fact that the mud is constantly an obscuring material makes the methods very difficult to apply.

f. The extreme difficulty of removing all microscopic organisms from the mud practically precludes the use of any gravimetric methods for measuring these populations.

Methods for Sublittoral Bottoms

Methods outlined previously for use of the Ekman dredge and the core sampler on profundal bottoms apply, with the same limitations, to soft and medium hard bottoms of sublittoral areas. For harder bottoms, the Petersen dredge is required.

PETERSEN DREDGE METHOD

Suitability. Suitable for qualitative and quantitative work on macroscopic animals in and on hard bottoms; not practicable for quantitative work on microscopic fauna.

Equipment. Petersen dredge of convenient size, and weight; additional weights for dredge; hoist and wire cable for operating dredge; several large tubs or similar containers; 2 or 3 pails; several quart-size, wide-mouth jars with tops and with serial identification numbers, in field case; several large, circular, 30-mesh screens; forms for records.

Procedure

1. Inspect dredge for good working order; attach securely to wire cable on hoist; install hoist in convenient operating position.
2. Select sampling position and anchor boat over it; set dredge in open position, lower slowly to bottom; allow moment for dredge to sink into bottom materials; allow cable to go slack to release locking bar; raise dredge slowly to surface, swing inboard, and discharge load into tub.
3. Proceed with sampling in way similar to that described (p. 300) for use of Ekman dredge. Send accumulated samples to shore and screen as described on p. 300.
4. For analysis of samples and computation of results, proceed according to outline given on pp. 303–306.

Operation of a Petersen dredge may be interfered with from time to time by objects caught between the jaws, preventing closure. Such imperfect samples should be discarded.

At the beginning of a sampling program, one or more trial samples should be taken to determine whether additional weight should be attached to the dredge in order to make certain that it bites deep enough into a hard bottom.

SAMPLING ALONG TRANSECTS

The use of transects in sampling the sublittoral zone is of particular value because a sloping bottom is involved and because of the possibility that *concentration zones* occur in these areas. Unless sampling is done systematically and at relatively close intervals along transects which extend completely across the sublittoral zone, concentration zones may be missed entirely or be only partly sampled. It is unlikely that one transect will be adequate; several should be used and at such distances apart as to insure that the regions involved are properly sampled. During periods of calm water such transects present no serious obstacles; during rough water, they are very difficult or even impossible unless the work is done from a large boat. Various ways of keeping the boat at the proper intervals along a transect may be devised by the operator. In the smaller waters, the boat can be operated along a rope stretched from a heavy anchor sunk in deep water to some fixed object on shore. Or a set of shore signals may be so arranged that the courses of the boat and the sampling intervals are properly executed.

TOW NETS AND DREDGES

SUITABILITY. Suitable only for qualitative work with macroscopic bottom organisms.

EQUIPMENT. Tow dredges are made in a great variety of forms. The two described here are simple in construction, easy to handle, and ordinarily effective in performance.

TOW NET ON RUNNERS. This tow net, originally designed by Reighard, is very useful for collecting organisms just above soft bottoms. To an iron ring constituting the mouth of the dredge (Fig. 81) are attached four strips of band iron which extend radially for about 3 in., then make a rounded right-angle bend and extend parallel to each other for a distance of two or three feet where they make a round right-angle bend toward each other, meeting at the center. There they are riveted together. The dredge net, composed of strong materials and of whatever size of mesh is desired, is attached to the iron ring and extends backward within the framework of the iron strips. From the ring also extend three or four cords which are brought together to form an attachment for the towline.

Such a dredge sinks to the bottom and comes to rest horizontally, the four runners insuring that it will always assume an operating position. When hauled along the bottom by means of the towline, the runners act as a sort of sled, holding the mouth of the net about at the surface of, or slightly above, the soft bottom materials. In the sublittoral region such a dredge must be drawn either up the slope or parallel to it. When hoisted to the boat the catch is deposited in appropriate containers by turning the net inside out.

TRIANGLE BOTTOM DREDGE. An equilateral triangle, with sides 12–15 in. long, constructed from a piece of heavy band iron (Fig. 82) about 3 in.

FIG. 81. (*Left*) Tow net on runners, designed by Reighard for use on littoral and sublittoral bottoms.

FIG. 82. (*Right*) Triangle bottom dredge.

broad, constitutes the mouth of the dredge. One edge of each of the sides of the triangle is cut into a continuous set of large saw teeth. These teeth are bent outward so that when the dredge is hauled on the bottom they tend to dig into it. A loop at each corner serves as attachment for the tow-line. At each corner there is fastened a stout iron rod, about 3 ft. long, which extends backward and ends freely. To the smooth edge of the iron triangle is attached the net which consists of (1) an outer supporting and protecting cover of burlap, coarse-mesh netting, or some other coarse, strong material, and (2) an inner lining of muslin at the bottom.

When attached to the towrope and lowered to the bottom, the rods insure that the dredge will always come to rest in the proper operating position. Drawn along the bottom, the toothed edge digs shallowly into the deposits and the catch thus contains those animals which are more or less imbedded in the uppermost layers of the latter. When hoisted to the top, the catch is delivered into containers by turning the net inside out. In the sublittoral region the dredge should be drawn either parallel to the slope or up the slope. Usually, it cannot be operated down the slope with any success without the use of complicated accessory arrangements.

Methods for Littoral Areas

Since, in general, the littoral zone shows the greatest diversity in bottom conditions of all lake regions, sampling methods must of necessity be chosen according to the kind of area examined. Qualitative work presents no special problems since any means of securing bottom samples will probably serve the purposes, but the requirements for quantitative work are more exacting. The first problem is the proper choice of a bottom sampler.

SAMPLING EQUIPMENT

Unit-area Sampler. A simple but effective unit-area sampler is made from a strip of spring sheet brass, 1/16 in. thick, about 6 ft. long, 3–4 in. wide, with rounded ends. The length of the brass strip is such that it can easily be made into a circle which contains an area of exactly 1/5 sq. m. At one end is a short bolt supplied with a wing nut; in the opposite end is a longitudinal slit the width of which is slightly greater than the diameter of the short bolt at other end. This slit fits over the bolt and the wing nut is tightened down to hold the two ends together. A circle of fixed area is thus produced; loosening the nut permits easy storage of the flat strip. At each interval of about 1.5 ft. along the brass strip there is riveted at the middle line a slender brass tooth, about 6 in. long, 1/8 in. thick and 1/2 in. wide, tapered gradually to a point at the free end. These teeth are fastened loosely on the rivets so that when not in use they can be turned parallel to the strip, simplifying storage. When ready for use the teeth are turned at right angles to the brass strip, the sampler is lowered into the water and the teeth are pushed into the bottom until the lower edge of the brass strip is buried a short distance in the deposit. One unit-area of bottom is thus definitely included. Such a device is best suited for barren sandy exposed shoals, although it may have a limited use on certain other littoral areas where conditions permit.

This unit-area sampler can be used only during periods of calm water. When it has been sunk into the bottom at any location, the operator, using a strong pint dipper of the usual type, removes carefully the bottom materials included within the unit area to a depth of about 1–1.5 in. Such materials are transferred immediately to a sieve of the appropriate mesh and screened. Often this screening can be done by a helper standing only a short distance away. Screenings are removed from the sieve and stored in the usual way (p. 301) until analyzed.

PETERSEN DREDGE. The Petersen dredge, described on p. 178, is particularly suited to those littoral regions having hard bottoms, but with proper care and an understanding of limitations, it may be used on certain other areas.

CORE SAMPLER. The core sampler may be used on the following littoral regions: barren sandy exposed shoal; protected sandy shoal; marly shoal; mud bottoms. It is not recommended for rock, stony or gravel bottoms which prevent its proper penetration, or for bottom materials containing so much water that the cores will not be retained.

SQUARE-FOOT SAMPLER. A bottom sampler (Figs. 83 and 84) described by Wilding (1940) consists essentially of two brass cylinders, one inside the other. The outer cylinder, 30 in. long, is open at both ends. On the lower margin are fastened 13 large saw teeth by means of which the cylinder, when rotated, cuts its way into the bottom. The diameter of the outer cylinder is such that exactly 1 sq. ft. of area is enclosed. The inner cylinder fits closely within the outer one and is perforated with minute holes, each $\frac{1}{64}$ in. in diameter and so spaced that there are 23 holes per linear inch or 529 holes per square inch. At the bottom of the inner cylinder is a valve composed of two sets of overlapping horizontal blades so constructed that the inner cylinder can be closed or opened at the bottom. The lower part of the valve is stationary; the upper part rotates on a shaft, the two parts forming a watertight bottom when in closed position. The valve is opened and closed by means of a special handle. When the valve is open the blades are directly superimposed; when closed, upper set closes the openings in the lower one.

When ready to use, the outer cylinder alone is rotated, toothed edge downward, to sink it into the bottom for a short distance. All coarse materials (rocks, vegetation, debris) are taken out by hand picking, placed in a container partly filled with water and washed; the more obvious bottom organisms are collected and sorted by direct examination. The water in which they were washed is poured through a piece of muslin loosely supported by a frame in order to secure the macroscopic animals which remain in the water. After the gross materials on the surface of the bottom area within the outer cylinder have been removed to a depth of 3–4 in., the remaining bottom materials are stirred thoroughly. The inner cylinder,

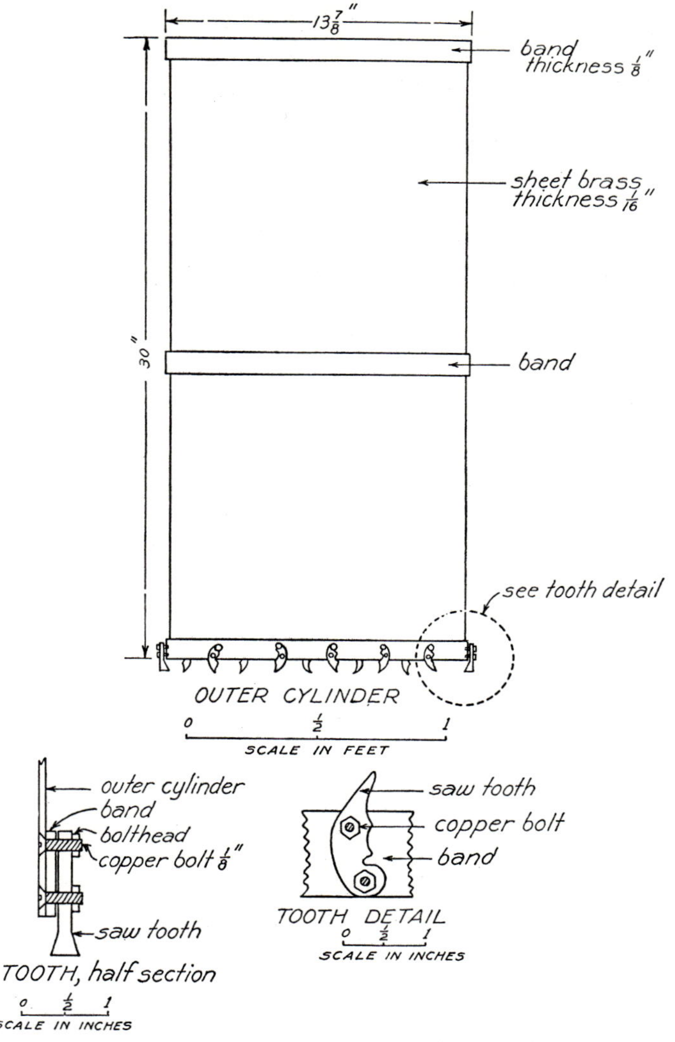

Fig. 83. External view of Wilding's square-foot aquatic sampler. (From Wilding, 1940.)

with the valve set in open position, is then inserted into the outer cylinder and plunged to the bottom, forcing the water contained in the outer cylinder through the openings of the valve. When the inner cylinder is pushed as far as it will go, the valve is closed by turning the handle which controls it. Then the inner cylinder is lifted out and the contained water allowed to strain out through the perforations, leaving the organisms thus collected inside the cylinder. This catch is washed out into a large receptacle and the water containing the organisms is then poured through the muslin cloth previously mentioned to complete the capture.

This sampler provides positive capture of practically all bottom organisms within the area selected, with the exception of those minute ones which pass through the perforations in the inner cylinder. It can be used

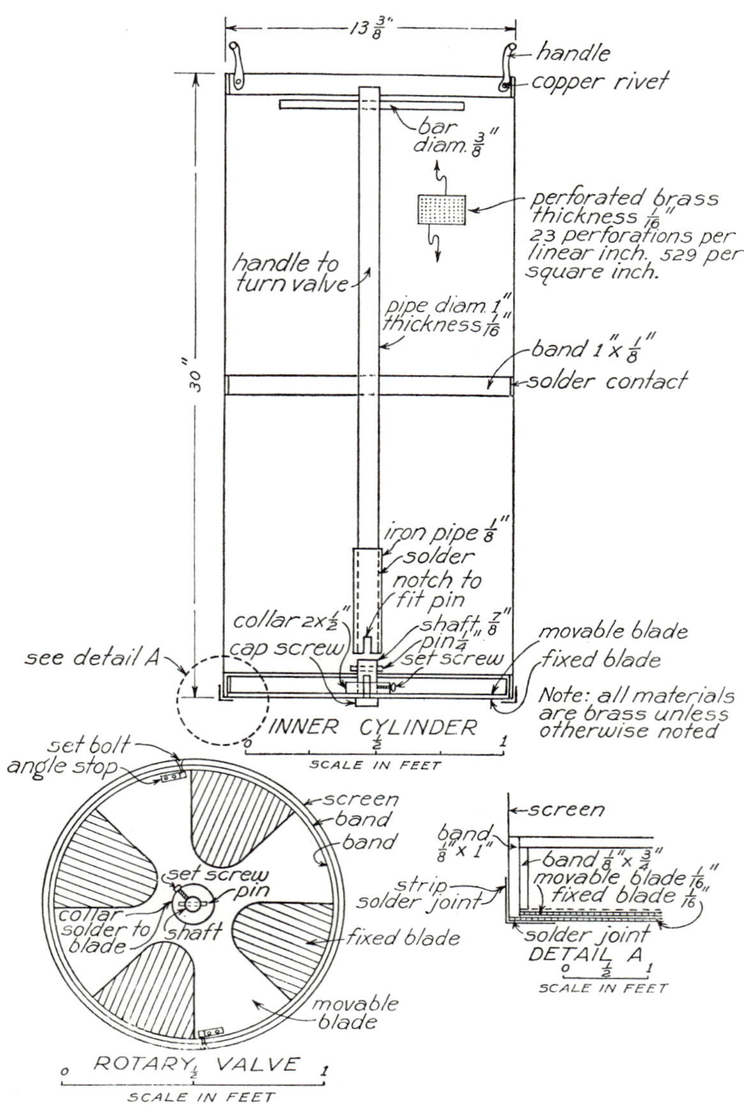

FIG. 84. View of internal construction of Wilding's square-foot aquatic sampler. (From Wilding, 1940.)

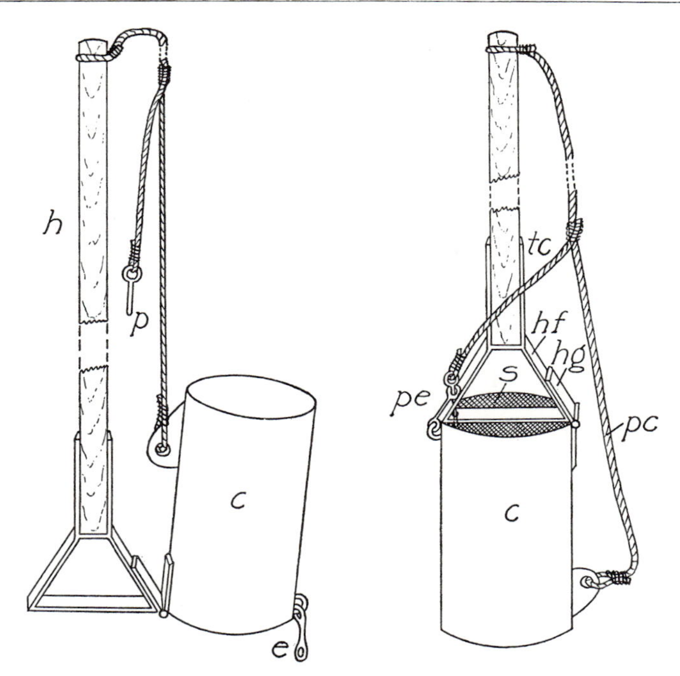

FIG. 85. Dendy inverting sampler. (*Left*) Sampler inverted. (*Right*) Sampler in operating position. (*c*) Cylinder. (*e*) Eye. (*h*) Handle. (*hf*) Handle frame. (*hg*) Hinge. (*p*) Pin. (*pc*) Pull cord. (*pe*) Pin-and-eye lock. (*s*) Screen. (*tc*) Trip cord. (From Dendy, 1944.)

in various kinds of littoral regions of lakes, in weed beds, and in gravel and rubble areas of streams. All parts are made of brass or copper.

EKMAN DREDGE. The Ekman dredge described on p. 176 can be used only in the soft, finely divided littoral bottoms which are free from vegetation, intermixtures of sand and miscellaneous coarse debris.

DENDY INVERTING SAMPLER. A very useful and convenient sampler of the inverting type (Fig. 85) was devised by Dendy (1944) for work on littoral areas. The main feature of the sampler consists of a brass cylinder 7.8 cm. in diameter and 13 cm. long. The top end of this cylinder is covered with brass screen the size of mesh of which can be varied to suit the character of the work. No. 30 mesh is a good average size. A handle frame is fastened to the top of the cylinder by a hinge on one side and by a pin-and-eye lock on the other. A wooden handle of suitable length is fitted into this frame. From the free end of the handle a pull cord extends to the bottom of the cylinder on the hinge side where it is tied into a projecting wing provided for that purpose. A trip cord, extending from the pin to the pull cord, is so adjusted that when tension is exerted on the pull cord the pin is withdrawn, releasing the lock, and then additional tension inverts the cylinder.

To secure a sample, the sampler is set in the position shown at the right in Fig. 85. By means of the handle the cylinder is pushed vertically into the stream or lake bottom to a depth about equal to its height. At the end of the downward push on the handle, the operator inverts the cylinder by further pull on the cord. This inversion of 180° insures that the full sample is secured and that on the trip to the surface of the water the only loss of material is that which may drain through the brass screen. When the sampler reaches the surface its contents are discharged into a screen and treated as described elsewhere (pp. 300–302), stored temporarily in an ice box, or preserved.

This sampler is suitable for use on bottoms composed primarily of mud, sand, marl, fine gravel, and similar materials into which the collecting cylinder can penetrate readily. Its capacity is small enough so that the entire sample may be analyzed without subdivision. This size also makes it possible to secure a larger number of samples spread more extensively over the area involved. It is easily handled by the operator while wading in shallow water and samples are taken rapidly. It may also be operated from a boat. Within limits, the handle and pull rope may be extended to a length which makes possible sampling in deeper water. The size of the mesh of the brass screen covering the upper end of the cylinder may be varied according to the needs of the work. This sampler will collect samples from bottom materials which contain an unusual component of water or which, because of some other feature, will not remain in the tube of a core sampler.

DISTRIBUTION OF LITTORAL SAMPLES

In some instances, the manner in which littoral bottom samples are spread over the whole area under consideration may be of little consequence; in others, special reasons may require a careful choice. The principal types of distribution are as follows:

Dispersed Random Sampling. By dispersed random sampling is meant the completely irregular selection of sampling positions. The essential requirements of such sampling are (a) samples should be scattered over the entire area involved; (b) conscious choice in the selection of any sampling position must be rigidly excluded; and (c) a sufficient number of samples must be taken to meet the precision requirements of the work.

Transect Sampling. Transect sampling has an important advantage over dispersed sampling, namely, that the position and distribution of samples are known and can be definitely mapped. In addition it may be easier to provide the distribution of samples over the area concerned and still maintain the requirements of random sampling. Transects may be transverse, parallel, or diagonal, to shore line depending upon circumstances and the character of the work.

Transect sampling is of two general types: (a) line transects; and (b) zonal or belt transects.

By a *line transect* is meant either the regular or irregular scattering of samples along a predetermined line which extends across the area concerned. It results in a single row of sampling positions in which the essentials of the random sampling method are met. If a sufficient number of samples is taken, it may be of little consequence whether the samples are all the same distance apart, or whether there is no regularity of intervening distance just so long as no conscious choices are involved and so long as obviously abnormal groupings of samples are avoided. Any substantial program of bottom study will almost certainly require several transects the position of which must be determined by the special features of the area. The position of transects should be located by permanent objects on shore, by measurements and sets of marked stakes, or by data secured with surveying instruments. When properly referenced to permanent shore features, a set of line transects can be accurately drawn on maps.

By *zonal* or *belt transects* is meant those which have a definite *width* as well as length. Such transects may be used in many ways. For example, zonal transects may be laid out and samples confined to them although *within* them the samples may be of the dispersed random type. In some situations zonal transects are of particular value for measures of the standing crop of the larger, visible animals, such as clams and snails. Dependable surveys of the clam population may be made by two observers who hold between them a light pole of known length and traverse an area, back and forth, counting all individuals which occur in the space between the observers. From several such zonal transects distributed over the area as a whole, very satisfactory computations of the standing population may be made.

ANALYSES OF LITTORAL SAMPLES

The general procedure for analysis of screened samples of the profundal region (p. 300) is suitable for analyzing the screenings from the littoral zone. Unfortunately no means has yet been devised for all littoral materials which eliminates the tedium and time consumption inherent in the necessary hand-sorting methods. However, for formalin-preserved screenings in which there is a preponderance of sand or fine gravel and minimal amounts of flocculent materials, the calcium chloride *flotation method* is very useful. This technique consists essentially of the following steps: (a) drain off excess formalin from each sample; (b) divide sample into convenient portions; (c) place each portion in large evaporating dish; (d) cover portion with a saturated solution of calcium chloride and stir vigorously. The high density of the calcium chloride solution causes all arthropods, excepting caddis fly larvae in sand-covered cases, to rise to the surface while the sand and gravel quickly settle to the bottom. The specimens are picked or skimmed off the surface of the liquid. Annelids and mollusks may be poured off after vigorous stirring and before settling can occur. The supernatant fluid should be poured through a piece of fine marquisette stretched over the mouth of a funnel to secure animals which are missed. The materials remaining behind in the evaporating dish should be stirred and washed 3–5 times in the manner just described and then placed in an 8-in. culture dish for further examination, first over a dark and then over a white background, with a large-diameter 10$\times$ reading glass for the capture of any organisms missed in the chloride treatment. The residue on the marquisette is washed into a finger bowl and examined under a binocular microscope. In some materials more than 95 per cent of the arthropods, caddis fly larvae and annelids may be so collected. Mollusks are effectively removed in this way only when very small. The calcium chloride can be used repeatedly but the full strength must be maintained otherwise loss of buoyancy results. The method as outlined here is a modification and simplification of one described by Beak (1938).

Increased speed in collecting and sorting littoral animals in screenings has been claimed for the sorting trough (Fig. 86) designed by Moon (1935) which partially grades and arranges the materials examined. It consists essentially of a trough, 1.3 m. long, with a bottom made of plate glass. This trough is divided into three compartments by two incomplete partitions *p* so arranged that the gap between the end of one partition and the side of the trough is on the side opposite that of the next one. Water is admitted into the upper end of the trough through a lead pipe *w*, 1.3 cm. in diameter, in which linear series of holes provides that the water enters the first compartment in the form of a diffuse jet. Entering water is controlled by a valve and leaves the trough at the lower end through a hole, 2.5 cm. in diameter, into which a short pipe is fitted. This pipe discharges into a small trough which in turn empties through a series of sieves *s* or filters and thence into the final drain. A strip of metal *st* 1.3 cm. high is installed across the front end of the pipe. The plate-glass bottom should be painted white on the outside surface, or the whole apparatus set over a white background. When in use, the original screenings are placed in the first compartment and water admitted by adjusting the tap. As water passes through the trough, a swirl is produced in front of each partition and before the lower end of the trough, thus producing a sorting effect. The heavier materials remain in the upper compartment while the lighter ones collect in the lower two compartments. The lighter debris and the smaller animals pass on through the pipe and smaller trough into the screens. Rate of current through the apparatus may be partly controlled

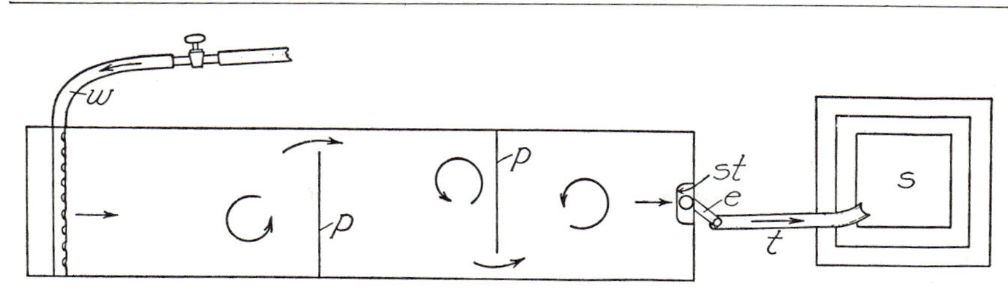

FIG. 86. Diagram of Moon's sorting trough for littoral bottom materials. Arrows indicate flow and performance of water. (*e*) Effluent pipe. (*p*) Partial partitions. (*s*) Sieves. (*st*) Strip of metal about opening into effluent pipe. (*t*) Drain trough. (*w*) Water pipe. (Redrawn with slight modifications from Moon, 1935.)

by tilting the main trough. After sufficient sorting effect by the water current has been effected, the water is turned off and the exit is closed by inserting a cork into the pipe. The accumulations of materials in the compartments are then carefully raked apart and the animals removed by hand picking. Material collected on the screens is washed into a white enamel sorting pan divided into numbered compartments. These subdivisions focus the attention upon one compartment at a time thus making sorting easier and more accurate. Since the partitions in the main trough are removable, the apparatus is adaptable to various modifications of the procedure. If, for example, the sample is small, the partitions may be removed and the light materials washed directly through to the screens. If the original screenings are too light for sorting, a small amount may be put through the trough, worked over, washed away, and replaced by the next portion.

METHODS FOR DIFFERENT TYPES OF LITTORAL BOTTOMS

1. SANDY AND FINE-GRAVEL SHOALS. For *qualitative* sampling, various kinds of scoops, dredges, drag nets, and other devices which collect surface layers of such a bottom are generally satisfactory. In the less accessible or deeper situations some use may be made of the *tray method*. A collection tray consists essentially of a metal-rod frame of convenient size and shape to which is attached a very shallow bag of strong sacking or similar material. The frame is provided with a bridle by means of which the tray is lowered and raised. The tray is loaded with clean bottom materials of the kind present in the area to be sampled; then lowered to the bottom, allowed to remain there for some days or weeks; then raised and the contents analyzed for the bottom organisms which have colonized the materials in the tray. Various accessories may be added to the tray to increase the ease and effectiveness of its use.

For *quantitative* work, the Petersen dredge and the core sampler are recommended. In the shallower portions of the area the unit-area form (p. 311) has some advantages. Trial samples must be run with Petersen dredge and the core sampler to determine the amount of weight necessary to cause these samplers to function properly.

2. Stony and Coarse-gravel Shoals. For *qualitative* sampling, heavy drag dredges or scoops, heavy hand-dip nets; rake dredges; and similar heavy-duty devices are necessary. Some limited use may be made of the tray method (p. 319). For *quantitative* work, some form of the unit-area method (p. 311) appears to be the most satisfactory. In the deeper areas, quantitative sampling may be very difficult and sometimes uncertain. The tray method may have some possible use in such areas but great care must be exercised that the tray samples are truly representative of both the kinds and quantities of bottom animals.

3. Rock Surfaces. Rock surfaces are very difficult to sample adequately, especially for quantitative results. *Qualitative* collections may be made by brushing the surface materials into the open end of a dredge or dip net. For *quantitative* work, results for some purposes may be obtained, in shallows and under conditions of calm water, by marking off a known area and then, by direct observation, counting the organisms. In deeper water some limited use may be made of the method of scrubbing a known area with a brush and sweeping the materials into the open mouth of a dredge provided with a flat lip. Some workers have attempted to make use of the method of transferring rocks from the bottom into a submerged container, bringing them to the surface, and determining the quantity of fauna in terms of some selected unit of rock-surface area.

4. Soft Mucky Bottoms. Soft mucky bottoms present fewer sampling problems than any other littoral areas. The Ekman dredge, Petersen dredge, and core samplers, usually provide adequate means of securing both qualitative and quantitative samples. Many of the simpler dredges, nets, scoops, and similar devices are useful for making qualitative examinations.

5. Marl Bottoms. Bottoms containing very large amounts of marl will probably require a Petersen dredge or a core sampler. Only on the soft marl bottoms having considerable intermixture of organic matter and silt is there any chance that an Ekman dredge might be used for quantitative purposes.

Methods for Stream Channels

Most of the methods already described as suitable for lake bottom are adaptable to stream-bottom conditions and need not be repeated. However, certain methods particularly suitable to bottom conditions in streams have been developed, two of which (use of Surber's stream-bottom sampler and Hess' circular sampler) are described here.

STREAM-BOTTOM SAMPLER

A stream-bottom sampler (Fig. 87), originally described by Surber (1937), is now widely used. It consists primarily of two square frames of equal size hinged together. One frame carries the net and the other, when in working position, encloses the sampling area (1 sq. ft.). Two braces provide means of locking the two frames into working position at right angles to each other. The net is attached to the vertical frame by means of four brass strips screwed onto the sides of the vertical frame. The net is composed of No. 000 XXX extra heavy silk bolting cloth and is 27 in. in length. It may be attached directly to the frame, or if added durability is desirable, a section about 12 in. long composed of muslin or light canvas may be attached to the metal frame and the bolting cloth net of correspondingly reduced length sewed to the opposite end. Two triangular wings, either of canvas, muslin, or bolting cloth fill the space between the vertical and horizontal frames on each side. All metal parts should be made of brass to prevent rusting. When not in use the whole sampler folds into a compact, flat form convenient for carrying in a field kit. A short handle installed in a vertical direction on top of the net-bearing frame may be convenient but is not a necessity.

Suitability. Especially suitable for sampling in stony or gravelly bottoms in regions of streams which are shallow and possess current enough to hold the net in an open position and wash the dislodged organisms into it. For other types of stream bottom its use is limited and in some instances impracticable. Only macroscopic organisms are collected.

Procedure

1. Set net in open position and fasten wing braces securely in place; select sampling position in stream, approaching it from the side or from a downstream position.

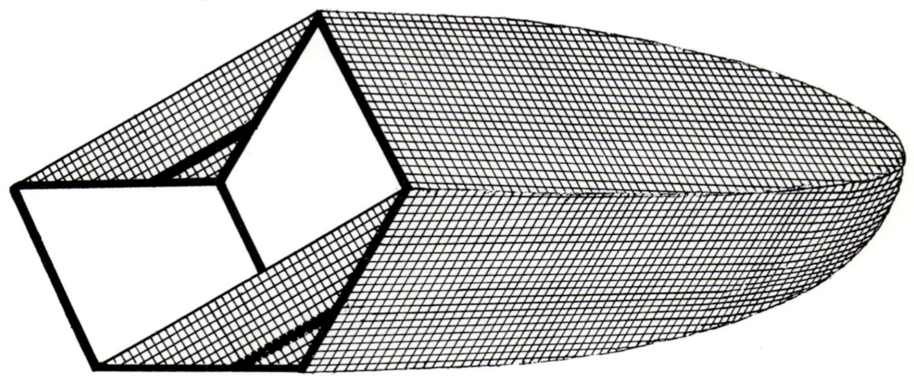

Fig. 87. Surber's stream-bottom sampler.

2. Grasp net by handle or, in absence of handle, by vertical frame; lower into water with open end upstream; work lower edge of horizontal frame into bottom until one-square foot area is completely enclosed; if spaces still occur under edge of horizontal frame, close them by pushing outside bottom materials against side; make sure that net is floating out in operating position.

3. Turn over and stir stones, gravel and other coarser materials enclosed by horizontal frame, allowing dislodged materials to float into net; make certain that all materials within frame are stirred thoroughly to a depth of at least one inch.

4. When mixing of materials within frame is completed, keeping open end upstream, lift net from water into vertical position and allow water to drain out of net. Transfer organisms collected in net to appropriate receptacle by turning net inside out; receptacle should contain water if temporary live storage is planned.

General Considerations

a. If the triangular wings fail to present sufficient area, rectangular wings may be installed on folding metal arms.

b. The form of the net is not a matter of prime importance although a rounded terminal section may be convenient.

c. Size of mesh in the net may depart from that indicated (p. 321), especially in direction of larger dimensions, if the purposes of the work warrant. Smaller meshes are subject to increasing inconvenience due to clogging.

d. Since the net is subject to considerable pressure when in use, all seams must be broad and substantial.

e. In stirring materials enclosed within the horizontal frame, care must be taken to avoid undue bringing of fine sand into suspension since it will accumulate in the net and cause much difficulty in handling samples; also damage to net may result from overloading its contents.

f. Ordinarily it is not practicable to collect from more than one square-foot area before emptying the net.

g. This type of apparatus is impracticable for work in the quiet sections of streams since its performance depends upon current. It is also impracticable for work in water deeper than arm's length.

h. Some workers have partially overcome the difficulty of making the horizontal frame seat completely on rough bottoms by padding the frame with burlap or some other coarse cloth which is sufficiently compressible to pack into uneven spaces.

i. At the conclusion of work, the net must be thoroughly rinsed out in clean water and then allowed to dry in open air but protected from direct sunlight.

WILDING'S SAMPLER

Wilding's aquatic sampler, described on p. 312, is suitable for use in gravel and rubble bottoms of shallow streams.

HESS' CIRCULAR SAMPLER

Recently Hess (1941) has described another form of circular, square-foot sampler suitable for gravel and rubble bottoms of streams which overcomes some of the difficulties sometimes encountered in the use of Surber's square bottom sampler. The body of the sampler is in the form of a cylinder, about 18 in. tall and somewhat streamlined in cross section. The frame of this body is composed of a top and bottom hoop of ⅛ × ¾-in. strap iron, both horizontal in position. These hoops are welded to two vertical, 0.5-in. iron rods which are opposite each other. Projecting ends of these rods are bent at the top into a horizontal position to form handles. The inside diameter of the bottom hoop is such as to enclose exactly 1 sq. ft. of area. A ⅙-in. mesh galvanized wire netting covers the front half of the cylinder. The back half is covered with heavy canvas in the rear of which is an opening, 1 ft. square, over which the collecting net is sewed. The net, about 24 in. long and conical in shape, is constructed of heavy grit cloth, No. 24 (23 meshes to the inch). A finer mesh may be used if desired. The circular form of the lower hoop and the strong handles make it possible to turn the lower hoop into the bottom even in the presence of rocks of considerable size; the screen guard on the front surface prevents escape of larger organisms and keeps rocks from rolling into the sampler; and the streamlined form eliminates swirling.

The method of taking samples is essentially the same as that described for the stream-bottom sampler (p. 321).

Analyses of Stream-bottom Samples

Methods of analysis of samples taken from bottoms of stream channels need not differ from those already described (pp. 299–308) for samples taken by the Ekman dredge, the ooze sucker, the Petersen dredge, and core sampler. All of these instruments may be used effectively on stream bottoms just so long as the limitations of each are observed. Samples taken with Surber's, Wilding's and Hess' samplers may be analyzed by the same methods described (pp. 308–321) for those samples taken in different kinds of areas on the littoral and sublittoral zones. Calculations are made, and results are expressed, in the ways already described (pp. 303–305).

Methods for Psammolittoral Areas

It is now an established fact that sandy beaches, particularly that zone sometimes designated as the inner beach (that area extending from the water's edge, during periods of calm, up the slope to the place where the surface of the sand ceases to be saturated with water and shows first traces of drying), maintain a population of organisms, often of considerable diversity and magnitude. Special methods are required for these situations. The following method was developed into its present form in the writer's laboratory by Neel (1948). Certain of its features are based upon earlier procedures devised by Pennak (1940).

Equipment. Several pieces of seamless brass tubing, 12 in. long, 1.5 in. in diameter, with walls 0.02 in. thick; at least 2 brass tubes of same dimensions but 18 in. long. Piston, 2.5 in. long, fitting snugly inside of brass tubes, and composed of 3 sections of good quality cork, each 0.6 in. thick, separated from each other by a thin metal disk, all closely fastened together into compact plunger. Piston is mounted on brass rod with diameter, 0.2 in.; length, 1.5 ft. Good grade No. 8 rubber stoppers: top diameter, 1.5 in.; bottom diameter, 1.25 in.; thickness, 1 in. Field kit provided with supports which hold tubes in vertical position.

PROCEDURE FOR SAMPLING

1. At position of sampling, thrust open tube into sand to desired depth; remove carefully and insert rubber stopper in each end of tube without disturbing core of material; keep tube in original vertical position; transfer to field kit and continue to keep in vertical position.
2. Secure subsequent samples in accordance with predetermined sampling program which may be according to dispersed random method or in accordance with some plan of transect sampling; transfer to laboratory for analysis.

GENERAL CONSIDERATIONS

a. Depth of sampling must be determined by the object of the work and by trial tests before the formal program is begun. If a complete survey of the psammolittoral organisms is contemplated, sampling may require cores at least one foot or more in length since some organisms occupy depths much greater than was formerly supposed.

b. Transverse transects across the beach are likely to be the most useful plan of sampling.

c. It may not be practicable to sample by this method those beaches which are composed of light, finely divided muds with a large component of water, since cores may not remain unchanged in the tubes.

Analysis of Psammolittoral Samples

1. Remove stoppers from sampling tube and insert piston into lower end; push core toward upper end and remove segment of core of required

thickness by cutting off with spatula or thin knife; place sample so secured in large flat dish.

2. By proper methods of subdivision (or use whole segment of core if of small volume), remove 10 cc. of segment of core, place in 3-in. diameter evaporating dish, and cover with about 50 cc. of tap water.

3. Introduce air jet of near capillary size into sand under water and adjust air pressure so that continuous stream of fine bubbles is produced, having force enough to bounce sand grains about but not enough to blow water over rim of dish. Allow air jet to stir sand vigorously for 3 min., separating sand masses into individual grains and providing thorough washing.

4. At end of 3-min. period of agitation, transfer water to two hand-centrifuge tubes and centrifuge.

5. After centrifuging, pour supernatant water back into 3-in. evaporating dish containing sample; stir concentrate in bottom of centrifuge tube and transfer to concentrate bottle.

6. Repeat this process (paragraphs 3–5) three or more times for each sample, using the same water over and over. After last centrifuging, rinse tubes with tap water and add this wash water to concentrate.

7. For identification and enumeration of concentrate, follow directions given for counting plankton (pp. 285–295); also make computations in same way.

GENERAL CONSIDERATIONS

a. Since some of the psammolittoral biota are very small, thin counting cells permitting use of higher magnifications of microscope may be required.

b. Volumetric or gravimetric measurements are often impracticable because of the difficulty of separating the organisms from the residual fine sand.

Selected References

(For Chapter 16)

Beak, T. W.: Methods of making and sorting collections for an ecological study of a stream, *Avon Biol. Research, Annual Rept. 1936–1937*, Progress Rept. No. III, pp. 42–46, 1938.

Beauchamp, R. S. A.: A new dredge, *Intern. Rev. ges. Hydrobiol. Hydrog.*, **27:** 467–469, 1932.

Dendy, J. S.: The fate of animals in stream drift when carried into lakes, *Ecol. Monographs*, **14:** 333–357, 1944.

Hess, A. D.: "New Limnological Sampling Equipment." 5 pp. Spec. Pub. No. 6, Limn. Soc. Am., 1941.

Leonard, J. W.: Comments on the adequacy of accepted stream bottom sampling technique, *Trans. Fourth N. A. Wildlife Conf.*, pp. 289–295, 1939.

Moon, H. P.: Methods and apparatus suitable for an investigation of the littoral region of oligotrophic lakes, *Intern. Rev. ges. Hydrobiol. Hydrog.*, **32:** 319–333, 1935.

Naumann, E.: "Einführung in die Bodenkunde der Seen." 126 pp. Vol. 9 in Thienemann's "Die Binnengewässer," Stuttgart, 1930.

Neel, J. K.: A limnological investigation of the psammon in Douglas Lake, Michigan, with especial reference to shoal and shoreline dynamics, *Trans. Am. Microscop. Soc.*, **67:** (in press), 1948.

Pennak, R. W.: Ecology of the microscopic Metazoa inhabiting the sandy beaches of some Wisconsin lakes, *Ecol. Monographs*, **10:** 537–615, 1940.

Surber, E. W.: Rainbow trout and bottom fauna production in one mile of stream, *Trans. Am. Fisheries Soc.*, **66:** 193–202, 1937.

Wilding, J. L.: "A New Square-Foot Aquatic Sampler." 4 pp. Spec. Pub. No. 4, Limn. Soc. Am., 1940.

METHODS FOR PLANT-INHABITING ORGANISMS

METHODS FOR PERIPHYTON

Periphyton is that assemblage of organisms which commonly forms upon surfaces of submerged plants, wood, stones, and certain other objects, forming a more or less continuous slimy coat. It may develop from a few tiny gelatinous masses into a woolly, felted coat that is slippery to the touch or crusty with included marl or sand.

QUALITATIVE METHODS. Collection of periphyton for qualitative study requires no special equipment or approaches. Material may be scraped

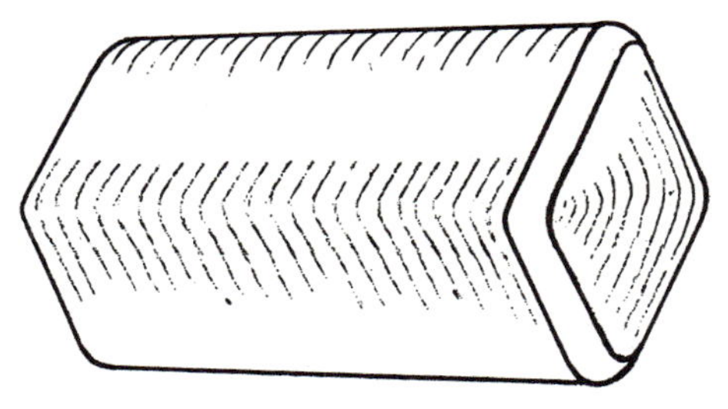

FIG. 88. Hollow-square instrument designed by Young for securing quantitative samples of periphyton from hard objects having large areas.

from rocks, petioles of aquatic plants, or other supports and brought into the laboratory. There is every advantage in the use of fresh, living material since preservation produces changes in many of the organisms. Usually, material should be collected under water, although some of the larger rooted plants, such as the bulrushes, can be cut off just above the bottom, allowed to rise to the surface, and pieces of the submerged portions introduced, with a minimum of disturbance, into bottles containing water.

QUANTITATIVE METHODS. Quantitative samples can be secured from aquatic vegetation by carefully cutting off sections of stems, petioles, and leaves, under water and introducing them into bottles also under water. Sections of roots, dead brush and similar harder materials may be cut off with tin shears and introduced into bottles in the same way. Quantitative samples of materials on stones, logs and similar surfaces can be secured by the use of a hollow-square instrument (Young, 1945) (Fig. 88) made by bending a piece of sheet brass so that the inside dimension of each side of the square is 1 or 2 cm. depending upon how large a sample is desired. One edge of the instrument is beveled and sharpened so that when it is set down upon a surface covered with periphyton it cuts through and encloses a measured area. The material in a zone outside the instrument is scraped away leaving the periphyton within the sampler as an isolated island. The sampler is removed and the enclosed sample is scraped off into a container.

ANALYSES OF PERIPHYTON SAMPLES

Numerical Methods

EQUIPMENT. Large counting cell (Fig. 89), made by cementing together pieces of plate glass to form cell 30 × 4 × 0.6 cm. inside measure-

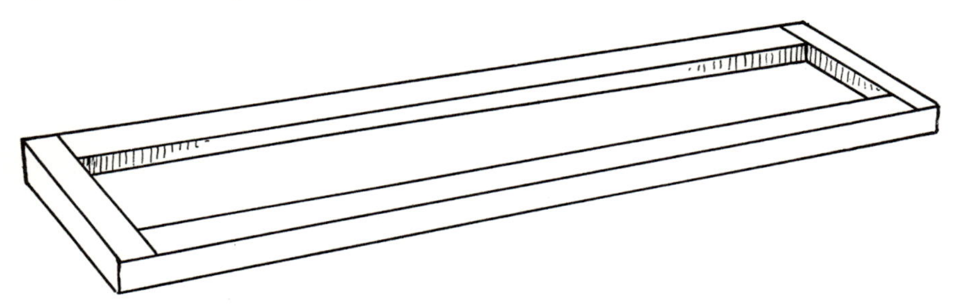

FIG. 89. Large glass counting cell designed by Young for examination and counting of larger periphyton organisms.

ments and having a capacity of 72 cc.; binocular microscope; compound microscope; Sedgwick-Rafter counting cell; Whipple ocular micrometer; wide-mouth pipet.

Procedure

To count *larger* periphyton organisms, proceed as follows:

1. Deliver sample, or known portion of sample, into large counting cell.
2. Tease apart all compact fibrous masses.
3. Place cell under binocular microscope and make total count of all larger organisms in cell.
4. Compute and express number of each of different kinds or organisms per square decimeter of original surface and enter on record.

To count *smaller* organisms, such as diatoms, rotifers, and nematodes, proceed as follows:

1. Tease apart all masses in sample and mix thoroughly.
2. With wide-mouth pipet, remove amount sufficient to fill Sedgwick-Rafter counting cell; examine and count organisms under compound

microscope equipped with a Whipple ocular micrometer, counting all rotifers and nematodes in entire cell, and all diatoms, desmids, and organisms of smaller size in ten 1-cu. mm. areas selected at random.

3. Compute number of each of different kinds of organisms per square decimeter of surface from which sample was taken.

Volumetric Method

Equipment. Small stender dish; hard-surface filter paper; narrow graduate, or small graduated centrifuge tube.

Procedure

1. Add convenient amount of water to sample; stir thoroughly and filter through small piece of filter paper, allowing all parts of filter paper used to become saturated with water; allow to drain completely.

2. Place carefully measured, convenient volume of distilled water in stender dish; carefully transfer piece of filter paper bearing periphyton to water in stender dish and carefully wash off all periphyton; remove filter paper, allowing all excess water on it to drain back into dish.

3. Transfer water containing periphyton to graduate or centrifuge tube and measure volume; volume of original water subtracted from volume after periphyton is added yields volume of periphyton.

4. Compute and express results as number of cc. per square decimeter of original surface supporting periphyton.

Gravimetric Method

Equipment. Filter paper; balance of appropriate sensitivity; drying oven with heat control.

Procedure

1. Select filter paper; weigh carefully.

2. Filter sample through filter paper; drain fully.

3. Transfer filter paper and periphyton to drying oven and evaporate to dryness, temperature controlled at 60° C.

4. Weigh dried sample; subtract weight of filter paper from total weight; difference represents dry weight of periphyton.

5. Calculate and express results in terms of grams per square decimeter of original periphyton area.

General Considerations

a. Volumetric and gravimetric measures of periphyton samples may sometimes be subject to sizable errors due to the presence of sand or marl. Separation of sand and marl from periphyton is difficult and sometimes impracticable.

b. Periphyton samples may be concentrated by the use of a centrifuge.

c. Pieces of silk bolting cloth may be substituted for filter paper in volumetric measurements if *net* periphyton will suffice for the results desired.

d. Certain plankton methods (pp. 385–395) may be adapted for quantitative studies of periphyton.

Methods for Vegetation-inhabiting Animals

WISCONSIN TRAP METHOD

Trap Construction

A frame (Fig. 90), approximately square, with inside dimensions of 36 × 37 cm., made from brass strip 2.5 cm. wide and 3 mm. thick; two halves of frame independent and attached to each other with double hinge so that when doubled upon each other the edges meet evenly; one row of closely spaced small holes parallels edges opposite closing ones, affording means of attachment for net; an elongated net 85 cm. long with transverse dimensions approximately same as brass frame, sides made of light canvas or drilling and bottom of No. 72 extra heavy grit gauze.

Fig. 90. Trap for plant-inhabiting animals, Wisconsin design.

Suitability. Suitable only for work in shallow water; for macroscopic animals living on exterior of plants; and for work with submerged and floating aquatic plants.

Procedure

1. Hold trap in vertical position, jaws downward and wide open; lower carefully over plants to be examined; loosen plants from bottom and close jaws of trap; invert trap into position with jaws at top; bring to surface and pull entirely out of water; allow water to drain out through grit gauze at bottom of net.

2. Transfer plants to large tub containing tap or filtered water; wash

plants in this water thoroughly to remove all animals; also turn net of trap inside out and rinse in same water to secure any animals dislodged in process of operation of trap.

3. Pour contents of tub through sheet of No. 72 grit gauze to concentrate the catch.

4. Sort, identify and enumerate caught animals according to methods described on p. 303. Express results in terms of number, weight, or volume of animals per square meter of bottom on which plants grew.

General Considerations

a. By use of a diving outfit this trap can be used to depths of 5–6 m.

b. Ordinarily two individuals are required for operation of this trap, one in the water to make the catches, and one to haul in apparatus and remove collection.

c. If desirable, the plants can be weighed (wet weight) after the water has been drained off, and the results of the enumeration of animals can be expressed in terms of number, weight or volume of animals per unit of weight of the plants concerned.

STREAM-BOTTOM SAMPLER METHOD

EQUIPMENT. Surber's square stream-bottom sampler (p. 321); accessories.

SUITABILITY. Suitable for low-growing bottom plants in shallow streams; only for macroscopic animals living on outside of plants.

Procedure

1. Set sampler carefully over plants to be examined; settle horizontal frame firmly on bottom.

2. As stream water flows over horizontal frame and into net, wash plants thoroughly dislodging animals and allowing them to be carried by current into net.

3. Secure catch and treat as described elsewhere (p. 322) for other samples taken with sampler.

4. Express results in terms of number, weight or volume of animals per square meter of stream bottom.

General Considerations

a. Care must be taken that animals not related to the plants are prevented from being drifted into the net by the current.

b. If desirable, the plants may be saved from the area examined, drained of excess water, weighed (wet weight), and the results expressed in terms of the number, weight or volume of animals per pound of plants involved.

HESS METHOD FOR FLOATING VEGETATION

Equipment. Cylinder of heavy galvanized iron, 6 in. high, having top area of 1 sq. ft., and bottom covered with No. 40 brass screening; wooden paddle.

Suitability. Suitable for securing samples from floating mats of aquatic vegetation; for macroscopic animals living on outside of plants.

Procedure

1. Slip sampler under mat of vegetation to sampling position; bring sampler up against lower surface of mat and lift slightly above water.
2. Cut out circular sample by striking all around edge of sampler with wooden paddle.
3. Wash animals from vegetation so collected into sampler, and transfer them to collecting jar; transfer plants to large paper bag; transport to laboratory for analyses.
4. Sort, identify, enumerate, and express in terms of number, weight, or volume per square meter of water surface; also if desired, drain plants and express results in terms of number, weight or volume of animals per unit of wet weight of plants involved.

WASHING METHOD

Suitability. Suitable only for general purposes in which results of but approximate accuracy are acceptable.

Procedure

1. By any means suitable and convenient for the situation involved, sever connections of plants from bottom and lift slowly and carefully to surface but *not out of water.*
2. Submerge a 30-mesh screen of type described on p. 186 and slip under plants as they approach surface; lift screen and contained plants and discharge all contents into large tub.
3. Repeat this form of collection until desired quantity of plants is secured. Transfer to shore for washing and weighing.
4. Pile plants on a 30-mesh screen and allow to drain for 5 min. Then transfer all contents of screen to large tub partly filled with clean tap water and wash plants *thoroughly* displacing all adhering organisms and extraneous materials.
5. Lift plants from water, pile on screen and allow to drain for 5 min.; weigh plants to secure their *wet weight.*
6. Pour contents of tub through a 30-mesh screen. With additional amounts of tap water, wash residue in screen until all extraneous materials which will pass through meshes have been eliminated.

7. Sort, isolate and count organisms in screen residue by method described on p. 303.

8. Compute number of organisms per pound of wet weight of plants; or if preferred, allow excess water to drain off organisms by method described on p. 304, measure volume, and compute volume of total organisms (or of various kinds of organisms) per pound of wet weight of plants; or, if desirable, weigh drained (p. 304) organisms, in total or by different kinds, and compute number of selected weight units of organisms per weight unit of wet weight of plants.

General Considerations

a. In this method no account can be taken of any organisms *within* the plants concerned.

b. The size of the screen employed determines the size of invertebrates lost through the meshes.

c. Plants must be so collected that samples are representative of the plant beds involved.

d. The operator must be on guard against losses of invertebrates which may occur while plants are being hauled up through the water. Such loss may be detected and measured by the use of a fine-mesh "umbrella net" swung under the plants when first detached from bottom, then brought to the surface immediately below the plant mass.

Selected References

(For Chapter 17)

Juday, C.: A third report on limnological apparatus, *Trans. Wisconsin Acad. Sci.*, **22**: 299–314, 1926.

Hess, A. D.: "New limnological sampling equipment." 5 pp. Spec. Pub., No. 6. Limn. Soc. Am., 1941.

Krecker, F. H.: A comparative study of the animal population of certain submerged aquatic plants, *Ecology*, **20**: 553–562, 1939.

Young, O. W.: A limnological investigation of periphyton in Douglas Lake, Michigan, *Trans. Am. Microscop. Soc.*, **64**: 1–20, 1945.

Appendix

LINES

So numerous and varied is the need for lines in limnological work that careful choices are necessary if dependable results are to be obtained.

Graduated Ropes

For work in inland waters graduated ropes have many uses. To be satisfactory a rope must be strong, flexible, closely plaited or twisted, durable, free from undue initial stretching, and must lend itself to some simple system of permanent graduation. Companies dealing in cordage supply information and samples on application. Rarely will graduated ropes of diameter greater than 3/8 in. be needed; smaller sizes are sometimes required for special purposes. Good linen or hemp ropes are usually acceptable. Certain kinds and sizes of "tiller rope" may be useful. Cotton cord must be carefully inspected and tested. Best-grade sash cord (window-weight cord), sizes No. 6, 7, and 8, is satisfactory. However, since not all sash cord is of equal quality, samples should be secured and tested in advance. White sash cord is preferable to colored or spotted since it does not obscure graduation marks. Ordinary clothes-line rope and similar loosely woven cotton lines should be avoided. All ropes, even under the best of care, deteriorate because of wear and aging and must be replaced from time to time.

Since all ropes will stretch and shrink, new ones, prior to graduation, should be treated as follows: Stretch the line tightly between two trees or posts, or wrap tightly about an oil drum or other cylindrical object of convenient size; wet thoroughly and allow to dry; restretch or rewind to take up slack and again wet and allow to dry; repeat treatment until the stretch seems to be out, but use care not to overstretch.

When a rope is ready to be graduated it should be laid out on the floor of a long corridor or on a clean concrete walk, pulled straight and stretched with a tension slightly in excess of that representing the weight of the instruments to be used on it, and then securely fastened at each end. Begin at one end to measure off the desired intervals with a previously tested steel tape or some other instrument of linear measure, using the desired units (feet or meters), and install the permanent markers of the selected graduation system. The zero mark should be at least one foot from the end of the rope in order to provide a tie-on margin.

Various systems of marking graduations may be used. Whatever the scheme of graduation, good visibility and ease of reading are prime requisites. If ropes are to be used in operations which do not require the use of metal messengers, the intervals may be marked in one of the conventional ways by knotting and threading pieces of cloth (white at all intervals except the fifth which is red and red at every fifth thereafter) through the meshes or strands of the rope. However, this method can not be used for ropes on which messengers must slide and since many instruments require messengers some other effective method of graduation must be provided. The writer uses the following scheme: the desired intervals are marked by a thorough application, with a small brush, of waterproof black ink (Higgins Eternal, or Higgins India), making certain that the ink soaks deep into the rope. When used on white sash cord, this affords an easily recognizable mark, and while its distinctness is dimmed somewhat in a dry rope which has had considerable use, it becomes distinct again when wet. One thorough treatment with the ink will usually suffice to last through the life of the rope. Renewal may be needed when ropes have an unusual amount of messenger or handling wear. A simple system of long marks, short marks, and groups of long and short marks can be devised which will enable the operator to read directly the various levels. In the case of marks, or of groups of marks of considerable length, the interval should be correct at the middle point.

It must not be expected that ropes, constructed as described above, will remain accurate without further attention. There is always the possibility that they will shrink in length when coiled and stored for long periods of time. They may even show some shrinkage if coiled and dried after ordinary use in the field. Therefore, graduated ropes should be frequently inspected—wet them, pull them out straight on a long flat surface and test them for accuracy along the entire length. Usually, if a rope is found to be short of the expected value, the shrinkage has been uniform throughout the entire length and correction can be made by pulling the rope until the shrink is removed, after which the rope will probably remain correct for that day's work or perhaps longer. However, when careful work is required, it will be much safer to test the ropes both at the beginning and at the end of the day's work. This checking may be done with a tape or meter stick, but a more convenient way is to mark off permanent intervals with copper tacks on the boat deck, dock, or laboratory floor.

During a period of active use, keep graduated ropes submerged in water. If for any reason this procedure cannot be followed, thoroughly soak them in water for about an hour and then test for accuracy before using.

Every graduated rope should bear some means of identification, preferably a serial number. Such identification number should be attached to the end opposite the zero end of the rope. The writer uses 3×3-cm. squares of heavy canvas on which the identification number is written boldly in black waterproof ink. The square is then soaked in melted paraffin and attached by a strong, paraffined cord sewed into the rope.

Metal Lines

Metal lines have both advantages and disadvantages. Their principal point of superiority is in the fact that they do not stretch or shrink (temperature effect usually negligible). They may also combine small bulk with greater length of life if properly cared for. Obvious disadvantages are the following points: (1) ordinarily they must be used on a reel or drum in order to prevent kinking or twisting; (2) they should be of nonrust metal; (3) it is difficult to attach graduation marks if graduation has not been built into the line by the maker; (4) use by hand is slow, uncomfortable, and sometimes impracticable; and (5) they are likely to be expensive.

Probably the best of all metal lines are the stranded wire and the piano wire which have been used extensively for sounding in deeper waters, but they can be operated only on a revolving drum and, in lieu of graduations on the line, depths must be determined by a recording dial geared to the drum mechanism. For soundings in depths of 200–500 fathoms, the U.S. Coast and Geodetic Survey uses a stranded wire composed of seven tightly twisted strands of double-galvanized wire, each No. 24 B. and S. gage, and with a breaking strength of not less than 500 lb. It is obtainable in sealed tins containing 300-fathom lengths. For greater depths, steel piano wire, No. 21 B. and S. gage, is employed. For some purposes graduated steel tapes may be used but they require much care to prevent rusting, and unless they are quite narrow, messengers cannot slide on them. Special tapes, suitable for limnological purposes, are expensive. Tapes can be used only by winding and unwinding on a reel. Metal fishing lines have been tried, but usually with little success since they also must be used on a reel, can be graduated with difficulty, and most, if not all, of them kink too readily. Small brass chains, graduated by solder marks in links or by soldering small circular stamped brass plates to links, have been used with some success but have the disadvantage of being uncomfortable to use by hand.

Small, flexible, multiple-strand wire cables of ⅛-in. diameter or less may be used on a hand reel. They should be galvanized to prevent rusting. The smaller sizes will probably be preferred, if they have adequate tensile strength, because of their greater flexibility and lighter weight. Such wire cables may be graduated by marking them with barn-red paint.

Combination Lines

Lines can be purchased from cordage and other supply companies which combine metal and rope in various ways, as for example, in some instances small copper wire strands are interwoven with the strands of the rope; in other kinds, the rope is woven about a centrally located copper wire. Such a combination increases the tensile strength of the rope and practically eliminates stretching or shrinking. For lead-line sounding, the

U.S. Coast and Geodetic Survey had adopted a line with wire center and finds that best results have been obtained with one known as Sampson mahogany tiller rope, size No. 8, a waterproof solid braided rope with phosphor bronze wire center ("Hydrographic Manual," 1928, p. 44). Such a line can be graduated in the usual ways. However, when the interwoven-strand types are used for limnological purposes, sooner or later the wire strands break, thus defeating the purpose of such lines and also making them very bad for handling because of the projecting broken ends of the wire. Combination metal-fabric lines last longer if used on drums or reels.

Lines for Heavy Equipment

Heavy samplers, dredges and similar equipment may demand stronger lines than those mentioned in the preceding section. If such apparatus does not require the use of a messenger, lines of larger size are usable. If messengers are required, a smaller line of relatively high tensile strength is necessary and must be specially selected if a rope is to be used, but a small wire cable will probably be preferable.

Heavy instruments carrying a heavy load should be handled with a drum lift if possible. Under these circumstances a well-chosen, small-diameter, nonrust wire cable is suitable. Manufacturers of wire cables will supply samples and full information. The variety of wire cables made for commercial purposes is so great that almost any desired type can be secured.

If heavier instruments are to be used by hand, metal cables are ill suited for such work. As a general utility article, a good-grade Manila rope of ½- to ¾-in. size will be found adaptable, durable, and easy to handle.

Wire cables of properly selected kinds and sizes are useful for many purposes, particularly for anchors, floats, buoys, and work rafts, but are expensive and handle with some difficulty. Ropes are in more general use. Good Manila rope of appropriate size is very serviceable, but its life in water is relatively short unless it has been impregnated with paraffin, linseed oil and paraffin, or some other waterproofing substance. The well-known tarring treatment employed by commercial fishermen and others is very effective but ropes so prepared are disagreeable to handle and possess other objections for scientific work.

Small Lines

The limnologist has innumerable needs for small lines. They may be chosen to suit his own preferences and uses from the great variety of cordage on the market. Linen twines and cords are very useful. The well-known fish-cord, in its various sizes, is convenient. When soaked in melted soft paraffin, its life is considerably prolonged.

Table 1

EQUIVALENT VALUES

Length

1 inch = 25.40 millimeters
= 2.54 centimeters
1 foot = 0.305 meter
= 30.5 centimeters
1 yard = 3 feet
= 0.914 meter
1 rod = 16.5 feet
= 5.5 yards
1 mile (statute) = 63,360 inches
= 5280 feet
= 1760 yards
= 320 rods
= 1609 meters
= 1.609 kilometers
= 0.867 geographic mile
1 millimeter = 0.0393 inch
1 centimeter = 0.393 inch
1 meter = 39.37 inches
= 3.281 feet
= 1.0936 yards
= 0.000621 mile
1 kilometer = 3281 feet
= 1000 meters
1 chain (Gunter's) = 792 inches
= 66 feet
= 4 rods
= 0.0125 mile
1 link (Gunter's) = 7.92 inches
= 0.04 rod
1 chain (engineer's) = 100 feet
1 link (engineer's) = 1 foot

Table 1—(Continued)

EQUIVALENT VALUES

Depth

1 fathom = 6 feet
= 1.829 meters

Area

1 square inch = 6.42 square centimeters
1 square foot = 929.03 square centimeters
1 square yard = 0.836 square meter
1 acre = 43,560 square feet
= 4840 square yards
= 160 square rods
= 10 square chains (Gunter's)
= 0.4047 hectare
1 section = 640 acres
= 1 square mile
1 square mile = 640 acres
= 259 hectares
= 2.59 square kilometers
1 square millimeter = 0.0015 square inch
1 square meter = 10.758 square feet
1 hectare = 10,000 square meters
= 2.5 acres (approximately)

Volume

1 cubic inch = 16.386 cubic centimeters
1 cubic foot = 28,316 cubic centimeters
= 7.48 gallons
= 0.0283 cubic meter
1 cubic yard = 0.7646 cubic meter
1 acre-foot = 325,850 gallons
1,000,000 cubic feet = 22.95 acre-feet
1 cubic centimeter = 0.061 cubic inch
1 cubic meter = 35.314 cubic feet
= 1.308 cubic yards

Table 1—(Continued)

EQUIVALENT VALUES

Capacity

1 U.S. pint = 473.18 cubic centimeters
1 U.S. quart = 2 pints
= 946 cubic centimeters
= 0.946 liter
1 U.S. gallon = 231 cubic inches
= 4 quarts
= 3784 cubic centimeters
= 3.784 liters
1,000,000 gallons = 3.07 acre-feet
1 liter = 61.027 cubic inches
= 2.11 pints
= 1.0567 quarts
= 1000 cubic centimeters

Miscellaneous

1 atmosphere pressure = about 15 pounds per square inch
= about 1 ton per square foot
= about 1 kilo per square centimeter

Angles

1 circumference = 360 degrees
1 degree = 60 minutes
1 minute = 60 seconds

Table 2

MATHEMATICAL FORMULAS

Given	*Sought*	*Formula*
Triangle		
1. Base (b) and altitude (a')	Area (a)	$a = \frac{ba'}{2}$, or $= b\left(\frac{a'}{2}\right)$
2. Area (a) and base (b) or altitude (a')	Base (b), or altitude (a')	$b = \frac{2a}{a'}$, or $a' = \frac{2a}{b}$
3. Three sides (d, d', d'')	Area (a)	Let s = sum of three sides, then $a = \sqrt{\left(\frac{s}{2} - d\right)\left(\frac{s}{2} - d'\right)\left(\frac{s}{2} - d''\right)\left(\frac{s}{2}\right)}$
4. Base (b) and perpendicular (p) of right-angle triangle	Hypotenuse (h)	$h = \sqrt{b^2 + p^2}$
5. Base (b) or perpendicular (p) and hypotenuse (h) of right-angle triangle	Base (b), or perpendicular (p)	$b = \sqrt{h^2 - p^2}$, or $p = \sqrt{h^2 - b^2}$
Trapezoid		
6. Sides (s and s') and altitude (a')	Area (a)	$a = a'\left(\frac{s + s'}{2}\right)$
Trapezium		
7. Diagonal (d) and perpendiculars (p and p') to diagonal drawn from vertices of opposite angles	Area (a)	$a = d\left(\frac{p + p'}{2}\right)$
Circle		
8. Radius (r)	Circumference (c)	$c = 2\pi r$
9. Circumference (c)	Radius (r)	$r = \frac{2\pi}{c}$
10. Radius (r)	Area (a)	$a = \pi r^2$
Sphere		
11. Radius (r)	Surface (s)	$s = r^2(4\pi)$
12. Radius (r)	Volume (v)	$v = r^3\left(\frac{4\pi}{3}\right)$

Table 2—(Continued)

MATHEMATICAL FORMULAS

Given	*Sought*	*Formula*
Cylinder		
13. Radius (r) and altitude (a')	Convex surface (s), or volume (v)	$s = a'(2\pi r)$, or $v = a'(\pi r^2)$
Cone		
14. Radius (r) and altitude (a')	Volume (v)	$v = \frac{a'}{3}(\pi r^2)$
15. Radius (r) and slant height (h)	Convex surface (s)	$s = \frac{h}{2}(2\pi r)$
Frustrum of Cone		
16. Areas of both bases (b and b') and altitude (a')	Volume (v)	$v = \frac{a'}{3}(b + b' + \sqrt{bb'})$
17. Circumferences (c and c') and slant height (h)	Convex surface (s)	$s = \frac{h}{2}(c + c')$

Table 3

FEET TO METERS*

Feet	*0*	*1*	*2*	*3*	*4*	*5*	*6*	*7*	*8*	*9*
0	0.000	0.305	0.610	0.914	1.219	1.524	1.829	2.134	2.438	2.743
10	3.048	3.353	3.658	3.962	4.267	4.572	4.877	5.182	5.486	5.791
20	6.036	6.401	6.706	7.010	7.315	7.620	7.925	8.229	8.534	8.839
30	9.144	9.449	9.753	10.058	10.363	10.668	10.972	11.277	11.582	11.887
40	12.192	12.496	12.801	13.106	13.411	13.716	14.020	14.325	14.630	14.935
50	15.239	15.544	15.849	16.154	16.459	16.763	17.068	17.373	17.678	17.983
60	18.287	18.592	18.897	19.202	19.507	19.811	20.116	20.421	20.726	21.031
70	21.335	21.640	21.945	22.250	22.555	22.859	23.164	23.469	23.774	24.079
80	24.383	24.688	24.993	25.298	25.602	25.907	26.212	26.517	26.822	27.126
90	27.431	27.736	28.041	28.346	28.651	28.955	29.260	29.565	29.870	30.174
100	30.479	30.784	31.089	31.394	31.698	32.003	32.308	32.613	32.918	33.222

*Length in feet expressed in left vertical column and in top horizontal row; corresponding lengths in *meters* in body of table.

Table 4

METERS TO FEET*

Meters	*0*	*1*	*2*	*3*	*4*	*5*	*6*	*7*	*8*	*9*
0	0.00	3.28	6.56	9.84	13.12	16.40	19.69	22.97	26.25	29.53
10	32.81	36.09	39.37	42.65	45.93	49.21	52.49	55.78	59.06	62.34
20	65.62	68.90	72.18	75.46	78.74	82.02	85.30	88.58	91.87	95.15
30	98.43	101.71	104.99	108.27	111.55	114.83	118.11	121.39	124.67	127.96
40	131.24	134.52	137.80	141.08	144.36	147.64	150.92	154.20	157.48	160.76
50	164.04	167.33	170.61	173.89	177.17	180.45	183.73	187.01	190.29	193.57
60	196.85	200.13	203.42	206.70	209.98	213.26	216.54	219.82	223.10	226.38
70	229.66	232.94	236.22	239.51	242.79	246.07	249.35	252.63	255.91	259.19
80	262.47	265.75	269.03	272.31	275.60	278.88	282.16	285.44	288.72	292.00
90	295.28	298.56	391.84	305.12	308.40	311.69	314.97	318.25	321.53	324.81
100	328.09	331.37	334.65	337.93	341.21	344.49	347.78	351.06	354.34	357.62

*Length in meters expressed in left vertical column and in top horizontal row; corresponding lengths in *feet* in body of table.

Table 5

TEMPERATURES—CENTIGRADE TO FAHRENHEIT*

Temp. ° C.	*0*	*1*	*2*	*3*	*4*	*5*	*6*	*7*	*8*	*9*
0	32.0	33.8	35.6	37.4	39.2	41.0	42.8	44.6	46.4	48.2
10	50.0	51.8	53.6	55.4	57.2	59.0	60.8	62.6	64.4	66.2
20	68.0	69.8	71.6	73.4	75.2	77.0	78.8	80.6	82.4	84.2
30	86.0	87.8	89.6	91.4	93.2	95.0	96.8	98.6	100.4	102.2
40	104.0	105.8	107.6	109.4	111.2	113.0	114.8	116.6	118.4	120.2
50	122.0	123.8	125.6	127.4	129.2	131.0	132.8	134.6	136.4	138.2

*Temperatures in degrees Centigrade expressed in left vertical column and in top horizontal row; corresponding temperatures in degrees Fahrenheit in body of table.

Table 6

TEMPERATURES—FAHRENHEIT TO CENTIGRADE*

Temp. ° F.	*0*	*1*	*2*	*3*	*4*	*5*	*6*	*7*	*8*	*9*
30	– 1.11	– 0.56	0.00	0.56	1.11	1.67	2.22	2.78	3.33	3.89
40	4.44	5.00	5.56	6.11	6.67	7.72	7.78	8.33	8.89	9.44
50	10.00	10.56	11.11	11.67	12.22	12.78	13.33	13.89	14.44	15.00
60	15.56	16.11	16.67	17.22	17.78	18.33	18.89	19.44	20.00	20.56
70	21.11	21.67	22.22	22.78	23.33	23.89	24.44	25.00	25.56	26.11
80	26.67	27.22	27.78	28.33	28.89	29.44	30.00	30.56	31.11	31.67
90	32.22	32.78	33.33	33.89	34.44	35.00	35.56	36.11	36.67	37.22
100	37.78	38.33	38.89	39.44	40.00	40.56	41.11	41.67	42.22	42.78

*Temperatures in degrees Fahrenheit expressed in left vertical column and in top horizontal row; corresponding temperatures in degrees Centigrade in body of table.

Table 7

EFFECT OF TEMPERATURE UPON DENSITY OF WATER*

From Birge

I *Temperature* °*C.*	II *Density*	III *Difference*	IV *Relative Difference for 1° C.*	V *Ergs*
0	0.999868	+0.000059	7.38	0.0491
1	0.999927	+0.000041	5.12	0.0342
2	0.999968	+0.000024	3.00	0.0200
3	0.999992	+0.000008	1.00	0.0067
4	1.000000	—0.000008	1.00	0.0067
5	0.999992	—0.000024	3.00	0.0200
6	0.999968	—0.000039	4.88	0.0325
7	0.999929	—0.000053	6.62	0.0441
8	0.999876	—0.000068	8.50	0.0566
9	0.999808	—0.000081	10.12	0.0675
10	0.999727	—0.000095	11.88	0.0791
11	0.999632	—0.000107	13.38	0.0891
12	0.999525	—0.000121	15.12	0.1008
13	0.999404	—0.000133	16.62	0.1108
14	0.999271	—0.000145	18.12	0.1208
15	0.999126	—0.000156	19.50	0.1299
16	0.998970	—0.000169	21.12	0.1408
17	0.998801	—0.000179	22.38	0.1491
18	0.998622	—0.000190	23.75	0.1583
19	0.998432	—0.000202	25.25	0.1683
20	0.998230	—0.000211	26.38	0.1758
21	0.998019	—0.000222	27.75	0.1849
22	0.997797	—0.000232	29.00	0.1993
23	0.997565	—0.000242	30.25	0.2016
24	0.997323	—0.000252	31.50	0.2099
25	0.997071	—0.000261	32.62	0.2174
26	0.996810	—0.000271	33.88	0.2257
27	0.996539	—0.000280	35.00	0.2332
28	0.996259	—0.000288	36.00	0.2399
29	0.995971	—0.000298	37.25	0.2482
30	0.995673			

*See note to table on facing page.

Note for Table 7

*Column I indicates temperature in degrees Centigrade; column II shows the density of pure water. Note that the numbers in columns III to V are not on a level with those in columns I to II but are opposite the spaces between these numbers. Column III shows the differences between the successive numbers of the previous column and indicates the change in density caused by a temperature change of 1°. The significant figures also show the difference in weight, in milligrams, between a liter of water at any given temperature and at a temperature 1° lower. Thus, a liter of water at 10° weighs 81 mg. less than one at 9°. These numbers express the differences in density and weight which for a temperature difference of 1° (1) enable a layer of water to set up convection currents if it lies above a warmer stratum and (2) enable a stratum of water, warmer above and cooler below, to resist mixture attempted by mechanical agents. Column IV, the convection capacity and the thermal resistance to mixture corresponding to the temperature difference of 1° at 4° to 3° or 4° to 5°, is taken as unity, and the relative value is given for the same difference at higher or lower temperatures. At 10°, its value is more than ten times greater than at 4°; at 15°, it has increased eighteenfold; and at 30°, it is more than thirty-seven times larger than at 4°. Column V contains a statement of work in decimals of an erg which would be required to mix a column of water 1 sq. cm. in area, 1 m. high, in which the temperature gradient is uniform and whose upper and lower surfaces differ in temperature by 1°. (Explanation largely quoted from Birge.)

Table 8

AMOUNT OF DISSOLVED OXYGEN IN WATER AT DIFFERENT TEMPERATURES WHEN EXPOSED TO AN ATMOSPHERE CONTAINING 20.9 PER CENT OF OXYGEN UNDER A PRESSURE OF 760 MM. INCLUDING PRESSURE OF WATER VAPOR*

Temp. ° *C.*	*Parts per Million*	*Cc. per liter (at 0° C. and 760 mm.)*	*Temp.* ° *C.*	*Parts per Million*	*Cc. per liter (at 0° C. and 760 mm.)*
0	14.62	10.23	16	9.95	6.96
1	14.23	9.96	17	9.74	6.82
2	13.84	9.68	18	9.54	6.68
3	13.48	9.43	19	9.35	6.54
4	13.13	9.19	20	9.17	6.42
5	12.80	8.96	21	8.99	6.29
6	12.48	8.73	22	8.83	6.18
7	12.17	8.52	23	8.68	6.07
8	11.87	8.31	24	8.53	5.97
9	11.59	8.11	25	8.38	5.86
10	11.33	7.93	26	8.22	5.75
11	11.08	7.75	27	8.07	5.65
12	10.83	7.58	28	7.92	5.54
13	10.60	7.42	29	7.77	5.44
14	10.37	7.26	30	7.63	5.34
15	10.15	7.10			

*Reprinted by permission from "The Microscopy of Drinking Water," by Whipple, Fair, and Whipple, published by John Wiley & Sons, Inc.

Table 9

GRADES AND SIZE RANGES OF SILK BOLTING CLOTH

Grade	*Range of Sizes*
Standard	Nos. 0000–25
X quality	Nos. 6–17
XX quality	Nos. 0000–16
XXX quality	Nos. 6–18
Grit gauze	Nos. 14–72
XXX Grit gauze	Nos. 14–72

Table 10

WENTWORTH'S CLASSIFICATION OF COARSER SEDIMENTS BASED UPON SIZE OF PARTICLES

Diameter of Particle in mm.	*Name Applied to Particle*
More than 256	Boulder
256–64	Cobble
64–4	Pebble
4–2	Granule
2–1	Very coarse sand
1–0.5	Coarse sand
0.5–0.25	Medium sand
0.25–0.125	Fine sand
0.125–0.062	Very fine sand
0.062–0.004	Silt
Less than 0.004	Clay

Table 11

WENTWORTH GRADE SCALE, $\sqrt{2}$ SCALE, $\sqrt[4]{2}$ SCALE, CORRESPONDING TYLER SIEVE OPENINGS AND MESH, AND CORRESPONDING MESH OF U.S. SIEVE SERIES*

Wentworth Grade Scale (mm.)	*The Openings Increase in the Ratio of*		*Tyler Screens*		*U.S. Sieve Series, Mesh*
	$\sqrt{2}$ or 1.414 mm.	*$\sqrt[4]{2}$ or 1.189 mm.*	*Mm.*	*Mesh*	
4 Granule	4.00	4.00	3.96	5	5
		3.36	3.33	6	6
	2.83	2.83	2.79	7	7
		2.38	2.36	8	8
2 Very coarse sand	2.00	2.00	1.98	9	10
		1.68	1.65	10	12
	1.41	1.41	1.40	12	14
		1.19	1.17	14	16
1 Coarse sand	1.00	1.00	0.991	16	18
		0.840	0.833	20	20
	0.707	0.707	0.701	24	25
		0.595	0.589	28	30
0.500 (½) Medium sand	0.500	0.500	0.495	32	35
		0.420	0.417	35	40
	0.354	0.354	0.351	42	45
		0.297	0.295	48	50
0.250 (¼) Fine sand	0.250	0.250	0.246	60	60
		0.210	0.208	65	70
	0.177	0.177	0.175	80	80
		0.149	0.147	100	100
0.125 (⅛) Very fine sand	0.125	0.125	0.124	115	120
		0.105	0.104	150	140
	0.088	0.088	0.088	170	170
		0.074	0.074	200	200
0.062 (1/16) Silt	0.062	0.062	0.061	250	230

*Reprinted by permission from "Methods of Study of Sediments," by Twenhofel and Tyler, published by McGraw-Hill Book Co.

Table 12

AVERAGE APERTURE SIZE OF STANDARD GRADE DEFOUR BOLTING SILK*

Silk No.	*Meshes per Inch*	*Size of Aperture (mm.)*	*Silk No.*	*Meshes per Inch*	*Size of Aperture (mm.)*
0000	18	1.364	10	109	0.158
000	23	1.024	11	116	0.145
00	29	0.752	12	125	0.119
0	38	0.569	13	129	0.112
1	48	0.417	14	139	0.099
2	54	0.366	15	150	0.094
3	58	0.333	16	157	0.086
4	62	0.318	17	163	0.081
5	66	0.282	18	166	0.079
6	74	0.239	20	173	0.076
7	82	0.224	21	178	0.069
8	86	0.203	25	200	0.064
9	97	0.168			

*Reprinted with permission from "The Oceans, Their Physics, Chemistry and General Biology," by Sverdup, Johnson and Fleming, published by Prentice-Hall, Inc.

Table 13

RELATION OF ANGLES OF WILD'S PLATE TO WIND VELOCITY*

Angle† in Degrees	0	4	15.5	31	45.5	58	72	80.5
Wind Velocity in Meters per Second	0	2	4	6	8	10	14	20
Beaufort Scale	0	2	3	4	5	6	7	8

*Reprinted with permission from "Physical Climatology," by Landsberg, published by Gray Printing Company.

†Angles measured from the perpendicular.

SOURCES OF HYDROGRAPHIC MAPS AND HYDROGRAPHIC INFORMATION

This list is suggestive only and claims no completeness. See elsewhere (p. 1) for caution on unsuitability for limnological purposes of some maps made by agencies whose interests lie elsewhere.

Federal

Coast and Geodetic Survey
Geological Survey
Lake Survey
Forest Service
Fish and Wildlife Service
Tennessee Valley Authority
Bureau of Reclamation
Mississippi River Commission

Aid may be secured concerning maps published by these and other federal offices by applying to the U.S. Map Information Office, Department of Interior Building, Washington, D.C.

State

Geological Surveys
Departments of Conservation
Biological Surveys
Highway Departments

Miscellaneous

City Engineers' Offices; Water Supply; Sewage Disposal
Water Power organizations
Irrigation organizations
Recreational systems
Sportsmen's preserves
Reservoir lake developments
Publications of inland biological stations

LOCATION OF LAKES AND STREAMS

In limnological descriptions, geographical location of streams and lakes must be expressed with certainty. Names alone cannot be depended upon save in the instances of large bodies of water. Names are sometimes wholly local and different ones may be used for the same water. Therefore some other means must be used. Ordinarily this can best be done through the use of conspicuous, permanent, and well-known physical features, an adopted system of land survey, an established system of political subdivisions, or a

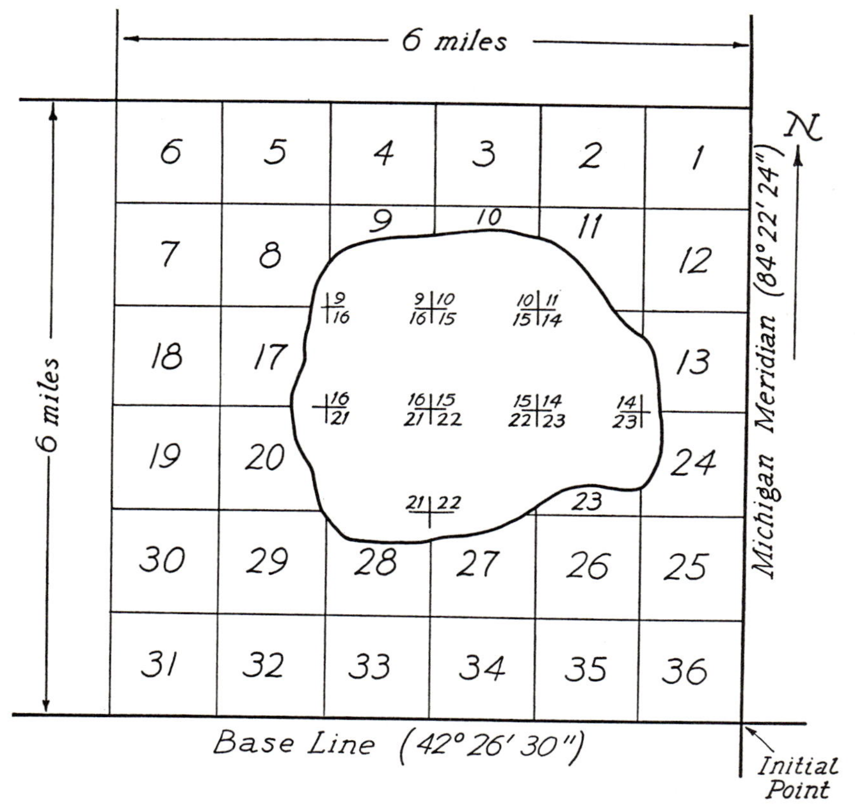

FIG. 91. Diagram of a congressional township showing method of numbering sections. Outline of lake within township shows numbering system for sections occupied wholly or in part by water.

judicious combination of these three. It is often preferable to base the description of inland lakes and streams upon the framework of the U.S. Public Land Survey in much the same way that tracts of land are specified for legal purposes. In this system, the congressional township is the basic unit. It is 6 miles square and contains 36 *sections* (approximately 36 sq. mi.). Each section is designated by a number and the system of numbering is shown in Fig. 91. Townships are numbered into *ranges* which extend north and south and into *tiers* which extend east and west. These ranges and tiers are related to an *initial point* formed by the intersection of the *principal meridian* and the *base line* of the whole area concerned. Principal meridians and base lines are often indicated on the larger geographic maps; ranges and tiers of townships are shown only on local, detailed maps. According to this system the imaginary lake shown in Fig. 91 is located as follows: Sections 15, 16, 21, 22, and portions of sections 9, 10, 11, 13, 14, 17, 20, 23, 24, 27, and 28, Township 1 North, Range 1 West of the Michigan Meridian (abbreviated T 1 N R 1 W Mich. M). Further details concerning the system of U.S. Land Surveys can be secured by consulting a standard text on surveying; also from the "Manual of Instructions for the Survey of the Public Lands of the United States," Government Printing Office.

ACCESSORY EQUIPMENT

Plankton Trawl

The plankton trawl is usually used for the capture of plankters which occur very near the bottom. Fig. 92 shows the construction of this device. The net is mounted on a metal collar which in turn is attached to a metal frame. A tow cord, tied to the frame, affords means of dragging this apparatus along the bottom. The horizontal metal sheet protects the net against damage.

The plankton trawl is very useful in connection with qualitative study of plankton in streams. In running water the trawl is not dragged but instead is anchored in position, mouth upstream, and left there during the collection period. When a coarser net is substituted for the fine-mesh bolting cloth of the plankton net, this device may be useful in the capture of those larger organisms which are carried downstream with the current and constitute much of what is called *stream drift*.

Field Kits

Portable field kits must be provided if equipment is to be properly guarded and if procedures are to be facilitated. Many different designs are required for diverse needs and the working limnologist must expect to plan and build cases for his own instruments since few such containers are available on the market.

Chemical Kits. Since in many chemical analyses water samples must be treated in the field to some safe point just as soon as they are secured, a

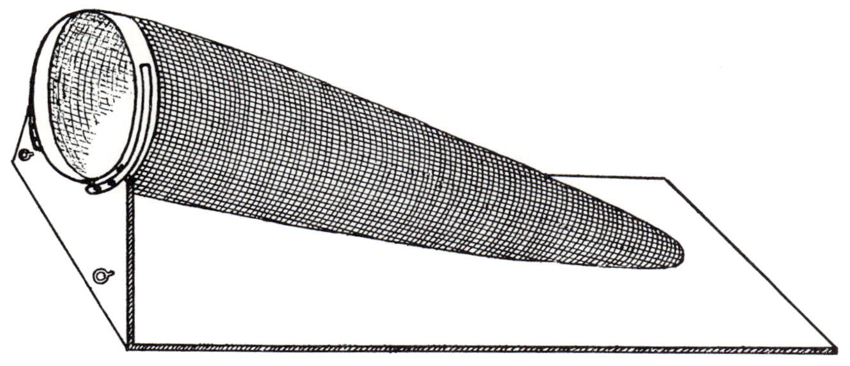

Fig. 92. Side view of plankton trawl.

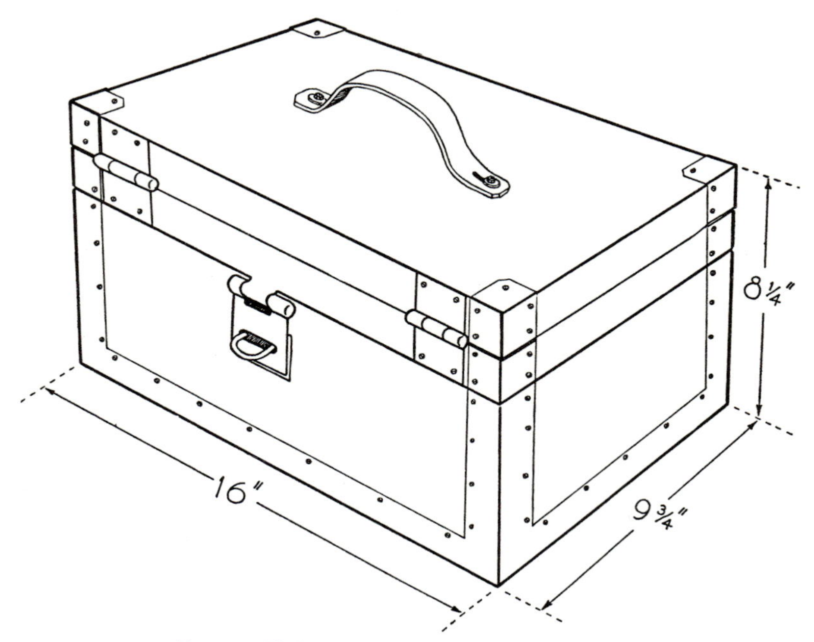

Fig. 93. Field chemical kit in closed form.

chemical kit containing the necessary reagents, pipets, and other equipment should be supplied. Figs. 93 and 94 represent the plan of a chemical kit developed in the writer's laboratory and used for many years. It is constructed of half-inch well-seasoned basswood, all edges are bound with metal (sheet copper or galvanized iron), and the whole kit is covered, inside and outside, with several coats of good "outside" paint. The double-hinged lid provides housing for the numerous pipets. Pipets are held in place by metal or rubber clips fastened to a removable board which fits into the lid space and is fastened to the walls of the lid by means of pegs and a spring catch. This board may be so constructed that pipets are supported on both surfaces.

The kit portrayed in Fig. 94 is designed to provide for equipment necessary for the routine field operations of analyses for dissolved oxygen, free carbon dioxide, alkalinities, hydrogen-ion concentration, and one or two others. The numerous circular holes of same diameter hold a full set of pH colorimeter tubes. Other compartments provide space for various bottles containing chemicals, indicators, necessary glassware, and general miscellany. Brown and Flaten (1943) recently described a chemistry kit of somewhat different pattern which incorporates some features of the one discussed here.

Instrument and Sample-bottle Kits. Kits for equipment accessory to chemical analyses, such as Kemmerer samplers, thermometers, graduated ropes, and miscellaneous articles, may be made of half-inch basswood to the following dimensions: 27 × 9 × 8 in. Corners should be metal bound and the whole covered, both inside and outside, with at least three coats of "outside" paint. The lid (top) is hinged along one side and fastened down on the other edge with two sturdy hasps, one near each end. The interior is divided lengthwise into two compartments by a partition about 3 in. high, thus making spaces for Kemmerer samplers. Above these spaces is a removable tray, 26 × 4¼ × 3¾ in., for other equipment. A strong leather handle is fastened to the top of the lid. If heavy equipment is to be handled, the lid may require reinforcement.

Kits for sample bottles may be made of the same materials and to the same dimensions as the instrument kit described above. The interior is

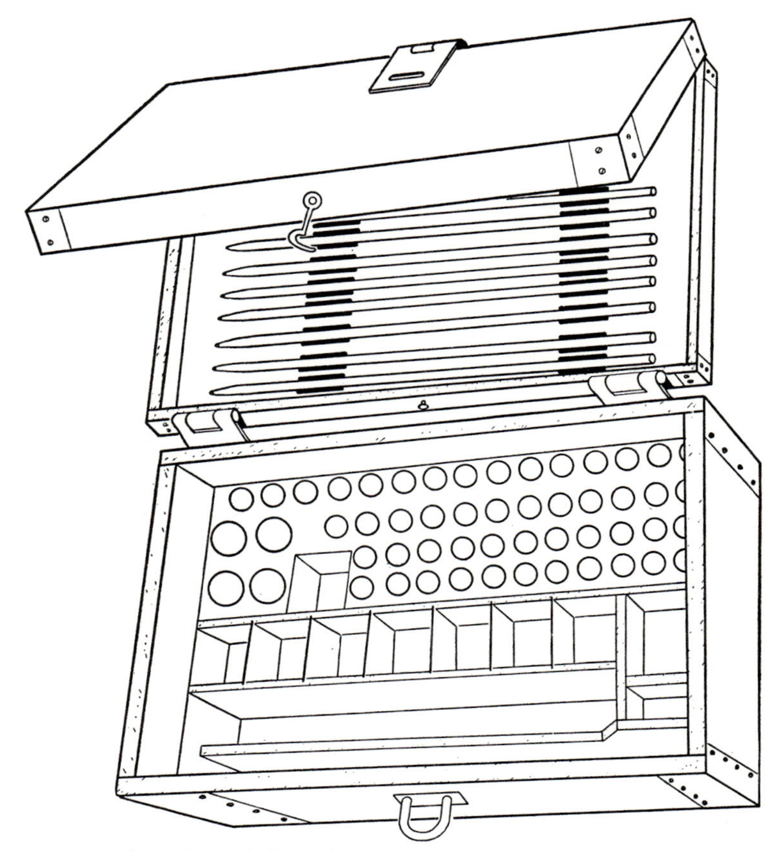

FIG. 94. Inside view of field chemical kit, showing pipets in holders on one surface of double lid; opposite surface of double lid also provided with holders (not shown in figure) for additional pipets.

divided into square compartments of size necessary to hold sample bottles. These kits may have lids as described above, or for some purposes they may be made without lids but equipped with fasteners at each end so that one sample-bottle kit can be attached to the lower surface of the instrument kit and both carried as one unit. Rubber feet at each corner of the lower surface reduce possible shock and breakage. Small holes in the bottom allow stray water to drain out, although they are no substitute for the necessary occasional complete drying of the interior. These kits must also be thoroughly painted on all surfaces.

Ice Drill

Cutting of holes for sounding through ice cover is slow and costly in effort unless performed by some power-driven device. Fig. 95 shows the chief construction features of an electric ice drill designed by Brown and Clark (1939) and used extensively by the Institute for Fisheries Research, Michigan Department of Conservation, in winter lake mapping. The principal specifications are as follows: *Motor*—1928 Studebaker starting motor (one which turns to the right); *drill bit*—1¾-in. wood bit; *shaft*—9⁄16-in. diameter steel rod (length determined by maximum depth of ice); *starting switch*—Universal starting motor switch; *battery*—regular 19-plate automobile battery. It is necessary to install a thrust bearing (*tb*) on the lower side of the motor. Battery cables may be of any convenient length. This drill will bore through 2–3 in. of ice per sec. and one operator can make

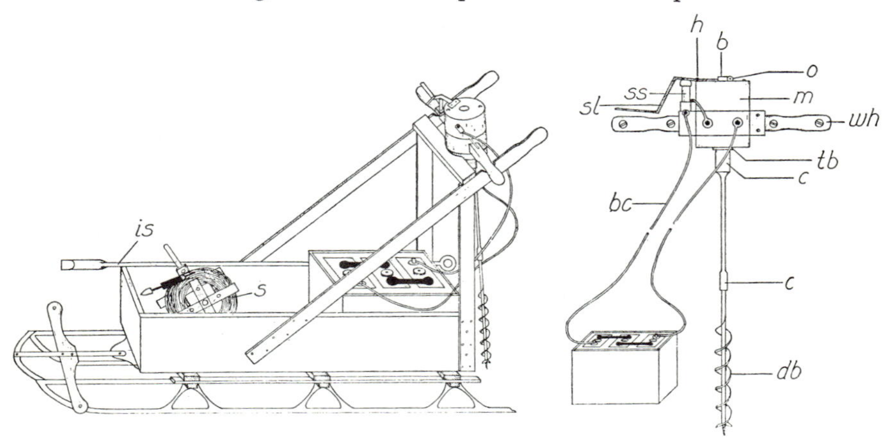

Fig. 95. Electric ice drill designed by Brown and Clark for cutting sounding holes through ice in winter hydrographic mapping. (*Left*) Transportation assembly. (*Right*) Drill unit. (*b*) Bearing. (*bc*) Battery cable. (*c*) Coupling. (*db*) Drill bit. (*h*) Hinge on starting lever. (*is*) Ice spud. (*m*) Motor. (*o*) Oil hole for upper bearing. (*s*) Sounding line. (*sl*) Starting lever. (*ss*) Starting switch. (*tb*) Thrust bearing. (*wh*) Wooden handle. (Drawings by William L. Cristanelli.)

300 or more sounding holes through thick ice in one day. At least one extra battery must be available if work is to be extended over two or more consecutive days since recharging will be necessary. Because of its relatively light weight the drill is easily handled. For ease in operation and transportation the whole outfit is assembled on a sled as shown in Fig. 95.

Water Telescopes

Examination of objects submerged in shallow water is greatly facilitated by the use of so-called water telescopes. Their essential feature is a plate of glass installed in some kind of frame which excludes water from its interior and is open at the top. When the lowermost, glass-bearing portion is shallowly submerged in water and the operator looks down into the open end, visibility of objects in the water is increased by the elimination of surface interference. Water telescopes may be made in many different forms to suit various needs. Fig. 96 represents three forms which serve general utility requirements: a very serviceable type made by the simple process of replacing the bottom of an ordinary pail with a piece of

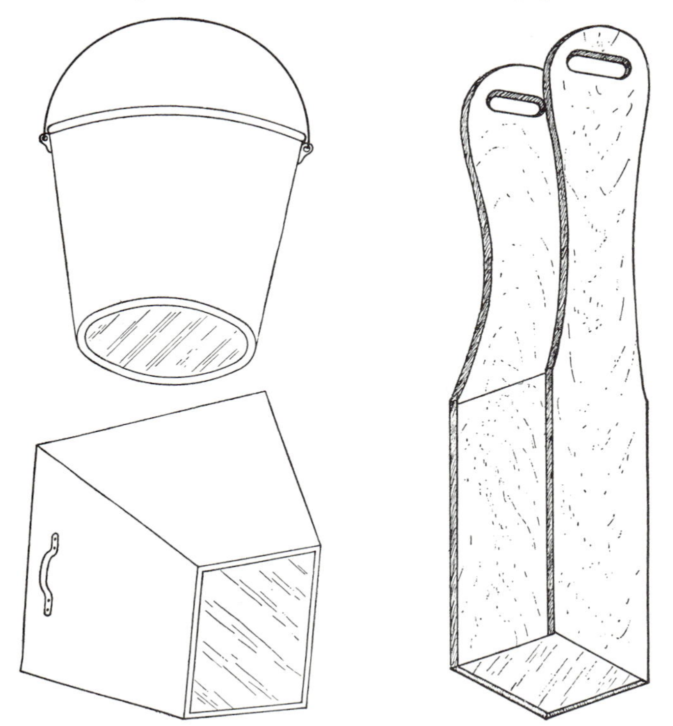

Fig. 96. Three types of water telescope. (*Top left*) Pail with glass bottom. (*Bottom left*) Rectangular metal box with glass bottom. (*Right*) Wooden frame with glass bottom.

plain glass in a watertight installation; watertight installation of glass in the smaller end of a flaring metal (brass or galvanized iron) frame, square in cross section; a sheet of glass sealed into the lowermost end of a special wooden frame of the form shown in the figure. Glass should always be installed in such a way that a projecting rim of the frame offers some protection against breakage when the telescope is set end down on a hard support. These types can be made in any metal shop and at small expense.

An elongated water telescope for use over the sides of boats may be made by sealing glass into one end of a tube having a diameter about that of ordinary stove pipe and putting handles on each side at the other end.

Selected References

Brown, C. J. D., and O. H. Clark: Winter lake mapping, *Michigan Conservation*, 8: 10–11, 1939. (Reprint of same paper issued to authors in enlarged form.)

———, and C. M. Flaten: "A Portable Field Chemistry Kit." 4 pp. Spec. Pub. No. 11, Limn. Soc. Am., 1943.

SOURCES OF LIMNOLOGICAL APPARATUS AND SUPPLIES

Limnological apparatus and supplies are so diverse that a working equipment must be assembled from many general sources. Certain items may be found in the catalogs of general scientific supply companies, but many others must be sought in highly specialized sources. In some instances it will be necessary to have instruments made to order. The Limnological Society of America published the extensive list of such sources indicated below.

Reference

"Sources of Limnological Apparatus and Supplies," rev. ed., 10 pp. Spec. Pub. No. 1, Limn. Soc. Am., 1939.

Correction Factors for Oxygen Saturation at Various Altitudes

Altitude Feet	Altitude Metres	Pressure mm.	Factor
0	0	760	1.00
330	100	750	1.01
655	200	741	1.03
980	300	732	1.04
1310	400	723	1.05
1640	500	714	1.06
1970	600	705	1.08
2300	700	696	1.09
2630	800	687	1.11
2950	900	679	1.12
3280	1000	671	1.13
3610	1100	663	1.15
3940	1200	655	1.16
4270	1300	647	1.17
4600	1400	639	1.19
4930	1500	631	1.20
5250	1600	623	1.22
5580	1700	615	1.24
5910	1800	608	1.25
6240	1900	601	1.26
6560	2000	594	1.28
6900	2100	587	1.30
7220	2200	580	1.31
7550	2300	573	1.33
7880	2400	566	1.34
8200	2500	560	1.36

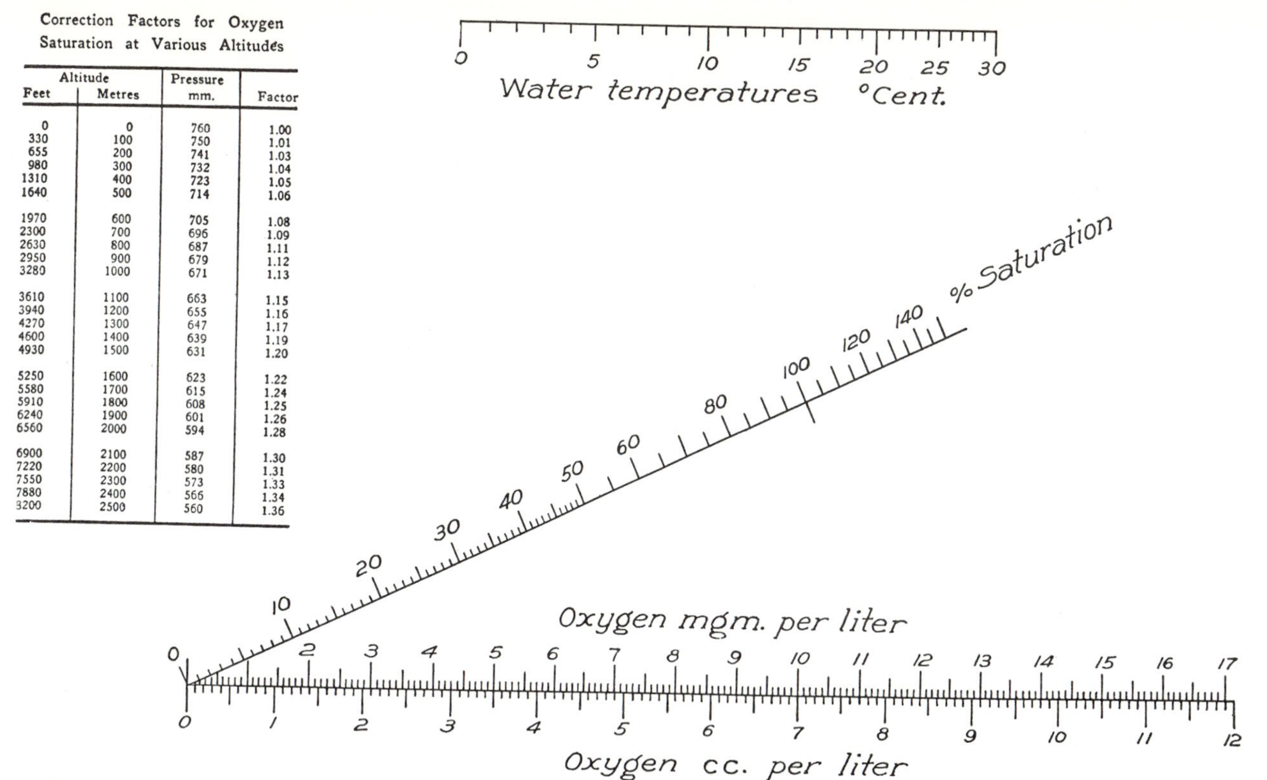

FIG. 97. Rawson's nomogram for obtaining oxygen-saturation values at different temperatures and at different altitudes; also for transforming oxygen values from one kind of unit to another. When a ruler, or preferably a dark-colored thread, is held so as to join an observed temperature on the upper scale with the observed dissolved-oxygen value on the lower scale, the values or units desired are read at points where the thread or ruler crosses the other scales. The associated table supplies correction values for oxygen saturation at various altitudes; for example: if 6.4 cc. per liter of oxygen is observed in a sample from a lake having an altitude of approximately 500 m. (1640 ft.), the amount of oxygen which would be present at sea level under the same circumstances is found by multiplying 6.4 by the factor 1.06, giving the product 6.8; then the percentage saturation is determined by connecting 6.8 on the lower scale with the observed temperature on top scale and noting point of intersection on middle (diagonal) scale. (Redrawn from Rawson, Spec. Pub. No. 15, Limn. Soc. Am., 1944.)

LITERATURE AIDS IN THE IDENTIFICATION OF PLANKTON

Selected List

References included here have been selected from the tremendous literature dealing with the taxonomy of plankton organisms. Items chosen are those which are general in nature and which seem suited to the needs of the worker who is making an entry into the field of plankton identification. It is believed that these references will also serve as a basis from which to secure leads into the more extensive literature. For the most part, only recent works have been listed. In a few instances certain unusually valuable old monographs have been included and are indicated by an *asterisk* (*). The inexperienced worker must not overlook the fact that these old works, while very helpful, are out-of-date in certain ways, as for example, in some details of nomenclature. Completeness is not attempted in any group.

GENERAL WORKS

Needham, J. G., and P. R. Needham: "A Guide to the Study of Fresh-water Biology," 4th ed., 90 pp. Ithaca, N. Y., Comstock Pub. Co., 1938.

Schoenischen, W.: "Einfachste Lebensformen das Tier- und Pflanzenreiches," 5th ed., Vol. 1, 519 pp.; Vol. 2, 522 pp. Berlin, Hugo Bermühler, 1925–1927.

Ward, H. B., and G. C. Whipple: "Fresh-water Biology." 1111 pp. New York, John Wiley and Sons, 1918.

Whipple, G. C., G. M. Fair, and M. C. Whipple: "The Microscopy of Drinking Water," 4th ed., 586 pp. New York, John Wiley and Sons, 1927.

PHYTOPLANKTON

Boyer, C. S.: "The Diatomaceae of Philadelphia and Vicinity." 143 pp. Philadelphia, J. B. Lippincott Co., 1916.

———: Synopsis of the North American Diatomaceae. Part 1, *Proc. Acad. Nat. Sci. Phila.*, Vol. 78, Suppl., pp. 1–228, 1926; Part 2, *Ibid.*, Vol. 79, Suppl., 229–583, 1927.

Fritsch, F. E.: "The Structure and Reproduction of the Algae." 792 pp. Cambridge University Press; The Macmillan Co., 1935.

Huber-Pestalozzi, G.: "Das Phytoplankton des Süsseswassers," in Thienemann's "Die Binnengewässer," Vol. 16, Part 1, 342 pp., 1938; *Ibid.*, Vol. 16, Part 2, 365 pp., Stuttgart, 1941.

Hylander, C. J.: "The Algae of Connecticut." 245 pp. Bull. 42, State Geol. and Nat. Hist. Survey, Connecticut, 1928.

Johnson, L. P.: Euglena of Iowa, *Trans. Am. Microscop. Soc.*, **63:** 97–135, 1944.

Pascher, A.: "Die Süsswasserflora Deutschlands, Osterreich und der Schweiz." Parts 1–15 (No. 8 not issued), 1913–1936. (2nd ed. issued under name "Die Süsswasserflora Mitteleuropas"; only Parts 9, 10, and 14 published.) Jena, G. Fischer.

Prescott, G. W.: "Iowa Algae." 235 pp. Vol. 13 in "Studies in Natural History," Iowa City, University of Iowa Press, 1931.

——— and A. M. Scott: The fresh-water Algae of southern United States. III, The desmid genus Euastrum, with descriptions of some new varieties, *Am. Midland Naturalist,* **34:** 231–357, 1945.

Smith, G. M.: "Phytoplankton of the Inland Lakes of Wisconsin." Part I. Wisconsin Geol. and Nat. Hist. Survey, Bull. 57, 243 pp., 1920. Part II. Wisconsin Geol. and Nat. Hist. Survey, Bull. 57, 227 pp., 1924.

———: The plankton Algae of the Okobojii region, *Trans. Am. Microscop. Soc.*, **45:** 156–233, 1926.

———: "The Fresh-water Algae of the United States." 716 pp. New York, McGraw-Hill Book Co., 1933.

Snow, Julia W.: The plankton Algae of Lake Erie, *Bull. U.S. Fish Comm. for 1902,* pp. 369–394, 1903.

Sparrow, F. K.: "Aquatic Phycomycetes." 785 pp. Vol. 15 in Scientific Series, Ann Arbor, University of Michigan Press, 1943.

Tiffany, L. H.: The filamentous Algae of northwestern Iowa with special reference to the Oedogoniaceae, *Trans. Am. Microscop. Soc.*, **45:** 69–132, 1926.

———: "The Plankton Algae of the West End of Lake Erie." 112 pp. Franz Theodore Stone Lab., Contr. No. 6, Columbus, Ohio State University Press, 1934.

Tilden, Josephine E.: "The Algae and Their Life Relations." 550 pp. Minneapolis, University of Minnesota Press, 1935.

West, G. S., and F. E. Fritsch: "A Treatise on the British Freshwater Algae," rev. ed., 534 pp. Cambridge, University Press, 1927.

West, W., and G. S. West: "A Monograph of the British Desmidiaceae," 5 vols., London, The Ray Society, 1904–1923.

ZOOPLANKTON

General

Brauer, A.: "Die Süsswasserfauna Deutschlands," No. 1–19, Jena, G. Fischer, 1909–1912.

Pratt, H. S.: "A Manual of the Common Invertebrate Animals," rev. ed., 854 pp. Philadelphia, The Blakiston Company, 1935.

Rylov, W. M.: "Das Zooplankton der Binnengewässer." 272 pp. Vol. 15 in Thienemann's "Die Binnengewässer," Stuttgart, 1935.

Protozoa

Ahlstrom, E. H.: Studies on variability in the genus Dinobryon (Mastigophora), *Trans. Am. Microscop. Soc.*, **56:** 139–159, 1937.

Blockman, F.: "Die Mikroscopische Tierwelt des Süsswassers." Part I, "Protozoa." 134 pp. Hamburg, Lucas Gräfe & Sillem, 1895.

Cash, J., G. H. Wailes, and J. Hopkins: "The British Freshwater Rhizopoda and Heliozoa," 5 vols., London, The Ray Society, 1905–1921.

Conn, H. W.: "A Preliminary Report on the Protozoa of the Fresh Waters of Connecticut." State Geol. and Nat. Hist. Survey, Connecticut, Bull. 2, 1905, pp. 5–69.

Eddy, S.: The fresh-water armored or thecate Dinoflagellates, *Trans. Am. Microscop. Soc.*, **49:** 277–321, 1930.

Kahl, A.: "Urtiere oder Protozoa." I. Wimpertiere oder Ciliata (Infusoria). 1–4. Die Tierwelt Deutschlands, Parts 18, 21, 25, 30. 886 pp. 1930–1935.

*Kent, W. S.: "A Manual of the Infusoria," 3 vols. London, D. Bogue, 1880–1882.

Kudo, R.: "Protozoology," 3d. ed., 778 pp. Springfield, Charles C. Thomas, Publisher, 1946.

*Leidy, J.: "Fresh-water Rhizopods of North America." 324 pp. Vol. 12, U.S. Geol. Survey Repts., 1879.

*Stokes, A. C.: A preliminary contribution towards a history of the fresh-water Infusoria of the United States, *J. Trenton Nat. Hist. Soc.*, **1:** 71–365, 1888.

Walton, L. B.: A review of the described species of the order Euglenoidina Bloch. Class Flagellata (Protozoa) with particular reference to those found in city water supplies and in other localities of Ohio. *Ohio State Univ. Bull.*, Vol. 19, No. 5. Ohio Biol. Survey Bull. 4, 1915, pp. 341–457.

Wenyon, C. M.: "Protozoology," 2 vols., 1563 pp. London, Baillière, Tindall and Cox, 1926.

Rotifera

Ahlstrom, E. H.: A revision of the rotatorian genus *Keratella* with descriptions of three new species and five new varieties, *Bull. Am. Museum Nat. Hist.*, **80:** 411–457, 1943.

Edmondson, W. T.: The sessile Rotatoria of Wisconsin, *Trans. Am. Microscop. Soc.*, **59:** 433–459, 1940.

Harring, H. K.: Synopsis of the Rotatoria, *U.S. Nat. Museum Bull.*, 81: 5–226, 1913.

Harring, H. K., and F. J. Myers: The rotifer fauna of Wisconsin, I, *Trans. Wisconsin Acad. Sci.*, **20:** 553–662, 1922.

——— and ———: The rotifer fauna of Wisconsin, II, *Ibid.*, **21:** 415–549, 1924.

——— and ———: The rotifer fauna of Wisconsin, III, *Ibid.*, **22:** 315–423, 1926.

——— and ———: The rotifer fauna of Wisconsin, IV, *Ibid.*, **23:** 667–808, 1928.

*Hudson, C. F., and P. H. Gosse: "The Rotifers or Wheel Animalcules," 2 vols., London, Longmans, Green and Co., 1889.

Jennings, H. S.: Rotatoria of the United States with special reference to those of the Great Lakes, *Bull. U.S. Fish Comm. for 1899*, 1900, pp. 67–104.

———: Synopses of North American invertebrates: XVII, The Rotatoria, *Am. Naturalist*, **35:** 725–777, 1901.

Myers, F. J.: The rotifer fauna of Wisconsin, V, *Trans. Wisconsin Acad. Sci.*, **25:** 353–413, 1930.

*Weber, E. F.: Faune rotatorienne du bassin de Léman, *Rev. suisse zool.*, **5:** 263–785, 1898.

Crustacea

General

Dodds, G. S.: A key to the Entomostraca of Colorado, *Univ. Colo. Studies*, **11:** 265–298, 1915.

Harrick, C. L., and C. H. Turner: "Synopsis of the Entomostraca of Minnesota." 525 pp. Geol. and Nat. Hist. Survey of Minnesota, Zool. Ser. II, 1895.

Weckel, Ada L.: Free-swimming fresh-water Entomostraca of North America, *Trans. Am. Microscop. Soc.*, **33:** 165–203, 1914.

Cladocera

Fordyce, C.: The Cladocera of Nebraska, *University of Nebraska Studies from the Zool. Lab.*, **42:** 119–174, 1901.

Hoff, C. C.: The Cladocera and Ostracoda of Reelfoot Lake, *Rept. Reelfoot Lake Biol. Sta.*, **7**: 49–107, 1943.

Scourfield, D. J., and J. P. Harding: A key to the British species of freshwater Cladocera with notes on their ecology, *Sci. Pub. Freshwater Biol. Assoc. Brit. Emp.*, **5**: 1–50, 1941.

Woltereck, R.: Races, associations and stratification of pelagic daphnids in some lakes of Wisconsin and other regions of the United States, *Trans. Wisconsin Acad. Sci.*, **27**: 487–522, 1932.

Copepoda

Ewers, Lela A.: "The Larval Development of Freshwater Copepoda." 43 pp. Franz Theodore Stone Lab., Contr. No. 3, Columbus, Ohio State University Press, 1930.

*Forbes, E. B.: A contribution to a knowledge of North American fresh-water Cyclopidae, *Bull. Illinois State Lab. Nat. Hist.*, **5**: 27–83, 1897.

Gurney, R.: "British Fresh-water Copepods," 3 vols., London, The Ray Society, 1931–1933.

*Marsh, C. D.: "Cyclopidae and Calanidae of Lake St. Clair." 24 pp. Bull. Mich. Fish Comm. No. 5, 1895.

———: A revision of the North American species of Diaptomus, *Trans. Wisconsin Acad. Sci.*, **15**: 381–516, 1907.

———: Distribution and key of the North American Copepods of the genus Diaptomus, with description of a new species, *Proc. U.S. Nat. Museum*, **75**: 1–27, 1929.

———: Synopsis of the calanoid crustaceans, exclusive of the Diaptomidae, found in fresh and brackish waters, chiefly of North America, *Proc. U.S. Nat. Museum*, **82**: 1–58, 1933.

Pesta, O.: "Krebstiere oder Crustacea." I, Ruderfüsser oder Copepoda (1. Calanoida, 2. Cyclopoida). 136 pp. Vol. 9 in Dahl, "Die Tierwelt Deutschlands," 1928.

Pine, Rose L.: Metamorphosis of Cyclops viridis, *Trans. Am. Microscop. Soc.*, **53**: 286–292, 1934.

Yeatman, H. C.: The American cyclopoid copepods of the viridis-vernalis group, *Am. Midland Naturalist*, **32**: 1–90, 1944.

Ostracoda

Furtos, Norma C.: The Ostracoda of Ohio, *Ohio Biol. Survey*, **5**: 411–524, 1933.

———: Fresh-water Ostracoda from Massachusetts, *J. Wash. Acad. Sci.*, **25**: 530–544, 1935.

———: Fresh-water Ostracoda from Florida and North Carolina, *Am. Midland Naturalist*, **17**: 491–522, 1936.

Hoff, C. C.: The Ostracods of Illinois, *Ill. Biol. Monogr.*, **19**: 1–196, 1942.

*Sharpe, R. W.: Contribution to a knowledge of the North American fresh-water Ostracoda included in the families Cytheridae and Cyprodidae, *Bull. Ill. State Lab. Nat. Hist.*, **4**: 414–484, 1897.

Bryozoa

*Allman, G. J.: "A Monograph of the Fresh-water Polyzoa, Including All the Known Species, Both British and Foreign." 119 pp. London, The Ray Society, 1856.

Davenport, C. B.: Report on the fresh-water Bryozoa of the United States, *Proc. U.S. Nat. Museum*, **27**: 211–221, 1904.

Rogick, Mary D.: Studies on freshwater Bryozoa: II, The Bryozoa of Lake Erie, *Trans. Am. Microscop. Soc.*, **54**: 245–263, 1935.

Index

INDEX

Q

R

U

V

W

Y